The Fabled Coast

The Fabled Coast

LEGENDS & TRADITIONS FROM AROUND THE SHORES OF BRITAIN & IRELAND

Sophia Kingshill &
Jennifer Westwood

arrow books

To Jane Kingshill (1923–2012)

Published by Arrow Books 2014

2 4 6 8 10 9 7 5 3 1

First published in Great Britain in 2012 by
Random House Books
Random House, 20 Vauxhall Bridge Road,
London SW1V 2SA

www.randomhouse.co.uk

Addresses for companies within The Random House Group Limited can be found at: www.randomhouse.co.uk/offices.htm

The Random House Group Limited Reg. No. 954009

A CIP catalogue record for this book is available from the British Library

ISBN 9780099551072

The Random House Group Limited supports the Forest Stewardship Council® (FSC®), the leading international forest-certification organisation. Our books carrying the FSC label are printed on FSC®-certified paper. FSC is the only forest-certification scheme supported by the leading environmental organisations, including Greenpeace. Our paper procurement policy can be found at www.randomhouse.co.uk/environment

MIX
Paper from
responsible sources

[illegible]
Design by Richard Marston

Typeset in Fayon Pro by Palimpsest Book Production Limited,
[illegible], Stirlingshire

Printed and bound in Great Britain by CPI Group (UK) Ltd., Croydon CR0 4YY

Extract fro[illegible] by Craig Williamson,
reproduced by kind permission of the University of Pennsylvania Press.
Material from the article 'The Tobermory Treasure' by Sophia Kingshill,
reproduced by kind permission of the journal Folklore.

ACKNOWLEDGEMENTS

This book is based very largely on material compiled by the late Jennifer Westwood, outstanding folklorist and my much-missed friend. She was generous enough to bequeath me her wide-ranging research into sea-lore, and the process has been truly a collaboration, to the point where I would find it hard to sort out what is my writing from what is hers. I hope the result falls not too far short of her breathtakingly high standards.

Neither Jennifer nor I could have got anywhere without the pioneering work of other writers. Fletcher Bassett's *Legends and Superstitions of the Sea and of Sailors in all Lands and at all Times* (1885), although not perhaps so literally comprehensive as its title suggests, is nonetheless an astounding compendium, and Wilbur Bassett, Fletcher's son, examines several legends in meticulous and absorbing detail in his *Wander-Ships* (1917). *The Broad, Broad Ocean* (1871) and *Credulities Past and Present* (1880), by William Jones, provide fascinating material, while Walter Gill's *Manx Scrapbook* (1929) and *Second Manx Scrapbook* (1932) have interest far beyond the shores of Man.

Of more recent works, David Thomson's *People of the Sea* (1954), which provided a majority of the seal legends, is remarkable literature in its own right. Gwen Benwell and Arthur Waugh's *Sea Enchantress* (1961) is the most wide-ranging of investigations into the mermaid legend, and no book involving sea-monsters, Krakens and the like could be attempted without consulting Bernard Heuvelmans's *In the Wake of the Sea-Serpents* (translated by Richard Garnett, 1968). Peter Anson's *Fisher Folk-Lore* (1965) has been a wonderful resource. Horace Beck's *Folklore and the Sea* (1973) is compulsively readable and endlessly entertaining, while Peter Kemp's *Oxford Companion to Ships and the Sea* (1976) is indispensable.

Many newer and much older authors have gone into the mix, far too many to list here, but I have to credit two in particular whose scholarship and imagination have revived me when I felt I couldn't stand another mermaid whispering in my ear. John Livingston Lowes's *Road to Xanadu* (1927) was a revelation – not just for a Coleridge-lover – and Dáithí Ó hÓgáin's *Lore of Ireland* (2006), as well as several of his other books, opened up a world of poetry, history, implication and elucidation.

Friends and colleagues have helped me in more ways than I can mention. Brian Chandler, Jennifer Westwood's husband, has been endlessly supportive, as have my editors at Random House, Gemma Wain (who has put in much detailed and extremely valuable work on the text) and Sophie Lazar, and Nigel Wilcockson, to whom Jennifer owed her association with the company.

Jacqueline Simpson has indexed the entire book, as well as providing wise encouragement. Janet Bord and Tristan Gray Hulse put me right about several Welsh questions, and any mistakes that remain are my own alone. Jeremy Harte was his usual inspiring self, and also put me in touch with Richard Freeman of the Fortean Society, who was kind enough to read through a whole draft, and commented most helpfully. Paul Cowdell's article on cannibal ballads contributed very largely to my own essay on survival cannibalism, and on the topic of monkey-hanging I have called on the expertise of Sandy Hobbs. I am grateful to the journal *Folklore* for permission to use material from my article on the Tobermory galleon, and in particular to Deborah East, Jessica Hemming, and Caroline Oates for their help in sorting this out.

For miscellaneous assistance and advice, I should like to thank particularly Frances Collinson, Chris Denton, Nick Dyer, Alan Griffiths, Gisela Holfter (and her helpful and knowledgeable friend Tadhg Ó hIfearnáin), Suzi Hopkins, Martin Morgan, Sophie Reissner, June Rose, Geoffrey Sample, Ann Shearer, Chris Sheffield, and my own family. Graeme Rosie has been forbearing and discreetly appreciative throughout.

Most importantly, I am grateful to Kristina Rapacki, picture researcher of genius, and to Clare Graydon-James, who tirelessly and cheerfully pursued queries with vicars, museum curators, and other authorities, and produced wonderful results. Not all the entries she investigated could ultimately be included in the book, for reasons of space, and I have to thank several kind informants whose contributions do not appear, as well as those whose do. They are, in alphabetical order: Reverend Peter Blackett, St Michael's church, Bowness-on-Solway; Sara Brown, Cambridge and County Folk Museum; Dr David Caldwell, Edinburgh Museum; Christina Carson,

St Just Library; Lionel Clauzon, Newport Reference Library; the staff of the Cornish Studies Library; Rebecca David, Boscastle Tourist Information Centre; Peter Drummond, Orkney Museum; Sam Ellis, Scilly Isles Tourist Information Office; Katherine Fetherstonhaugh, Colwyn Bay Library; Howie Firth, Orkney Science Museum; Julian Gaisford-St Lawrence, Howth Castle; Graham Groom, Lockerby Library; Reverend Helen Griffiss, All Saints church, Mudeford; Father Milo Guiry, St Declan's church; Paddy Hodgins, Seamus Roe and Sandra & Bryan Rogers, Clogherhead Historical Society; Mairi Hunter and Ruth Airley, Ewart Library, Dumfries; Alison Kentuck, Receiver of Wrecks; Diana Leake, Bridport Library; Diane Leggett, Local Studies Centre, North Shields Library; Anne Lenihan, Dungarvan Central Library; Florence Luscombe, St Ives Library; Reverend Donald MacEwan, St Monans church; Clare McIntyre, National Museums of Scotland; Jessie McKerrow, Middlebie church; John Mudge, Steeple Woodlands Nature Reserve; Mike Murphy, St Ives Archive; David Muscat, Colchester Library; Micheal O Coileain, Kerry County Council; Norma O'Meara, Dungarvan Tourist Information Centre; Richard Parkinson, British Museum; Una Rowan, Cushendall Development Office; Reverend Ronald Seaman, formerly Dornock parish church, and Jeannette Seaman; Anne Sharp, Local Studies Centre, South Shields Library; the staff of the Shetland Library; Joe Smedley and Alison Burgess, Ewart Library, Dumfries; Reverend Hugh Steele, Dornock parish church; Reverend Robert Thewsey, Forrabury church.

After Jennifer's death, a fund was set up to help complete her unpublished work on sea legends. Heartfelt thanks to those who generously contributed: Alison Anderstrem; Geoff Ball; Caroline Baxter; Dr J. H. Belcher; Lin Bensley; Ruth Binney; Janet & Colin Bord; Elaine Bradtke; Celia Cannell; Veronica Cato; P. R. & M. R. Clarke; Mabel & Ken Colman; Grace Corne; Steve Corrsin; Linda & Mike Edwards & family; Peggy Everson; The Folklore Society; Sharon & Vikki Fulcher; family of the late Ted Fulcher (Joan, Valerie, Maureen & Susan); Terry & Melanie Goodwin; Bob & Beth Hardman; John Harris; Tristan Gray Hulse; Christine & Gareth Jones; Barbara & Tony Kelly; Leslie Leach; Joe Liddane; Barbara Littlewood; Kay, Dennis & Peter Loades; Sybil Loades; E. R. Loveday; Angus & Ann Mackenzie; Kirk & Julie Murdoch; Jackie & Jimmy Paterson; Mr & Mrs Roberts; Malcolm & Rita Rose; Graham Rowe; Jacqueline Simpson; E. M. F. Swale; Valerie Tate; John & Jennifer Ward; Valerie & Harry Wones.

PICTURE CREDITS

Colour illustrations are reproduced by kind permission of:

The Bridgeman Art Library: *The Fisherman and the Syren: From a Ballad by Goethe* by Frederic Leighton (© Bristol City Museum and Art Gallery), Woodcut of Noah's Ark from the Nuremberg Bible (Private Collection/The Stapleton Collection), *St. Nicholas rebuking the Tempest* by Bicci di Lorenzo (Ashmolean Museum, University of Oxford), Carving of Zennor mermaid (The Marsden Archive), Jonah and the whale by J. Vernet (Private Collection/© Look and Learn), Illustration of St. Cuthbert from 'Life and Miracles of St. Cuthbert' by Bede (© British Library Board. All Rights Reserved), *Kidd on the Deck of the 'Adventure Galley'* by Howard Pyle (Delaware Art Museum, Wilmington), A *Sea Ghost* by George Frederic Watts (© Trustees of the Watts Gallery, Compton, Surrey), *Smugglers on the Eastern Coast of Sweden* by Josef Wilhelm Wallander (Bibliotheque des Arts Decoratifs, Paris /Archives Charmet); British Museum: Photograph of the excavations at Sutton Hoo (© The Trustees of the British Museum); Getty Images: Photograph of Lerwick's *Up Helly Aa festival* (VisitBritain/John Coutts); Greenwich Hospital: *Battle of Trafalgar, 21 October 1805* by J. M. Turner (Greenwich Hospital Collection); Lawrence University: Illustrations of Blemmyae and Sciapods from Hartman Schedel's *Nuremberg Chronicle* (Lawrence University Special Collections, Appleton, Wisconsin); Leen Helmink: Scan of *Septentrionalium Regionum Descrip.* map by Abraham Ortellius; The Natural History Museum: *Dactylioceras commune* ammonite (© The Natural History Museum, London); Tate Collection: *The Boyhood of Raleigh* by John Millais (© Tate, London 2011).

CONTENTS

INTRODUCTION

Nay, come up hither. From this wave-washed mound
Unto the furthest flood-brim look with me;
Then reach on with thy thought till it be drown'd.
Miles and miles distant though the last line be,
And though thy soul sail leagues and leagues beyond, –
Still, leagues beyond those leagues, there is more sea.

from 'The House of Life' (1881), by Dante Gabriel Rossetti

A picture that used to appear regularly in school history books is Millais's *Boyhood of Raleigh*. It shows a weatherbeaten man pointing out to sea, spinning a tale to two spellbound boys. One of the listeners is the youthful Walter Raleigh, and this is his moment of epiphany, the awakening of a spirit of exploration that will take him across the wide Atlantic to the shores of the New World.

The scene of the old sea-dog telling his adventures could be set just about anywhere, in any century almost up to the present. Until very recent times, only sailors travelled widely, and so only they could describe the wonders of foreign lands, the perils and marvels of the open waves. Even now, we still know less about the ocean, its farthest reaches and lowest depths, than we do about mountains, jungles, deserts or ice-fields.

Our coast is more familiar. Certainly we have plenty of it: the shoreline of the UK mainland alone is more than five times as long as that of France, and once you add Ireland and all the islands of both countries, you have a lot more edge than middle. Legends flourish in these borders between land

and sea. All things supernatural favour the territory linking one state with another – twilight between day and dark, doorways and gates between in and out, bridges, dreams between sleep and waking. The shore is another liminal area, joining earth to water, known to unknown, and this is the setting for some of the most beautiful, terrible, and memorable tales of folklore.

Secrets of the sea

Landsmen love to hear sailors' yarns because, however fabulous, they could be true. Sailors love telling them to show off, and to convert grim experience into romance. There is no occupational group with more tales than sailors: none until quite lately spent more time in isolation from all but a group of their fellows; none had more startling truths to tell, nor more scope for lies. These are ideal conditions for storytelling.

Every maritime nation has its sea legends, and in our group of islands to the west of the European mainland, examples of just about every theme are found. By the nature of their profession, sailors form an international community, and their beliefs and traditions became common currency, translated into different idioms and languages, but recognisably similar from the Hebrides to the seaboard of America, from the Mediterranean to the Channel.

The prototype of seafarers is Odysseus, whose exploits, first told by Homer in the eighth century BCE, were a blueprint for all later WONDER VOYAGES (p. 458). As sailor's yarns go, the *Odyssey* set a high standard. It took the hero ten years to travel from Troy to Ithaca – a distance of three hundred miles or so as the crow flies – and when he finally walked in and his wife Penelope asked, as well she might, 'And where have *you* been?', she heard a tale of spectacular risks and stupendous freaks. Her husband had evaded Scylla, a monstrous squid or octopus like the vast Kraken that haunted the imagination of later Scandinavian seafarers (*see* BURRA, Scottish Highlands & Islands), and Charybdis, mother of all whirlpools (*see* CORRYVRECKAN, Scottish Lowlands). He had heard the magical song of the Sirens without being tempted to his death, and on enchanted islands ruled by seductive queens he had mysteriously forgotten to go home. Penelope might have had a thing or two to say about that, or she might have decided the less said the better. Odysseus had come up with a winning formula, the model for centuries to come. His are the same improbable but undisprovable yarns told by generations of sailors.

They are not pretty tales for children, but tough stories told by tough men, whose lives held real dangers. Whirlpools could and still can swallow up small craft, and although no octopus has ever capsized a vessel, or plucked a sailor from the rigging with its snake-like tentacles, whales are easily big enough to sink boats, and have done so. As for the islands, countless rocks and reefs, uncharted in Odysseus's day and for two thousand years or more after, are found even in a sea as small as the Mediterranean. Some islands, as if they were bewitched fairy realms, really *do* come and go. Vegetation clotted with earth can form masses large enough to count as 'floating islands', and there are seldom-exposed reefs of rocks or sandbanks, such as SCROBY SANDS (East Anglia), that can grow and shrink. Mirages including the famous 'fata Morgana' play their part (*see* RAMSGATE, South-East England), and then there is the whole volcanic box of tricks, from the 1963 appearance of Surtsey off Iceland – coming up – to Santorini, largely destroyed by eruption in the seventeenth century BCE and leading, perhaps, to the myth of ATLANTIS (p. 24) – going down.

Throughout the ancient world, there were tales of island paradises, attainable before death if you just sailed far enough. To the Greeks they were the Fortunate Isles or Hesperides; to the Celts, Tir na nÓg, the Land of Youth, or Hy Brazil (*see* INISHMORE, Northern Eire & Northern Ireland). For the early Celtic monks, it became a form of pilgrimage to get in their coracles and launch themselves on the sea to journey they knew not where, for the love of God. Whatever happened on the actual voyages, in poem and legend they are as extraordinary as the adventures of Odysseus – but the men were *saints*, the stories *must* be true. This kind of thinking kept mythic lands on the map for centuries. St Brendan's Isle, supposedly discovered on Brendan's journey from BRANDON CREEK (Southern Eire), continued to be marked on charts of the Atlantic for hundreds of years, and was seriously searched for until the eighteenth century. People are still looking for Atlantis.

Even when navigation and exploration had largely erased such elusive landmarks, tales of 'fairy islands' continued to haunt the imagination. Near MILFORD HAVEN in Wales, it was rumoured, lay 'the green meadows of the sea', from which fairy shoppers came to market on the mainland. The Scottish equivalent is Heather-Bleather, somewhere around Orkney (*see* EYNHALLOW, Scottish Highlands & Islands), and the seas around Ireland, according to popular report, are or were positively infested with magical islands (*see* for example RATHLIN O'BIRNE ISLAND, Northern Eire & Northern Ireland, and LAHINCH, Southern Eire).

Not only were there strange lands in the ocean, there was even another sea above the earth. Medieval ideas of this aerial ocean were called into play to explain sightings of 'sky-ships' (*see* BODMIN, South-West England & Channel Islands), vessels sailing in the air, a strand of folklore that has in fact grown more common in modern times, although today witnesses are more likely to interpret their visions, or illusions, as alien spacecraft. While the Unknown now is outer space, in ancient times it was the sea.

Unnatural natural history

Parallel to the imaginary geography that pictured heavenly and hellish lands somewhere just over the horizon was a fantastic zoology, populating the waves with hideous monsters and ravishing mermaids. Underwater, there was believed to be a 'duplicate creation' providing equivalents for everything found on land, not only sea-horses and sea-cattle (*see* ST BRIDES, Wales, and DURSEY ISLAND, Southern Eire), but even sea-bishops and other such STRANGE FISH (p. 144).

As scientific knowledge advanced, people stopped taking the word of classical authors as the undoubted voice of authority, and started really looking at the world around them. Authentic wonders began to replace fables in popular report, and the mermaid drifted out of the naturalist's sphere, where she had remained for centuries in one form or another (*see* SIRENS, SEA-NYMPHS, AND MERMAIDS, p. 428), and into fairy talc and fantasy, where she remains a vivid symbol of feminine allure and danger.

Some of the old theories, however, were remarkably tenacious. Barnacle geese – birds born from shellfish (*see* BANNOW BAY, Southern Eire) – disappeared from natural history books, but even in the twentieth century it remained acceptable among Catholics in some places to eat geese on fast days, since their flesh counted as a form of fish. Other prodigies too remained stock elements of sailors' travelogues. The immensity of the sea, its inconceivable depths, its ultimate mystery – coupled with the fact that we are able to justify our wish to believe with scientific discovery, for instance of the Colossal Squid (*see* ROTHESAY, Scottish Lowlands) – these make marine monsters and SEA-SERPENTS (p. 348) among the most viable of legendary ideas today.

The supernatural

Many stories told of the sea cannot be fitted into our modern world-view, but instead tap into such deep levels of the human psyche that we are never likely to outgrow them. These are our most potent legends, those that have inspired painters, poets, and musicians. Their common theme might be called 'the wrath of God'.

The ocean's terror has been largely forgotten in the western world today, among travellers who fly or take the Channel Tunnel if they want to go abroad, but every so often we get a rude reminder. At the beginning of the twentieth century, technology was king, but then along came the iceberg that sank *Titanic* (*see* BELFAST, Northern Eire & Northern Ireland). At the beginning of the twenty-first, we might have thought ourselves safe from anything similar, but the sinking of the *Costa Concordia* early in 2012 made us think again.

The human instinct when confronted with such awesome power is to propitiate (to call the sea a god and try to keep it pleased). From there it is a short step to SUPERSTITIONS (p. 360) – and no group of people in the world is so observant of taboos and omens as seafarers.

On the other hand, the ocean might be, not itself a deity, but God's instrument of retribution. Fear of the sea as the ultimate weapon of the Powers That Be, to be hurled against us if we misbehave, underlies the stories of Noah's Flood and other deluges of ancient myth, as well as the legend of Atlantis, and notions of impiety, blasphemy, and sin are bound up with tales of 'drowned cities'. On the east coast of England the losses are not only real but relatively recent: more than thirty places mentioned in the Domesday Book (1086) have disappeared by erosion from the Holderness coast alone. To the west, we have to go back long before the historical record begins, to the changes brought about by the last Ice Age, in order to justify the stories of kingdoms drowned in CARDIGAN BAY and CONWY BAY (Wales) and off LAND'S END (South-West England & Channel Islands). In the cases of genuine sunken towns such as DUNWICH (East Anglia) there is no suggestion that they deserved their fate – reality is enough, perhaps – but not so with the mythic lost lands of the western coast, overwhelmed by the sea in punishment for their pride or evil ways.

The apocalyptic tone of such tales is echoed on an individual level by stories of men doomed and damned because they defied destiny, or God. In the late eighteenth century, Coleridge created a literary myth from the legendary archetype of a cursed man, condemned to 'death in life' (*see* WATCHET, South-West England & Channel Islands), and like the Ancient Mariner, the captain of THE FLYING DUTCHMAN (p. 30) experienced the sea as an endless journey. Perhaps he still does. Phantom ships are reported from time to time even now, and although some, like those that haunt the GOODWIN SANDS (South-East England), are identified as ghostly reminders of historical tragedies, others sail nameless through the mist, and might be the *Flying Dutchman* herself.

Men and women of the sea

To the long-distance sailor, the ocean is not only a highway but itself a country, where only the landsman is 'all at sea'. The mariner knows its language and its sometimes bizarre customs (*see* for example ABERDEEN, Scottish Lowlands, for a description of the ceremony performed when crossing the equator). No matter what its dangers and hardships, it is home. To the coastal fisherman and his family, the sea has a more ambiguous status, for though it provides, its goods come at a high price. Countless men have gone out with their nets one night only to meet with a deadly storm; countless wives and children have watched in the morning for a sail that never came. Trying to second-guess fate, fisherfolk were alert to a variety of omens, making their forecasts from clouds, seabirds, fish, and sounds heard at sea and on shore (*see* for example PATERNOSTERS, South-West England & Channel Islands, and SALTBURN-BY-THE-SEA, North-East England), and to assert some notional control they developed a remarkable range of taboos (*see* ROSEHEARTY, Scottish Lowlands).

The hazards were no less for a smuggler or buccaneer, but the potential rewards were far greater. Some, like Isaac Gulliver of WIMBORNE MINSTER (South-West England & Channel Islands), achieved not only wealth but respectability, while others got the worst of both worlds: Captain Kidd of GREENOCK (Scottish Lowlands) gained little but the gibbet, and went down in history as a brute. More notable than either in her own territory was Gráinne Mhaol, or Grace O'Malley, as the English called her, the sixteenth-century pirate queen of Connaught (*see* CLARE ISLAND, Northern Eire & Northern Ireland), who became a goddess-figure in later legend. Sir Francis Drake too took on a mythical character, gaining the reputation of a wizard during his lifetime (*see* PLYMOUTH, South-West England & Channel Islands), and was said centuries afterwards to have been reincarnated as that other English champion Nelson (*see* SCAPA FLOW, Scottish Highlands & Islands).

Much earlier, the seafaring saints exerted hands-on control over their environment. Columba or Colm Cillé, according to his first biographers, could direct the weather (*see* IONA, Scottish Lowlands), and moreover had a humane relationship with creatures of the deep, his encounters with whales mediated by prayer rather than slaughter. Man's subsequent mastery of the sea, such as it is, has been largely a tale of conquest rather than co-existence, an imbalance for which we may yet pay the price. Tales of vengeful mermaids who curse their enemies are as much allegory as folk tale: molest nature at your peril, is the message.

We don't listen. Humankind is incorrigibly rash, cheeky, invasive, callous, daring. The sea poses its questions, and something in us has to rise to the challenge. In Longfellow's 'The Secret of the Sea' (1850):

> 'Wouldst thou' – so the helmsman answered, –
> 'Learn the secret of the sea?
> Only those who brave its dangers
> Comprehend its mystery!'

There speaks the spirit of Odysseus: the man who faces up to Creation, wrestles with it, fears it, wonders at it, and at the end of the day, for the benefit of the less bold-hearted, brings back tales.

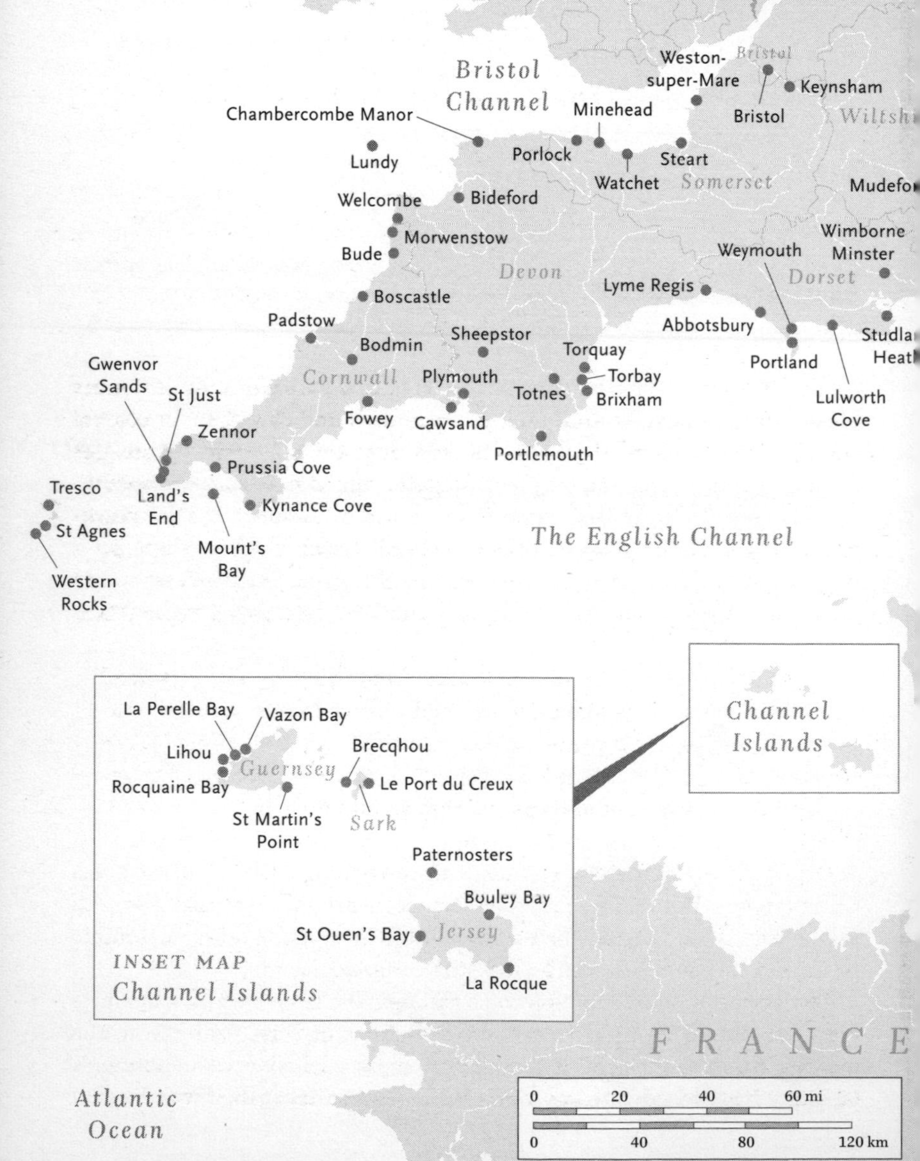

St George's Channel
WALES
Herefordshire
Worcs.
Gloucestershire
Brockweir
Lydney
Weston-super-Mare
Bristol
Keynsham
Bristol Channel
Chambercombe Manor
Minehead
Porlock
Watchet
Steart
Somerset
Wiltsh
Lundy
Welcombe
Bideford
Morwenstow
Bude
Devon
Mudefo
Wimborne Minster
Weymouth
Dorset
Lyme Regis
Boscastle
Padstow
Abbotsbury
Studla Heat
Sheepstor
Bodmin
Torquay
Torbay
Portland
Gwenvor Sands
Cornwall
Plymouth
Totnes
Brixham
Lulworth Cove
St Just
Fowey
Cawsand
Zennor
Portlemouth
Prussia Cove
Tresco
Land's End
Kynance Cove
St Agnes
Mount's Bay
Western Rocks
The English Channel
La Perelle Bay
Vazon Bay
Lihou
Brecqhou
Guernsey
Rocquaine Bay
Le Port du Creux
St Martin's Point
Sark
Paternosters
Bouley Bay
St Ouen's Bay
Jersey
La Rocque
INSET MAP
Channel Islands
Channel Islands
FRANCE
Atlantic Ocean
0 20 40 60 mi
0 40 80 120 km

SOUTH-WEST ENGLAND & CHANNEL ISLANDS

Channel Islands, Cornwall, Devon, Dorset, Gloucestershire, Scilly Isles, Somerset

ABBOTSBURY, DORSET

Garland Day

All over England, May Day used to be celebrated with garlands of flowers picked in the morning and hung up in houses and churches. In coastal communities, the wreaths were often cast into the sea, a springtime gift from the land to encourage a plentiful gathering of fish from the waters.

Most such rituals lapsed, often from clerical disapproval of their pagan implications, but in Abbotsbury near Chesil Beach they continue, in a modified form, to this day, despite (or perhaps because of) interference and opposition. This was the Abbotsbury Garland Day as described in the 1860s:

> The children belonging to the crew of each boat build up a large garland of handsome flowers upon a frame, and carry it from house to house, usually getting a few pence a-piece from those who can afford it. The people throng the beach, weather permitting, in the afternoon, when the garlands are taken out in boats and thrown into the sea.

Originally the ceremony was not an overtly religious or at least not a Christian one, but by the mid nineteenth century the vicar had begun to take part, accompanying the procession to the shore, reading a suitable passage from the Bible, and leading the congregation in prayers.

The occasion continued to evolve through the twentieth century, progressively losing its links with the sea. Before the First World War, the making and carrying of the garlands, once exclusive to fishermen's families, had extended to any local children, and after 1918, it was decided that it would be more suitable to lay the wreaths on the war memorial than to cast them into the water. In 1954, Garland Day nearly came to an end when a high-handed policeman, new to the district, took it upon himself to ban the parade altogether. He declared that the children who took the garland from house to house, collecting contributions, were breaking the law by begging, and the money was confiscated. Paradoxically, he may have helped keep the custom going, since not only the children but their parents, who had been losing interest in Garland Day, were furious. A protest was organised, and the Chief Constable of Dorset issued an apology and restrained the overzealous officer. Abbotsbury school continued to support the tradition, giving its pupils a day off on 13 May (Old May Day, by the pre-Gregorian calendar), but after the school was closed in 1981 the children no longer got a holiday for the occasion. Garland Day has survived, however, as a local custom. Flowers are still taken round the houses on the evening of 13 May to collect money for local charities, and the garlands are put on the war memorial in the churchyard.

BIDEFORD, DEVON

The last battle of the Revenge

The Tudor admiral Sir Richard Grenville of Bideford had a fearsome reputation. One of his contemporaries was told 'by divers credible persons who stood and beheld him, that he would carouse three or four glasses of wine, and take the glasses between his teeth and crush them in pieces and swallow them down', and in Britain's wars with Spain his 'intolerable pride, insatiable ambition' made such an impression that Spanish nurses used to terrify their charges with tales of his ferocity.

In his final battle he surpassed himself. Commanding the *Revenge*, in 1591 he helped lead an expedition to the Azores in the hope of capturing Spanish treasure ships. When the Spaniards came in sight, however, they greatly outnumbered the British fleet, which had also been struck by fever, leaving many men too sick to fight. The *Revenge* was intercepted by two Spanish squadrons, and with the choice of sailing on between them or turning to escape, Grenville declared that 'he would rather choose to die than to dishonour himself, his country, and her Majesty's ship' by showing his back to the enemy. The *Revenge* advanced, one ship against fifty-three.

Grenville's decision went down in history. Tennyson's poem 'The Revenge' (1878) imagines his suicidal courage:

> Sir Richard spoke and he laughed, and we roared a hurrah, and so
> The little Revenge ran on sheer into the heart of the foe . . .

The unequal encounter should have been over in minutes. In fact it lasted all night. By morning, around fifteen hundred Spaniards were dead, and the *Revenge* had sunk or repulsed fifteen ships. Her powder was all gone, her masts were down, her pikes broken, forty of her hundred able men dead and most of the rest injured. Nevertheless, the Spanish made no further attempts to board, since they were convinced that Grenville would destroy his vessel sooner than surrender, and they were quite right. Sir Richard, mortally wounded, had ordered his master gunner to blow up the ship. This, however, was one step too far for his men, who locked up the gunner and negotiated with the Spanish, reaching agreement that the English officers were to be ransomed and the rest of the crew returned to their native shores. Too weak to argue, Grenville was removed to the Spanish admiral's ship, where he died three days later:

> And they stared at the dead that had been so valiant and true,
> And had holden the power and glory of Spain so cheap
> That he dared her with one little ship and his English few;
> Was he devil or man? He was devil for aught they knew,
> But they sank his body with honour down into the deep.

After more than twelve hours' battle with the Spanish, Sir Richard Grenville's ship Revenge *was little better than a mastless hulk, but still her captain refused to surrender.*

It was commonly believed that some supernatural power beyond human courage had been at work, and was displayed in the frightful tempest that followed. More than a hundred Spanish ships were wrecked in the storm, while the *Revenge* herself now sank, as if disdaining to survive her commander. The English held that this was the work of God, the Spanish that it was infernal, and 'so soon as they had thrown the dead body of the Vice-Admiral Sir Richard Grenville overboard, they verily thought that as he had a devilish faith and religion, and therefore the devil loved him, so he presently sunk into the bottom of the sea and down into hell, where he raised up all the devils to the revenge of his death, and that they brought so great a storm and torments upon the Spaniards, because they only maintained the Catholic and Romish religion.'

In an 1881 essay on 'The English Admirals', Robert Louis Stevenson comments that although Grenville chewing wine glasses at table made no very pleasant figure, his achievement continued to inspire an age less heroic than his own:

> I contend it ought not only to enliven men of the sword as they go into battle, but send back merchant-clerks with more heart and spirit to their book-keeping by double entry.

BODMIN, CORNWALL

A sea above the earth

Riding from Bodmin towards Fowey, one evening in May 1580, a company of gentlemen saw a wonderful sight in the sky, described in detail in John Stow's *Chronicles of England* (1580). First in the north-east there appeared a great fog, 'much like unto the sea', and out of it a castle emerged. Then a fleet of ships came sailing across, looking very warlike, and a second battalion from the south-west. Three or four galleys followed, with smaller boats in their wake, a procession that continued for around an hour.

The sight of sky-ships was by no means unique, though it was always reported as a marvel. This example sounds like an unusual cloud display, and other more remarkable phenomena have been explained as mirages (*see* RAMSGATE, South-East England). Observers, however, sometimes interpreted such visions as proof that there was a sea above the earth, mirroring the one below. The idea of the world as a solid island, in the midst of waters over and under it, is common to many mythologies, and appears in Genesis 1:7, where we are told that when creating the world,

God 'divided the waters which were under the firmament from the waters which were above the firmament'.

In ninth-century France, many educated people thought that in the air was a country called Magonia, from which cloud-ships would sail to carry off the corn beaten down in storms (making a profit for sorcerers called *tempestarii*, who raised the gales and then sold the ruined crops to the sky-sailors). A more specific account by the medieval historian Gervase of Tilbury relates how some people leaving a church saw a cable stretching from sky to earth, attached at the bottom to a ship's anchor caught on a tombstone, while the top was invisible, disappearing into thick cloud. The rope was straining as if something above was tugging at it, and soon a man came climbing down it from the clouds. He freed the anchor, but was then seized by the onlookers and died in their hands, 'suffocated by the humidity of our dense air as if he were drowning in the sea'. About an hour later the rope tumbled to earth, cut from above. The anchor remained where it was, and the citizens converted it into fastenings for their church door.

Very similar events were reported in Ireland in the eighth and tenth centuries, at Clonmacnoise and TELTOWN (Northern Eire & Northern Ireland), while another tale from Gervase implies that the aerial sea was continuous with our own oceans, though far distant. A Bristol sailor on a long voyage happened to drop his knife overboard one morning.

> That very hour it fell through an open window in the roof of the citizen's own home – the kind of window which the English call a skylight – and stuck fast in a table which stood beneath it, before the eyes of his wife.

She recognised it immediately, and when her husband came home, they compared dates and found that he had dropped it and she had received it in the same hour of the same day, which Gervase takes as incontrovertible evidence of an ocean either 'in the air or above the air'.

BOSCASTLE, CORNWALL

Drowned bells

Long ago, the inhabitants of Boscastle, whose church had no bell, were jealous of the beautiful peal at neighbouring Tintagel, and raised money to buy bells of their own. In *Cornish Feasts and Folk-Lore* (1890), Miss M. A. Courtney relates how the newly cast bells were brought by sea. In line with usual practice, the ship had taken on a local man to navigate into the harbour, and this pilot happened to be from Tintagel. Hearing his own parish bells ring out for vespers, he piously crossed himself, and thanked God that he would soon be safe on shore.

> On this the captain grew very wroth, and said, 'Thank the ship and the canvas at sea, thank God on shore.' 'No!' meekly replied the pilot, 'we should thank God at sea as well as on land.' At this the captain grew still more angry, swore and blasphemed, and with an oath exclaimed, 'Not so, thank yourself and a fair wind.'

A storm at once arose. The ship was driven on to the rocks and sank, and all on board except the devout pilot were drowned. As the vessel foundered, the bells were heard tolling, and they still ring out before a gale, 'but woe to the unhappy ship's crew that hears them, for wreck, misfortunes, and deaths are sure to follow.'

This moral tale was widely repeated in the late nineteenth century, and the Reverend Robert Hawker, vicar of MORWENSTOW from 1834 to 1875, composed a popular poem on the subject, 'The Silent Tower of Bottreau' (Boscastle was once the seat of the De Bottreaux family and known as Bottreaux or Bottreau Castle). There is no consensus as to whether the bell-less church referred to was the medieval church of St James at Boscastle, demolished in the nineteenth century, or St Symphorian's at Forrabury, now the church for Boscastle parish, or St Merteriana's at Minster, and it is many years since anyone has reported hearing the spectral bells. Hawker may, in fact, have made the legend up, but if so, he was drawing on far older accounts of bells drowned in punishment for sacrilege, such as that told at ST OUEN'S BAY.

It was widely believed that phantom bells were heard by drowning sailors, although, of course, like other near-death experiences, this can only be learned from the reports of those who have survived. In *Credulities Past and Present* (1880), William Jones cites an account of an old man who was on the point of death but revived:

> He had the ringing of bells in his ears, which increased as consciousness was becoming less, and he felt as if 'all the bells of heaven were ringing him into Paradise!'

The person to whom the old man told his story added that 'I know the locality where the circumstance occurred, and there is no bell within a circuit of six miles, but one old cracked church bell.'

BOULEY BAY, JERSEY, CHANNEL ISLANDS

The Dog of Bouley

At Bouley Bay, the nocturnal apparition of a large black dog, with staring saucer eyes, was said to foretell a storm. In Channel Islands patois, the beast was called *Le Tchan du Bouôlé* ('The Dog of Bouley'), and it has been suggested that *Tchan* (*chien*) is a corruption of *Chouan*, the name applied to royalists who escaped to Jersey during the French Revolution. 'History tells us that they indulged in a good deal of horse-play, and were constantly playing pranks on the inhabitants,' wrote the Jersey folklorist John L'Amy in 1927. 'May it not be that one of them disguised himself as a dog?' – a superficially tempting but, when you think about it, rather improbable theory. Elemental beasts (animal apparitions linked to the weather) are very often big, black, and canine (*see* OVERSTRAND, East Anglia), and Bouley's dog is a classic example, reports of him almost certainly pre-dating the French refugees.

BRECQHOU, CHANNEL ISLANDS

The devil-fish

Many dreadful stories have been told about octopuses, which were believed to be aggressive and dangerous, seizing swimmers and holding them underwater, and capable even of sinking boats. An old legend of this kind was told of the river BOYNE (Northern Eire & Northern Ireland), and in the Channel Islands, an octopus was said to have caught and drowned a fisherman in a cave at Brecqhou. Victor Hugo described the horrid animal in his 1866 novel *Toilers of the Sea*:

> A grey form oscillates in the water; it is as thick as a man's arm, and about half an ell long; its outline is ragged; its form is that of a closed umbrella without a handle. This ragged outline gradually advances toward one. Suddenly, it opens, and eight radii dart, all at once, from around a face with two eyes; these radii are alive; they wave like a flame; when spread out they resemble a sort of wheel, four or five feet in diameter . . . This creature clings to its prey, covers it, and knots its long hands about it. Underneath, it is yellowish, on top, it is earth-coloured; this dusty shade is indescribable; it appears to be a creature made of ashes, living in the water. It is the shape of a spider with the colouring of a chameleon.

An illustration to Victor Hugo's novel Toilers of the Sea *(1866), showing the hero Gilliatt in combat with the 'devil-fish', a giant octopus. 'Suddenly, the creature detached its sixth tentacle from the rock, and, throwing it on Gilliatt, tried to seize his left arm.'*

Hugo wrote that he had himself seen the 'devil-fish' chase a swimmer in another cave, the Boutiques in Sark. It was four foot across, and had four hundred suckers, which, when dying, it thrust out convulsively.

This specimen was, however, nothing compared to the monster described by Pierre Denys de Montfort in his *Histoire Naturelle des Mollusques* (1802). That had arms forty-five foot long and two and a half foot thick at the top, with suckers as big as plates. Denys de Montfort supported his contention that cephalopods were 'the most enormous animals that exist on the globe' with a picture of a ship being attacked by a monster as big as itself, a drawing which defeats its object by being so obviously incredible.

The largest known species, the Giant Pacific Octopus, usually attains an arm span of around fourteen foot, although there are reports of specimens with a span of thirty foot. Tales of truly titanic creatures may relate not to the octopus but to Giant or Colossal Squid, which can apparently be as much as sixty foot long (*see* ROTHESAY, Scottish Lowlands). Octopuses found around the British coast are generally quite manageable, but Richard Carrington notes in *Mermaids and Mastodons* (1961) that 'anyone who has felt the strength in the suckers of even the smallest octopus will not be likely to underestimate the risk.' He records a well-authenticated nineteenth-century episode when several French soldiers mysteriously disappeared while swimming off the Algerian coast:

> On one occasion a group of soldiers were bathing when one of their number screamed for help. Rushing to his aid, the others found that he was firmly gripped by four tentacles of a large octopus, the remaining tentacles anchoring the creature to a rock.

They released their comrade with difficulty, and took their revenge by eating the octopus, boiled.

BRISTOL

Remoras

A tale current among Bristol sailors in the seventeenth century told how a ship of that city had been infested with witches. This was discovered when the quartermaster went down to the hold, where to his astonishment he saw a lot of women, 'his knowne neighbours, making merry together, and taking their cups liberally: who having espied him, & threatning that he should repent their discovery, vanished suddenly out of sight'. Their ominous words were soon carried into effect. The quartermaster became mysteriously lame, and on the approach to Bristol the ship stopped dead. Although the wind was favourable and strong, the vessel remained immovable, even with help from on shore, until one man shoved it off with his shoulder. (The poet George Sandys, telling this in a note to his 1632 translation of Ovid's *Metamorphosis*, suggests that the powerful assistant may have been 'one of those whom they vulgarly call Wise-men, who doe good a bad way, and undoe the inchantments of others'.) When at last the ship came to harbour, the quartermaster laid an accusation of sorcery against the women he had seen below decks, several of whom confessed and were executed.

Sandys also mentions an alternative explanation offered when ships became mysteriously motionless, that they had been trapped by the remora, a small fish said to have great powers. The Roman writer Pliny maintained that this fish, commanding the fury of the winds, could hold a ship fast against a tempest strong enough to wrench up anchors and part cables. The emperor Caligula was supposed to have been delayed by a remora, his galley alone being detained out of his whole fleet, although it had five men to every oar. A man dived in to investigate, and found the fish attached to the rudder.

Fabulous tales of the remora's ability to stop ships in their tracks, and even to pull them backwards, remained current for a long time. A late sixteenth-century Dutch book of voyages tells the 'short and true Historie of a Fish' that lurked under a vessel's keel for a fortnight, 'drawing the ship with it against wind and weather'.

Physical descriptions of the remora vary considerably. Pliny says that Caligula's fish was like a large slug. The Dutch traveller quoted above mentions 'a great broad taile', and says that the rest of the fish was wound round the ship's beakhead, conveying a notion of something quite large and serpentine. There is a real remora, which is not very like either of these: a slender fish, between one and three foot long, whose salient feature, which gave rise to the legends, is an adhesive disc on the head, by which it can attach itself to any flat surface, so strongly that it can only be dislodged with great difficulty. It does occasionally stick to ships, in order to eat whatever food may be thrown overboard, but prefers to make its journeys with whales, sharks, or turtles, living on the larger creatures' leavings – and it is certainly incapable of steering or holding still any boat.

See also BODMIN.

BRIXHAM, DEVON

Coins to pay Neptune

In the early twentieth century, the folklorist Edward Lovett made enquiries into the survival of superstitions, and found a particularly rich vein among fishermen. Among other customs, he noted that a coin was often pushed into one of the corks used to float nets, and an old man of Brixham told him that about thirty years earlier, his skipper declared that he had had enough of the recent poor catches, and put a half-sovereign into his trawling net. 'That night they had such a big catch that it literally filled the boat, whilst the other boats did no better than usual.'

Paying the gods or spirits of the sea for their favour was a time-honoured rite, and the related practice of placing money somewhere in the boat, often under the mast before it was 'stepped' (set up), also dates back centuries. A Roman trading vessel of the second century BCE, found in the Thames in 1962, had in her mast-step a coin showing the goddess Fortuna holding a ship's rudder, and a Spanish wreck salvaged in the Orkneys was found to have a coin dated 1618 wrapped in canvas under the keel, evidently placed there as a charm. The tradition continued well into the twentieth century. Commander W. Beckett, writing of naval customs in the 1930s, records that 'Up to the present day we find that coins are still put in a ship – often under the step of the mast when she is built. The present Royal Yacht is a case in point.' Offerings like this were still made in the 1970s, when fishermen from Lowestoft in Suffolk used to throw copper coins overboard when setting out to sea, 'to buy a good catch'.

BROCKWEIR, GLOUCESTERSHIRE

The mermaid's curse

A mermaid was once caught in the tidal stretches of the river Wye, as reported by Margaret Eyre in a letter to the authors of *Sea Enchantress* (1961), a book on mermaids and other sea-spirits. Two fishermen were trawling for salmon with a net stretched between their boats, and noticed that something exceptionally large was struggling in the mesh. One of the men, excited at the catch, steered quickly towards his partner to close the mouth of the net, and carelessly pinched the creature's tail between the sides of the two boats. At this the 'fish' wriggled upright in the water, showing the upper half of a mermaid, and broke free, angrily telling the culprit that 'very few of his descendants would die in their beds'.

This was said to have happened in the early twentieth century, and was clearly remembered in Brockweir (misprinted in the book as Brockwin). A local woman said that one of the fishermen's descendants had fallen to his death while mending his roof. 'How sad,' said Miss Eyre. 'Ah, yes,' the woman replied, 'that was the mermaid's curse.'

BUDE, CORNWALL

A fake mermaid

As a young student in the 1820s, Robert Hawker, later vicar of MORWENSTOW, played an elaborate trick on his neighbours at Bude. One moonlit evening, he swam or rowed out to a rock a little way off shore, draped his head in seaweed, wrapped oilskin round his legs for a tail, and (otherwise naked) sat on the rock with a hand mirror, singing and yelling until he got attention. People ran into Bude saying that a mermaid was plainly to be seen and heard, and everyone rushed to the beach. In *The Vicar of Morwenstow* (1876), Sabine Baring-Gould reports that the performance was repeated for several nights, with people coming from all the surrounding villages to see the marvel, until at last Hawker grew tired and hoarse from his attempts at siren song. 'He therefore wound up the performance with an unmistakeable "God save the King," then plunged into the waves.'

The hoax would not have worked, of course, if people had not been predisposed to believe in a mermaid. In early nineteenth-century Cornwall it was evidently quite credible, although still remarkable, that one should appear. A good deal later in his career, Hawker was told that the father of one of his Morwenstow parishioners, Tony Cleverdon, had seen two 'merry-maids', as they were locally known. Once was at night, when old Cleverdon heard music from the sea. He crept nearer, and at last saw the creature quite plainly, swimming and crooning.

> But my father said it was very sad and solemn to hear – more like the tune of a funeral hymn than a Christmas Carol by far – but it was so sweet that it was as much as he could do to hold back from plunging into the tide after her. And he an old man of sixty-seven, with a wife and a houseful of children at home!

The second mermaid was 'the bootifullest merry-maid that eye could behold', twisting her long hair up like a girl getting ready for her sweetheart. Cleverdon's father got very near, hoping to catch hold of her, but before he could do so she caught sight of him, dived off her rock, turned a somersault in the water, 'and cast a look at my poor father, and grinned like a seal!'

CAWSAND, DEVON

A quiet baby

There are many traditional tales of smugglers outwitting the revenue. Isaac Gulliver of WIMBORNE MINSTER fooled his pursuers by faking death, and a coffin was used to convey contraband at GORLESTON-ON-SEA (East Anglia), but infants as well as corpses could be brought into play. At Cawsand, according to Sabine Baring-Gould's *Book of the West* (1899), there lived a woman known in her old age as Granny Grylls. In her youth she had often walked to the beach and back carrying a baby that was never heard to cry. One day a customs officer said to her, 'Well, Mrs Grylls, that baby of yours is very quiet.'

'Quiet her may be,' came the reply, 'but I reckon her's got a deal o' sperit in her.'

And so she had, of course, for the baby was a jar of brandy.

CHAMBERCOMBE MANOR, DEVON

The skeleton in the closet

Chambercombe Manor near Ilfracombe retains much of its original eleventh-century architecture, and is worth a visit for its beautiful gardens alone, but it also has a splendid story attached.

According to legend, a seventeenth-century owner of the house, Alexander Otway, was a 'wrecker' who deliberately lured ships on to the rocks so that he could plunder them. One night his son William, following him on his wicked errand, rescued a beautiful girl from the waves. She became William's wife, and a daughter, Kate, was soon born to the couple. When Kate was grown up, she married an Irishman and moved to Dublin, and after this things began to go wrong for her parents. Finding himself short of money, William decided to follow his father's example, and took to engineering wrecks for his own profit.

One night a dreadful storm swept the Ilfracombe coast, hurling a vessel

The magazine The Leisure Hour *first printed the tragic story of Chambercombe Manor in 1865, with this image of William Otway rescuing his future wife from the waves.*

ashore. William hurried to see what he could salvage, but what he found first was a badly injured woman cast up by the waves, her face battered beyond recognition. He carried her back to his house, where she died. William robbed her body of its jewellery and said nothing about finding her, but he later learned from some sailors who had survived the wreck that she had been none other than his own daughter on her way from Ireland to see him. Tormented by remorse, he walled up Kate's body and left the house for ever.

The tragedy came to light in the nineteenth century, when a tenant noticed that the upper storey of the house contained one more window than he could account for. Breaking open the passage wall opposite the mysterious window, he found a room with ancient furnishings, including a four-poster bed whose mouldering curtains he nervously drew aside. Upon the bed lay a skeleton, all that remained of William Otway's daughter.

It seems that a secret room was genuinely discovered at Chambercombe Manor in the 1860s or thereabouts, and the saga of the Otways first appeared in the magazine *The Leisure Hour* in 1865. One investigator, more literal-minded than most, reported that he had examined the 'room' in question, and found that it was no more than a space in the roof, without a floor and more importantly without a window, but this has not stopped the legend appearing in many later histories of Chambercombe Manor, and naturally it is said in modern retellings that Kate's ghost haunts the house.

FOWEY, CORNWALL

John Dory

An old ballad tells of the prowess of the Cornish sailor Nicholl, who fought and overcame a boastful privateer, John Dory. Dory had offered his services to the king of France, promising that he would subdue all of England, but in a bloody battle he was captured by Nicholl and killed. The song was popular in the late sixteenth century, with its encouraging message of English superiority over French aggression, and in *The Survey of Cornwall* (1602), the antiquary Richard Carew tells the episode as a piece of history. Although he probably gets most of his information from the ballad, he adds that Nicholl was a Fowey man, a detail not found in the verses, suggesting that in Carew's time there was a local tradition of Nicholl and his triumph.

Unfairly, perhaps, the villain is better remembered than the hero, having given his name to a fish, which may have been originally called *jaune dorée*, 'yellow gilt' in French, from its colour. Other derivations for its name have been proposed, some claiming that it comes from French *adorée*, 'worshipped', since according to Fletcher Bassett's *Legends and Superstitions of the Sea* (1885), it was regarded as sacred in Greece until at least the nineteenth century. Others, however, prefer an Italian etymology, suggesting a corruption of *il janitore*, 'door-keeper', a reference to St Peter as keeper of the keys of heaven. This may not be quite as far-fetched as it seems. The John Dory is indeed sometimes known as Peter's fish, or in France *le poisson de St Pierre*, from a tale that the two black blotches on its sides are the marks of St Peter's fingers, left when he took money from its mouth (Matthew 17:27). Folk explanations of marks on fish often relate to saints, or otherwise to the Devil, as in the story told of the haddock at FILEY BRIG (North-East England).

GWENVOR SANDS, CORNWALL

Tregeagle's tasks

A Cornish phrase, 'to roar like Tregeagle', comes from the legend of Jan Tregeagle, who may have been a historical figure, perhaps a seventeenth-century steward who oppressed the peasantry, or, less convincingly, a Bluebeard-like serial killer of his wives. Whatever his sins in life, he achieved lasting fame in folklore for his punishment after death with a series of impossible tasks.

First he was set to work with a cockleshell to empty Bodmin Moor's Dozmary Pool, said to be unfathomable. In *Popular Romances of the West of England* (1865), Robert Hunt comments that 'Unfortunately for its bottomless character, in a recent hot and rainless summer, this little lake became dry,' but that was none of Tregeagle's doing, for he had been chased away from Bodmin by demons. At Padstow he was then set to weave ropes of sand, a hopeless job traditionally assigned to evil spirits to keep them out of mischief, but here he made so much noise, roaring and howling in frustration as his ropes were washed away, that the local residents begged the Cornish saint Petroc to get rid of him.

Petroc arranged for Tregeagle to carry sand from Bareppa to Porthleven, but this too made him bellow, for the tide brought the sand back as fast as he could remove it. With the aid of priests and prayers he was once more removed, this time to Gwenvor Sands near Land's End, where he remains to this day. His shrieks reach around the whole of Whitesand Bay, and in the late nineteenth century, a local man reported that 'I have often heard him howling before a westerly hurricane in the still of midnight at my house in Penzance, a distance of ten miles.'

KEYNSHAM, SOMERSET

Snakestones

In the sixth century, St Keyne resolved upon a solitary life, and chose a site for her cell at what became known as Keynsham. The place, however, was infested with snakes, which at the saint's prayers were turned into rock – since when, according to John Collinson's *History of Somerset* (1791), 'the stones in that country resemble the windings of serpents'.

Like the similar but further developed legend of St Hilda of WHITBY (North-East England), this refers to a local profusion of ammonites, fossilised shells of cephalopods that are distant ancestors of the Pearly Nautilus. Their distinctive coiled form reminded the ancient Greeks and Romans of rams' horns, sacred to the god Jupiter Ammon, and they were known in Latin as *cornu Ammonis*, 'horns of Ammon', from which the English name ammonites is derived. Their shape also suggested curled-up serpents, hence the name 'snakestones', and they were used by the Greeks as a remedy for snakebite.

In Britain they were thought to relieve muscular pain, as were many other fossil remains of shellfish from the primordial seas. Belemnites, an extinct group related to cuttlefish, left behind their internal shells, like daggers with the hilts broken off, which were widely known as 'thunderbolts', supposed to have been darts thrown from heaven during storms, and prized as cures for rheumatism and sore eyes. Fossilised sharks' teeth, known as 'tongue-stones', were considered effective against cramp and rheumatism (*see* WHITSTABLE, South-East England), and the 'Devil's toenail', a fossil oyster, was used against arthritis. The logic behind this folk medicine is a form of sympathetic magic, the clenched pain of cramp and other ills being thought of as transferable to the stony amulet.

KYNANCE COVE, CORNWALL

The Witch of Fraddam

One of the most powerful sorceresses of the West Country was the Witch of Fraddam. Her arch-enemy was the lord of Pengerswick Castle, who had learned his magic in eastern lands and had thwarted her spells many times. Determined to destroy him, she raised the Devil by her incantations and promised him her soul in return for his help. The site she chose for her spells was Kynance Cove, a wild and craggy place, with one prominent rock pierced by a cavern known as the Devil's Throat from the rumbling noise it made in stormy weather.

The Devil was not sure of his power over the Enchanter of Pengerswick, who was said to have potent sorcery at his command, but he played along, telling the witch that the Enchanter's mare must drink poisoned water, so that she would throw her rider. Pengerswick himself must be drenched 'with some hell-broth, brewed in the blackest night, under the most evil aspects of the stars', and would then be under the hag's control for ever.

During the wildest storms, the witch rode over the moors and mountains collecting her deadly ingredients. At length all was ready, and one dark night she set her tub of deadly water and her crock of deadlier potion in a lane where she knew Pengerswick must pass. Soon horse and rider approached, but when the mare saw the tub she snorted and her eyes flashed fire. The Enchanter whispered in the mare's ear and she bucked, sending the tub flying. The tub knocked over the crock, the crock struck the witch's legs, and she fell into the tub, which at once assumed the shape of a coffin. Raising his voice, the Enchanter uttered words in an unknown language, and a whirlwind arose, with the Devil in its midst. The witch in her coffin rose high into the air and the crock followed them, while the Enchanter laughed: 'She is settled till the day of doom!'.

According to her legend as recorded by Robert Hunt in *Popular Romances of the West of England* (1865), the Witch of Fraddam still floats along the Cornish coast in her coffin, the crock bobbing behind, and she still works mischief, stirring the sea up with her ladle and broom till the waves are mountain high and foam flies from their crests. The Enchanter of Pengerswick, however, retains his power over her: he has only to stand on his tower and blow three blasts on his trumpet to bring her peacefully to shore.

LAND'S END, CORNWALL

The drowned land of Lyonnesse

Tales of drowned places – churches, cities, even continents – are almost universal. The oldest example, in the Sumerian *Epic of Gilgamesh*, dates from around 2000 BCE, while in the Old Testament story of Noah's Ark, a cataclysmic flood wiped out the world, and classical legend tells how Deucalion and his wife Pyrrha were the sole survivors of a deluge sent by Zeus to punish mankind. In the fourth century BCE, Plato wrote of ATLANTIS (p. 24), a fabulous empire engulfed by the waters.

Medieval tales tend to be more localised, telling of the encroachment of the sea upon coastal towns and fields, and many such tales from along the shores of Britain and Brittany have a factual basis, for here indeed the

The realm of Lyonnesse, it was said, once stretched west of Land's End, but in the distant past the sea rose to cover it.

water has eaten at the land, and not only in the distant past. One quite recent example is given in Miss M. A. Courtney's *Cornish Feasts and Folk-Lore* (1890), where she records that no trace remained above the waves of some fields east of Newlyn, Cornwall, on which cricket had been played in living memory.

A Breton legend, current in various forms since the twelfth century, tells how Ker-Is, a great city off the coast of France, was drowned when the king's daughter treacherously opened the sluices and let in the sea. Of all its inhabitants, only the king himself escaped, outriding the rushing waves on his horse. By the fifteenth century, the tale had crossed to the other side of the Channel, and it became commonly reported that a great tract of country between Land's End and the Scilly Isles had been submerged. As in the French version, it was said that one man alone had been preserved, his white steed galloping fast enough to save its master from the rising tide; here the survivor was named as Trevilian. In support of the story, the arms of the Trevilians (later Trevelyans) were cited, showing a horse's head emerging from waves – a heraldic motif that appears on other coats of arms, and is less fancifully explained as symbolising the right of certain families to claim salvage from a wreck if they could ride on horseback into the water and touch the goods with their lance (*see* ISLE OF SHEPPEY, South-East England).

In the late sixteenth century, the undersea territory west of Cornwall became identified as 'Lyonnesse', home of the knight Tristan who fell fatally in love with his uncle's wife Isolde or Iseult, a tale linked to the Arthurian cycle. By the nineteenth century, detailed narratives were in circulation locating King Arthur's final battle with the rebel army of Mordred off Land's End, and concluding with a devastating tidal wave. An account in H. J. Whitfield's *Scilly and Its Legends* (1852) tells how Arthur's followers were fleeing from Mordred's troops when they saw between themselves and the advancing forces the gigantic ghost of Merlin. Mordred himself sensed an oppressive change in the atmosphere, and saw the phantom wizard raise his arm, at which the earth began to quake and the sea to rise.

> At evening there was nought from what was then first termed the Land's-end, to St Martin's head, but a howling and boiling wilderness of waves, bearing here and there upon its bosom a fragment from the perished world beneath or a corse tossed upon the billows, over which sea birds wheeled and screamed.

See also CARDIGAN BAY (Wales).

LIHOU, CHANNEL ISLANDS

Fishermen salute the rocks

On the tiny island of Lihou there once stood a chapel dedicated to the Virgin Mary. By the seventeenth century there was little of it left except the steeple, which continued to be saluted by sailors lowering their topsail as they passed.

This was explained as a religious observance in honour of the Virgin, but the custom may be more pagan than that. In *Guernsey Folk Lore* (1903), Edgar MacCulloch notes that there are other rocks around the Channel Islands 'which the fishermen are in the habit of saluting without being able to give any reason why they do so'. Jersey fishermen, for instance, used to row around a rock known as Le Cheval Guillaume, in St John's Bay, every Midsummer Eve, a practice noted by several twentieth-century writers without any further explanation.

Another rock honoured by the Guernsey mariners was Le Petit Bon-Homme Andriou ('Andriou the little fellow') in Moulin Huet Bay, a figure which looks rather like an old man wrapped in a monk's cloak and hood. MacCulloch records that fishermen passing the point would take off their hats to show respect, and would insist on strangers doing likewise.

> Formerly it was not unusual with them, before setting sail, to offer a biscuit or a libation of wine or cider to 'Le Bon Homme,' and, if an old garment past use chanced to be in the boat, this was also cast into the sea.

A tale current in the early twentieth century said that 'Andriou' had been turned to stone while trying to find hidden treasure among the rocks, while another told that he had been the last of the Druids, who had retreated to a cave when the islanders converted to Christianity. One day, seeing a ship in danger out at sea, he prayed in vain to his own deities to save it, and then in desperation sought help from the Christian God, vowing that if this ship were spared he would himself become a Christian and dedicate a chapel to the Virgin Mary. The waves became calm, the Druid was baptised, and the chapel he built, according to one version, was the one on Lihou.

The mingling or conflict of Christian and pre-Christian tradition in stories like this may have inspired Victor Hugo to invent a satirical legend in his Guernsey novel *Toilers of the Sea* (1866). He writes that old sailors often saw someone sat reading on Ortach, a rock near Les Casquets, and used to kneel as they passed the place, believing it to be St Maclou (Magloire). 'It has been discovered,' he goes on, 'and is now well known that what dwells on the rock of Ortach is not a saint, but a devil. This devil, called Jochmus, for several ages maliciously passed himself off for Saint Maclou.' Fletcher Bassett's survey of maritime folklore, *Legends and Superstitions of the Sea* (1885), records that Guernsey sailors once believed that St Maclou lived on Ortach, but his only source is Hugo. It seems likely that the novelist had heard the traditions about Lihou and refashioned them into his ambiguous saint/demon, his aim being to deflate one piece of superstition with another.

LULWORTH COVE, DORSET

Napoleon's visit

During the Napoleonic wars (1803–15), the British went in fear of invasion. The Dorset coast was thought to be particularly vulnerable, and beacons on hills and clifftops were ready to be lit as warnings in case of attack. Bonaparte's army was camped just a few miles away on the other side of the Channel, and even at the beginning of the twentieth century, locals could remember being scared as children by threats that 'Boney' would come and carry them away if they were naughty.

A circumstantial tale relates that Napoleon actually visited Lulworth Cove one night to see whether it was a suitable place to land his troops.

Having surveyed the narrow bay and rocky approach, he folded his map and announced regretfully to his companion that the plan was impossible.

> The whole incident was watched and the conversation overheard by a local farmer's wife who had learnt French as a young girl so that she might help her father (a china merchant) in his business.

This account comes from a collection of Dorset folklore compiled in the 1930s using reports from local Women's Institutes. It was backed up with the assertion that the lady who saw Napoleon 'was born in 1784, and lived to be 104, and was alive when the West Lulworth contributor first heard the story'.

The report seems, however, to have begun as literature. According to the memoirs of his second wife, in around 1882 Thomas Hardy was asked to write 'something of the nature of a fireside yarn', and invented Napoleon's trip, framing it as an old man's memory of his distant youth. The story was published in *Life's Little Ironies* (1894) as 'A Tradition of Eighteen Hundred and Four' and Hardy was amazed when people commented, 'I see you have made use of that well-known tradition of Napoleon's landing.' He wondered whether, unknown to him, such a report had already been in circulation, and whether indeed Napoleon *had* come to Lulworth, but when he made enquiries he was informed definitely 'that such a visit would have been fatuous, and wellnigh impossible' and 'that there had never existed any such improbable tradition'.

It certainly exists now, the 'tradition' sometimes reported as having inspired Hardy's fiction, sometimes the other way about.

LUNDY, DEVON

Invasion by the French

A ship flying the Dutch flag once anchored off Lundy, and the crew sent ashore for milk, saying that their captain was ill. This was in the reign of William and Mary (1689–94), when the Dutch were Britain's allies, so the islanders were happy to supply the sailors. After a few days, news came that the captain had died, and his men asked permission to bury him in the graveyard. This was granted, and the coffin was brought to the church, where the sailors asked to be left alone while they performed their service. Unsuspicious, the locals waited outside, until the doors were suddenly flung open. The friendly 'Dutch' sailors proved to be hostile French privateers, armed with cutlasses and muskets which they had hidden in the coffin. They behaved abominably, ill-treating the livestock, throwing the island's cannon off the cliffs, and stripping the inhabitants of all their possessions and even of their clothes, and after having devastated Lundy they departed as abruptly as they had arrived.

Folklore, or perhaps the predatory Frenchmen, may have borrowed the ruse from the occasion in the 1550s when the Flemish used a very similar ploy to capture Sark (*see* LE PORT DU CREUX).

LYDNEY, GLOUCESTERSHIRE

The sea-god Nodens

Near Lydney, about a mile from the Severn shore, deep glens skirt a ridge of the Forest of Dean that was known in the eighteenth century as the Dwarf's Hill, while some ruined walls there were called the Dwarf's Chapel. Dwarves, as such, feature quite rarely in British folklore, and here they can be considered as synonymous with the fairies or 'little people'. The ruins were obviously deemed to have mystic significance, and identification of the site as supernatural territory may go back a long way. When the site began to be excavated in the nineteenth century, remains were found of a temple complex dating back to the third or fourth century CE, and memories of its sacred nature probably contributed to a continuing feeling that this was enchanted ground.

The archaeologists who first uncovered the building in 1805 found a remarkable mosaic pavement dedicated to the god Nodens, two sea-monsters at its centre, rather like plesiosaurs, with flippers and elongated serpentine necks, and around them a frieze of fish, possibly salmon. Another find was a bronze plaque showing a figure crowned with sun-like rays and brandishing a sceptre, his chariot drawn over the waves by four horses, and flanked by winged spirits and reclining tritons with fish tails, one holding an anchor and a conch shell, the other with two oars or paddles, while a smaller panel shows a fisherman hooking a salmon.

All this sea imagery indicates that Nodens – or Nudons, as his name appears on other inscriptions in the Lydney temple – was a marine deity, although he clearly had other aspects too, probably a role as sun-god as implied by the chariot-rider with his rayed crown, and certainly a healing function. The Lydney sanctuary had baths and a big hall for visitors who brought votive offerings and sought cures from the god.

The name Nodens is etymologically linked with that of the ancient Irish hero Nuadhu or Nechtan, husband of the goddess Bóinn or Bóand (*see* ROCKABILL, Southern Eire), who gave her name to the river Boyne. The root of both names may be a Germanic word meaning 'catch', reinforcing the suggestion that Nodens was a fishermen's deity. Much of this, however, is necessarily guesswork. Hard evidence for early British religion is very scarce, and the Lydney complex, one of the most important finds in this context, was carelessly treated by its first excavators, so that the sea mosaic no longer exists. Other artefacts from the site, however, can be seen at the Lydney Park Museum.

LYME REGIS, DORSET

The Bermudas and The Tempest

In 1609, the retired admiral Sir George Somers, mayor of Lyme Regis, was summoned back to sea to command a fleet taking supplies to the new colony of Virginia. A hurricane overtook his flotilla in the Atlantic, driving Somers's flagship, the *Sea Venture*, far from the other vessels, and after four days it was wrecked on a reef off the Bermuda islands. The crew were able to salvage most of their stores and equipment, and built two small boats in which all but two men continued their journey, arriving in Jamestown, Virginia, in May 1610.

Two accounts of the wreck and the time on the island, written by eye-witnesses Silvester Jourdain and William Strachey, were soon in circulation, and became major sources for Shakespeare's *Tempest* (1611). Strachey's in particular, a vivid piece of description, can be seen to have inspired many details in the play. 'During all this time,' he writes of the storm (in a version with revised spelling and punctuation), 'the heavens looked so black upon us that . . . nor a star by night nor sunbeam by day was to be seen.'

> Only upon the Thursday night Sir George Somers, being upon the watch, had an apparition of a little, round light, like a faint star, trembling and streaming along with a sparkling blaze, half the height upon the main mast and shooting sometimes from shroud to shroud.

For several hours it continued its dance, 'running sometimes along the main yard to the very end and then returning', and towards morning it disappeared. Compare the spirit Ariel's report to his master Prospero in *The Tempest*, Act I scene 2:

> I boarded the King's ship; now on the beak,
> Now in the waist, the deck, in every cabin,
> I flamed amazement. Sometime I'd divide
> And burn in many places – on the topmast,
> The yards and bowsprit would I flame distinctly,
> Then meet and join.

Atlantis

The lost continent of Atlantis, swallowed by the waves as punishment for the sins of its citizens, is one of our most enduring romantic symbols, and over the thousands of years since its legend was first told, many people have tried to demonstrate that its existence is a historical truth. In the fifth century BCE, Plato described an island in the Atlantic 'larger than Libya and Asia combined', where a great civilisation once thrived, but later became degenerate, awaking the gods' anger, and was destroyed by earthquake and flood. Plato attributed the tale to the ancient Egyptians, but he may simply have invented it to illustrate his philosophy. On the other hand, it has been argued that such a land did once exist, and that rumours of its vanishing had genuinely survived into classical Greek times.

One theory popular in the sixteenth century was that Atlantis could be equated with America. According to this hypothesis, the

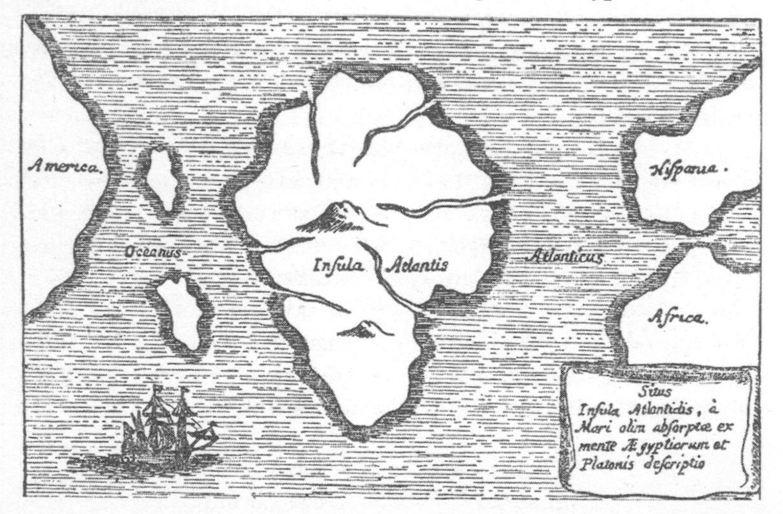

The seventeenth-century scholar Athanasius Kircher marked Atlantis midway between Africa and America. The Latin caption translates roughly as 'The site of the Island of Atlantis, swallowed by the Sea, from the memory of the Egyptians and Plato's description'.

island had not been drowned, but archaic knowledge of its whereabouts had been forgotten, so that it was remembered only as a dim and fabulous realm. Another idea was that the Mid-Atlantic Ridge, between America and Africa, was the remains of sunk Atlantis, as shown on a seventeenth-century map drawn by Athanasius Kircher, and as maintained over two hundred years later by Ignatius Donnelly, whose book *The Antediluvian World* (1882) was a prime example of Atlantean myth-making, proposing the drowned continent as an explanation for everything from ancient legends of the Deluge to the modern distribution of plant and animal life.

The discovery in 1968 of the 'Bimini Road', an underwater stone structure about half a mile long off the Bahamas, led to excited claims that this was a remnant of a prehistoric city, but it was later proved to be a natural feature. More persuasive is a suggestion that Atlantis may have represented the Greek island of Thera (Santorini), where a volcano erupted with devastating effect in the seventeenth century BCE, burying the flourishing Minoan town of Akrotiri. This eruption, and the tidal wave that followed, might well have provoked legends of divine wrath.

Whatever its origins, the tale of Atlantis has resonated down the centuries. The number of books written about it extends into the thousands, taking in the scientific, the poetic, and the lunatic, and in folklore it is the prototype for tales of Lyonnesse and Ker-Is (*see* LAND'S END, South-West England & Channel Islands), Cantre'r Gwaelod and Tyno Helig (*see* CARDIGAN BAY and CONWY BAY, both Wales), and all drowned lands. In 'Fragments' (1921), John Masefield imagines how the 'green and greedy seas' now roll above the courts of old Atlantis, but as the tide falls, its gold yet shines:

> The Atlanteans have not died,
> Immortal things still give us dream.

What Ariel is impersonating, and what Strachey describes, is the corposant, or St Elmo's fire, a phenomenon well known to sailors and the subject of many superstitions (*see* CROMER, East Anglia).

On landing, the sailors found that they were on 'the dangerous and dreaded island, or rather islands, of the Bermudas', known as 'the Devil's Islands', yet they found the place fertile and temperate, and particularly enjoyed a drink they made from steeping berries in water – like the 'Water with berries in't' that Prospero gives Caliban (Act I scene 2).

When, after several months, the improvised vessels finally departed for the colonies, the two men left behind were Bermuda's first settlers, and the islands were claimed as an English possession. For a while in the seventeenth century they were known as the Somers Islands in honour of Sir George, who died there in 1610 on his way back from Virginia, 'of a surfeit in eating of a pig', but the original name (which commemorates their sixteenth-century discoverer, a Spanish captain called Bermudez) was later restored.

In the nineteenth century, Somers's descendants preserved a lodestone, mounted in iron, which had been one of the admiral's most precious possessions. Before going to sea, he had always touched his compass needle to that particular magnet, presumably for luck. Magnetite or magnetic iron ore, commonly called lodestone, was used in the navigation of ships from at least the twelfth century, and was considered to have magical powers. It was sometimes said that it was alive, or at least capable of speech. In *Credulities Past and Present* (1880), William Jones reports an ancient belief that a magnet washed in spring water would be seen to breathe, and would reply to questions in 'a voice like that of a sucking child'. The stone was also considered to cure gout, if worn next to the skin.

MINEHEAD, SOMERSET

Mrs Leakey's whistling ghost

The story of Mrs Leakey, the whistling ghost of Minehead, appears as a note to Sir Walter Scott's poem *Rokeby* (1813). Before her death, Mrs Leakey knew or suspected that she would return in a less pleasant guise:

> This old gentlewoman was of a social disposition, and so acceptable to her friends, that they used to say to her and to each other, it were pity such an excellent good-natured old lady should die; to which she was wont to reply, that whatever pleasure they might find in her company just now, they would not greatly like to see or converse with her after death, which nevertheless she was apt to think might happen.

She was right. Soon after she died she began her haunting, appearing all over Minehead and particularly down at the quay, where she used to stand and call for a boat. Her son was a merchant with several ships which traded between Minehead and Waterford in Ireland, and as soon as one of his vessels approached, Mrs Leakey's ghost would give a piercing whistle that invariably called up a tempest. When she had wrecked all her son's ships, she turned on his family, tormenting his wife and even strangling her own grandchild.

Scott used the story to illustrate the common piece of sailors' lore forbidding whistling at sea, since it would summon a wind, and usually a storm (*see* FISHGUARD, Wales). His source was John Dunton's *Athenianism* (1710), where the legend appears with the title 'The Apparition-Evidence', and is introduced as an 'Original Manuscript' sent by 'a Citizen of great Integrity', although it is quite possible that Dunton invented most of his account. An inquiry had been held in 1637 into reported appearances of Mrs Leakey at Minehead, but the official record tells a commonplace tale of poltergeist activity, reaching the conclusion 'that there was never any such Apparition at all'. Dunton (or his anonymous informant) made a much more sensational tale of it, including the old lady's warnings that she might return after death, the ghost's pranks on the quayside, and the murder of her granddaughter. Finally, in this version, the ghost accuses Dr John Atherton, Bishop of Waterford, of getting her niece pregnant, but when invited to repent of his sin, the bishop's only reply is 'That if he were born to be hang'd, he should not be drown'd' (*see* KINGS CAN'T DROWN, p. 212).

The slander of Atherton must be the point of Dunton's anti-Catholic fabrication, but it was the whistling wraith of Mrs Leakey that took Scott's fancy and entered popular folklore. William Jones wrote in *Credulities Past and Present* (1880) that the sailors of Minehead continued to believe that in storms they could hear 'the whistle of the horrible old lady', and the Somerset folklorist Ruth Tongue reported in 1965 that local tradition was still lively about Mrs Leakey.

MORWENSTOW, CORNWALL

Reverend Hawker and the wreckers

Robert Hawker was one of nineteenth-century Cornwall's more notable eccentrics. As a twenty-year-old student, he married his godmother, who was more than twice his age, and although his reasons were partly financial (she had enough money to let him continue his studies at Oxford), the union seems to have been a perfectly happy one. Hawker liked to provoke public opinion, his youthful pranks including an impersonation of a mermaid at BUDE, and after he became vicar of Morwenstow in 1834, he used to go on his parochial rounds accompanied by a pet black Berkshire pig named Gyp. When he came to church, according to his biographer Sabine Baring-Gould, he was generally followed by ten or more cats, which used to sport about in the chancel during the service. As kind to humans as to animals, on several occasions he helped rescue sailors from shipwrecks on the rocky Cornish coast, and let one destitute survivor stay for weeks at the vicarage, eventually sending him on his way with a generous supply of money. He was also a scholar with a taste for folklore and tradition, and may have invented what became a famous legend about BOSCASTLE.

In his memoirs, he describes his parishioners at Morwenstow as a mixed bag of 'smugglers, wreckers, and dissenters of various hue'. His own servant, Tristram or Trim Pentire, had himself been a smuggler, and initiated his master into various secrets of the trade, including 'The Gauger's Pocket', a 'gauger' being a customs officer or exciseman, and the 'pocket' in question a crevice in a large rock near the sea. When the smugglers wanted the coast to themselves, they put money in the hole and then murmured to the local officer, 'Sir, your pocket is unbuttoned.' If they got the reply 'Ay! ay! but never mind, my man, my money's safe enough,' they would know he had taken the hint and the cash, and that he would keep well out of the way.

Local feeling was strongly on the side of the smugglers, and the customs men were regarded as pestilent creatures. Trim mentioned to Hawker a belief that no grass would grow on the grave of a man unjustly hanged, and as an example pointed to a bare grave in Morwenstow churchyard. Hawker asked what the dead man had done to deserve execution. 'Done? Nothing whatsoever but killed the exciseman!'

It was often said, as for instance at PORTLEMOUTH, that clerics took an active part in smuggling or wrecking, and Hawker tells an entertaining though probably apocryphal tale of a nearby parish where a stranger saw some smuggled goods being landed on the beach. Laden boats were passing between ship and shore, and a crowd surrounded a keg of cognac, drinking from whatever they could find. One man was using his shoe for a cup. Others were fighting and cursing over the spoils, by the light of a lantern held aloft by a man on the outskirts of the scene. Horrified, the stranger cried out for a magistrate or justice of the peace, and was told phlegmatically that there was none within eight miles. 'Is there no clergyman hereabout?' the newcomer persisted.

'Aye! to be sure there is.'

'Where is he?' asked the indignant stranger.

'That's he, sir,' – pointing to the man holding the light – 'yonder, with the lanthorn.'

MOUNT'S BAY, CORNWALL

Sarah Polgrain and Yorkshire Jack

Fishermen out in Mount's Bay after dark, it is said, used to be frightened by the ringing of ghostly wedding bells and a cry of 'I will, I will!' These sounds, echoing down the years, signalled the unholy union of Sarah Polgrain and Yorkshire Jack.

Sarah was a farmer's wife who lived in Ludgvan in the early nineteenth century, and had an adulterous affair with Jack, a handsome horse dealer. Her husband died, the cause announced as cholera, but there was strong suspicion that the death had not been natural. Sure enough, when he was dug up, Mr Polgrain's corpse was found to contain enough arsenic to have poisoned three men. Sarah was convicted of murder, and sentenced to be hanged. Her lover accompanied her to the scaffold, where he kissed her, and they were seen to whisper to each other. Those standing near heard Sarah say to Jack, 'You will?' and his reply, 'I will!' Soon after she was dispatched, ghastly rumours began to circulate that her ghost had been seen, the black marks of the rope plain on her swollen neck, dressed in her shroud and digging at her husband's grave.

Yorkshire Jack, meanwhile, had joined the merchant service and gone to sea, but his bold, dashing air had deserted him. He crept about, always glancing nervously over his shoulder, and at last confided his terrible secret, that he had vowed to marry Sarah, 'living or dead', after a certain number of years, and now he was in constant dread that she would claim him. Disasters followed wherever he went, until his shipmates considered him a 'Jonah' who brought bad luck (*see* JONAH AND THE WHALE, p. 164), and would have been glad to see the back of him. Soon enough their wish was granted. Returning from a voyage to the Mediterranean, the ship was pursued by a frightful storm, in which Sarah and the Devil were howling, arms outstretched for their prey. An immense wave swept Jack from the deck, and the tempest passed onward, now with a third figure riding in a black cloud amid the thunder and lightning.

The nineteenth-century folklorist Robert Hunt, one of the first to record this legend, found it in common currency among the local people. An old lady from Ludgvan, when he commented on the bad weather, replied, 'It's all owing to Sarah Polgrain,' evidently considering her as a sort of elemental or nature spirit.

The Flying Dutchman

Tales of a phantom ship are told among sailors throughout the world, a vessel that sails forever across the wide ocean with her crew of spectres or damned spirits, and brings destruction on anyone who sees her. Her captain goes by many names, and many crimes have been laid at his door to account for the curse on him and his vessel. Sometimes he is Dahul, an Arab pirate, and sometimes he is identified as Bernard Fokke, a seventeenth-century Dutch sailor said to have been taken by the Devil. Another legend calls him Falkenberg, and tells how he murdered his brother and his brother's bride out of jealousy, since when a demon and an angel have cast dice eternally to win his soul.

Most often, however, he is called Vanderdecken, and his ship is famed and feared as the *Flying Dutchman* (because the Netherlands produced so many sailors, English seamen called almost any north European a Dutchman). The story goes that Vanderdecken sailed round the Cape of Good Hope with the wind against him, his crew imploring him to put in to land. He laughed at their fears and the weather, and even when God appeared to him, he cursed and fired his pistol at the vision. For this most terrible of sins, he was condemned to travel on and on, never coming to land, his ship appearing to doomed sailors as an omen of disaster.

This maritime slant on the Wandering Jew (who would not allow Jesus to rest when carrying the cross, and can therefore take no rest himself until Judgement Day) was used by Captain Marryat in his novel *The Phantom Ship* (1839), and by Richard Wagner, whose 1843 opera allows his anti-hero to find salvation through love. These versions helped popularise the name 'Flying Dutchman', which in the nineteenth century became the generic title for all ghost ships.

A typical account comes from a journal kept in 1881 by the young princes Albert Victor and George (later George V), who were serving as midshipmen on board the *Bacchante*:

At 4 A.M. the *Flying Dutchman* crossed our bows. A strange red light as of a phantom ship all aglow, in the midst of which light the masts, spars, and sails of a brig 200 yards distant stood out in strong relief as she came up on the port bow.

A spectral ship looms on the horizon. She is sailing against the wind, her sails bellying out in the opposite direction from that of the small craft whose bows she crosses.

The red light was observed by the crews of two other ships, and the vessel was seen by thirteen people on the *Bacchante*, but it then completely disappeared, in calm and clear conditions. At 10.45 the same morning, the sailor who had first reported the strange craft 'fell from the foretopmast crosstrees . . . and was smashed to atoms', while at the next port they came to, the admiral died.

The princes seem quite calm about the whole thing, mentioning that everyone felt sad at the sailor's death, then going straight on to an account of their algebra exam.

MUDEFORD, DORSET

Blessing the sea

Rogation Sunday (from Latin *rogere*, 'to ask' or 'to supplicate'), five weeks after Easter, is traditionally the time when God is asked to bless the land, and in many rural areas of Britain the boundary of the parish is visited on this day, a ceremony called 'beating the bounds'. At Mudeford, Dorset, a similar rite is practised, but because the parish is on the coast, it is the sea rather than the land that is blessed.

The participants gather at All Saints church and then walk in procession, accompanied by the Salvation Army band, down to the pier, where the congregation joins in prayers for the fishermen and the Royal National Lifeboat Institute. The vicar is then rowed out into the Run, a stretch of rough water where the cross-currents are sometimes so boisterous that a lifeboat rather than an ordinary boat is used. She (or he, but since 2006 All Saints has had a woman vicar) then beseeches God to bless the sea, and drops a cross into the water. In good weather, up to two hundred people take part in the service.

PADSTOW, CORNWALL

The mermaid's revenge

In about 1610, the historian John Norden described Padstow as 'the beste haven in the north parte of the Shyre, being capable of manie Ships', although he added that a skilful pilot was needed, since the harbour was rocky to the west and barred with sand on the east. By the early nineteenth century the situation had drastically deteriorated. According to Davies Gilbert's *Parochial History of Cornwall* (1838), 'the harbour is so surrounded by rocks and obstructed by sand, that vessels even of a small size are unable to find shelter there when the wind blows on shore.'

The rocks had not changed since the seventeenth century, but the sandbank, known as Doom Bar, had built up considerably, and a legend told the reason why. Long ago, the port had sheltered many a large vessel, and had been under the care of a 'merry-maid', as mermaids were known in Cornwall. One day, however, a man with a gun had taken a shot at her. She had dived for a moment, then risen and raised her right arm with a vow that from that day the harbour should be desolate, a curse fulfilled when sand blocked the entrance.

See also GWENVOR SANDS.

PATERNOSTERS, JERSEY, CHANNEL ISLANDS

Laments of the Sea

The noises made by the wind along the coast, sounding like wailing women or children, were known in Jersey as *Les Cris d'la mé* ('The Laments of the Sea'). Understood as warnings of a coming storm, they were believed to be the voices of the drowned, and local stories associated the sounds with particular tragedies.

Around Sorel, the howls were thought to come from the Pierres de Lecq, a dangerous reef which came to be known as the Paternosters from the prayers fishermen said as they passed it ('*Pater Noster*' being the opening words in Latin of the Lord's Prayer, 'Our Father'). Some said that the crying souls were those of five children drowned in 1565 during the settlement of Sark, when the colonists' boat was wrecked on the rock.

On the west coast, the wind howling through the rocks called Les Tombelènes, near Bouley Bay, was explained by a detailed legend first recorded in 1880 by a French writer, Carolus Pipon. His tale is dated in the mid fifteenth century, and tells of a young man, Raulin de l'Ecluse, who was seized one night by the Moutonniers (French sheep stealers, also referred to as 'cavalry' since they were always on horseback). Raulin was engaged to Jeanne du Jourdain, whose dog Fidèle alerted the girl, and she followed him to La Creux Bouanne ('The Dark Cave'), where she saw the Moutonniers in the act of hanging Raulin. She seized a knife and cut her lover down, but the brigand chief stabbed the young man and then, inflamed with lust, threw himself upon Jeanne. She thrust her knife into her assailant, broke free, and ran to the top of L'Islet, at the end of a chain of rocks stretching into the sea. As the Moutonniers followed, they heard unearthly screams, and a flash of lightning revealed the girl with arms upraised, her long hair streaming in the wind, before a huge wave arose from Vicart Point and swept her away. Her body was washed up a few days later, and buried with that of Raulin, but her shrieks were heard whenever a storm passed over the rocks, and were known as *Les Cris des Tombelènes*.

LA PERELLE BAY, GUERNSEY, CHANNEL ISLANDS

Sailing on a bone

An animal's bladebone or scapula, the flat bone at the back of the shoulder, was said to have magical qualities. In the hands of a seer it could give information of future or distant events, but a witch could use it as a raft to cross the sea. A Guernsey story tells of a man who went out to La Perelle

Bay one morning and saw a ship approaching. While it was a long way off it looked very large, but as it came nearer it appeared to grow *smaller*, until when it reached the shore it looked like a toy. A tiny man stepped to land, and the islander then saw that the strange ship was nothing more than a sheep's bladebone tangled in seaweed.

The Guernseyman questioned the small visitor, who did not speak much of his language and said only that he was travelling. He then settled down in the island and built himself a house which the locals christened 'Le P'tit Casquet' because it used to show a light late at night and therefore looked like a smaller edition of the nearby lighthouse on Les Casquets. Descendants of the stranger were said still to live in the area in the late nineteenth century, and were known as skilled smiths, a trade which often carries a suggestion of the supernatural. Metalworkers were anciently considered to practise mysterious and magical rites, a belief that may go back to prehistoric times when the making of iron tools was a newly discovered art.

This is not the only case of real people said to have magical origins - in Ireland and Scotland, several families were supposed to be descended from seals or mermaids (*see* for example NORTH RONALDSAY, Scottish Highlands & Islands) - but here the tale seems to have mingled with legends of saints or sacred objects sailing the sea on rocks (*see* AARDMORE, Southern Eire). There is also a factual basis for the idea that bones could be used to cross water, since in the Middle Ages, people used to tie animal bones to their feet as ice skates.

PLYMOUTH, DEVON

Sir Francis Drake the wizard

Not many towns have a magician for their MP. Sir Francis Drake (c. 1540–96), admiral, explorer, and scourge of Spain, became Plymouth's representative in 1593, and it was said that he supplied the citizens with water by bewitching a spring that followed him from Dartmoor. In fact he had obtained an Act of Parliament authorising him to divert the stream through artificial channels, but popular opinion magnified his deeds into supernatural exploits.

His defeat of the Armada in 1588 was the triumph that made him a legend. Popular lore held that he had sat on Plymouth Hoe whittling a stick and throwing down the splinters, 'and all the chips that fell into the sea, they did turn into ships, to go fight the Spanish Armada.'

While a hero to the English, in Spain he became *El Draco*, 'The Dragon', a sorcerer in league with the Devil, and Spanish ill-feeling probably

England's victory over the Spanish Armada was a defining moment for the country and for the admiral Sir Francis Drake. This engraving of the battle scene was taken from a tapestry that once hung in the House of Lords, but was destroyed in a fire.

accounts for the tradition that he drowned a cabin boy. While in the Antipodes, according to the legend, he asked the lad what was directly opposite on the other side of the world. Drake's house, came the reply, which the Admiral knew by his own occult powers to be correct. Supposing this might have been a fluke, he waited a while and then asked again. London Bridge, said the youth, getting it right again, at which Drake exclaimed, 'Hast thou, too, a devil? If I let thee live, there will be one a greater man than I am in the world.' With that, he threw the boy into the sea.

In later centuries, it was said that Drake was reincarnated, most notably as Nelson (*see* SCAPA FLOW, Scottish Highlands & Islands), and also as Captain Frederick Walker, who led an anti-submarine escort group in the Second World War. The idea that Drake's spirit survived can be found as early as 1596, in a long poem by Charles Fitz-Geffrey that laments the hero's death. Its preface begins, 'Once dead and twice alive, thrice worthie Drake', and one of Fitz-Geffrey's stanzas runs:

> He who alive to them a dragon was,
> Shall be a dragon unto them again;
> For with his death his terror shall not pass,
> But still amid the air he shall remain
> A dreadful meteor in the eye of Spain;
> And, as a fiery dragon, shall portend
> England's success, and Spain's disasterous end.

PORLOCK, SOMERSET

Soul-birds

Collecting Somerset folklore in the mid twentieth century, Ruth Tongue was told around Porlock and Lynmouth that it was dangerous to feed a seagull and particularly to look it in the eyes: 'If you do, one day when you are clinging to a wreck, or perhaps only swimming, it will find you and peck out your eyes, and leave you to drown.'

More common is the idea that seagulls are soul-birds, embodying the spirits of the dead. On land, birds most often said to house souls are crows, ravens and choughs, and at sea, they include storm petrels, traditional harbingers of tempest also known as MOTHER CAREY'S CHICKENS (p. 264). It was forbidden for a seaman to shoot a storm petrel, since in doing so, it was believed, they would be murdering one of their former comrades, an act that would lead to the death of another sailor who would become a petrel in his turn. Superstitions prohibiting the killing of seabirds may have been influenced by Coleridge's *Ancient Mariner* (*see* WATCHET), although the poem contains no suggestion that the albatross has a human soul.

A storm petrel swoops low between the crests of two waves. The birds were thought to embody the souls of dead sailors, and to kill one of them was often said to bring bad fortune.

Though writers on sailors' superstitions generally prefer the soul-bird explanation of the ban on killing gulls, among fishermen there was good practical reason for it. When flocks of birds were active on the surface of the water, that was a sign that plenty of fish were below, showing where it was best to cast the nets. An old Manxman who first went out with the herring boats in the early nineteenth century remembered sixty years later that 'gulls were our best friends', and so it was law that they must not be shot.

Thus the two ideas - that it was unlucky to kill gulls, and that they embodied the souls of the dead - were not inseparable. The naturalist Peter Emerson found belief in reincarnation as gulls active in the nineteenth century, but without the taboo on killing the bird. He was astonished to learn that many old fishermen on the east coast believed that they would turn into gulls when they died, and asked one man if he did not dislike their being shot on that account.

> He replied philosophically, 'No! they hev been dead oncet, they hev been on earth oncet, and we hev got quite enough old men now.'

From Emerson's account, it is clear that such transmigration was not limited to fishermen and sailors, nor perhaps to the drowned. He asked his fisherman what became of the children, and was told that they were kittiwakes not gulls, being less 'artful'. 'And the women?' Emerson asked.

> 'The wives,' he replied, 'don't come back no more, they hev seen trouble enough; but the old women torturise the young 'uns.'

Another vessel for the soul was the gannet. Emerson gives a story about a man who went out in a North Sea fishing boat seeking his brother, recently drowned, among the seabirds. Before starting out on his last voyage, the brother had said to him:

> 'Jack,' he say, 'when I'll die, I'll be an owd gannet, and if I heave round you'll heave me a herrin', won't yow?' '*George*,' I say, 'how shall I know yow along with the other gannets?' and he say, 'I'll hev a pair o' black arm-sleeves, so yow'll know me.'

LE PORT DU CREUX, SARK, CHANNEL ISLANDS

The taking of Sark

A small chapel above Le Port du Creux was long ago the scene of a deadly stratagem. Sark at the time was infested by pirates, whom the English were determined to expel, and one day a Sussex ship anchored off the island, sending a message that their captain had died, and asking permission to bury him in consecrated ground. This was granted, on condition that he was accompanied only by unarmed men. The crew came defenceless, as requested – but the coffin they carried was full of weapons. Once they had set down their burden in the chapel, they unloaded the arms and sallied forth against the Sark pirates, killing many of them, and driving the rest from the island.

In *Recollections and Legends of Serk* (c. 1880), Louisa Lane cites a Latin manuscript as authority for this tale, adding the local tradition that an old woman, who was busy baking at the time, heard the sounds of slaughter and ran, carrying a loaf of bread, to the western side of the Coupée, a high, narrow neck of rock connecting the main island to Little Sark. There she hid in a cave, subsisting on her loaf, until she was rescued by a Guernsey fishing boat.

Lane dates the episode to the reign of Edward III (1327–77). She or her source may have conflated two bits of history, since although it seems that Sark was overrun by pirates in the fourteenth century, it was not until the sixteenth, according to other writers, that the coffin trick was carried out, and then it was by Flemish sailors displacing the island's French occupiers. The tale may, however, be legendary, since a similar ruse is attributed to earlier adventurers, including the Viking Harold Hardráda, and indeed to later ones at LUNDY. Nevertheless, the story is such a good one that it has passed into folk history. The old woman with her loaf of bread may be a tradition adapted from the Guernsey tale of VAZON BAY's invasion by the men of Aragon.

PORTLAND, DORSET

Magical beasts

What is known as the Isle of Portland is not actually quite an island, but until relatively recently the inhabitants thought of themselves as distinct from the mainland English, whom they regarded as foreigners. In the early nineteenth century, they were most reluctant to allow anyone to buy land there who was not Portland-born, and on the rare occasions this was done, they had as little as possible to do with the newcomers. It was said that they believed themselves to be descendants of the Phoenicians, an ancient maritime race that travelled widely from the second millennium BCE onward. Tradition states that they came from their Middle Eastern homeland as far as south-west England, where they traded for tin, and were evidently thought to have left settlers, the distant ancestors of the Portland people.

Whatever their racial origins, the Portlanders shared superstitions with many other coastal dwellers. A common apparition was a black hound known as the Roy or Row Dog – 'row' rhyming with 'cow', and meaning rough or shaggy, making this animal kin to the phantom Shuck (*see* OVERSTRAND, East Anglia). It was described as being 'as high as a man, with large fiery eyes as big as saucers', and was often seen at Cave Hole near Portland Bill. In wild weather, when waves sometimes spout up through the cavern, 'you have to stand on a stone, out of his reach, or he may come up out of the hole and seize you and drag you under the water,' said a local lady in 1967. Rumours of the Row Dog have been explained as having been put about by smugglers, wanting to scare people away from the shore after dark, but supernatural dogs are, or were, one of the most widely reported British spectres, and although smugglers may have used the story for their own purposes, that is not to say that they invented it.

A much more unusual, indeed unique, creature was the gigantic cockerel that emerged from the sea off Portland in 1457, as reported in the sixteenth-century *Chronicles* of Raphael Holinshed. He had 'a great crest upon his head and a great red beard, and legs of halfe a yard long: he stood on the water & crowed foure times, and everie time turned him about, and beckened with his head, toward the north, the south, and the west'. After this he vanished, but tales about this wonderful bird continued to circulate into the twentieth century. In 1965, the Dorset folklorist Edward Waring heard an oral version of the story that he thought made 'slightly better sense', that during a flood a great golden rooster appeared and crowed three times, after which the flood subsided. The original account, however, gives no reason for the bird's appearance.

PORTLEMOUTH, DEVON

The vicar and the Spanish galleon

One version of an anecdote widely repeated around the British shores is set in Portlemouth church. In *The Coasts of Devon and Lundy Island* (1895),

John Page writes that the parson was in the middle of a service when one of his parishioners entered hurriedly and whispered in his ear. The minister continued for a short while with his sermon:

> Then his pent-up excitement found vent, and shouting 'There's a ship ashore between Prawle and Pear Tree Point, *but let's all start fair!*' he tore off his gown, sprang from the pulpit, and, followed by his suddenly awakened congregation, raced across country to the scene of the disaster.

He, like his parishioners, and like the vicar of WALNEY ISLAND (North-West England & Isle of Man), was of course eager to plunder the wreck.

So far, so funny, but Page goes on to relate how the unfortunate crew of the wrecked ship, a Spanish galleon, were left to drown by the greedy looters, and begged in vain for ropes to pull themselves to safety (*see* SMUGGLERS AND WRECKERS, p. 56). Page reports that to his own day, coastguards heard voices crying from the spot as they patrolled the cliffs.

> One of them, indeed, solemnly assured an acquaintance that he had heard the words 'More rope, more rope!' in tones of agony, and that his dog heard it, too, for its hair bristled with terror.

PRUSSIA COVE, CORNWALL

John Carter the King of Prussia

In the eighteenth century, the coast of western Cornwall swarmed with smugglers, whose illegal trade enjoyed a great deal of popular support. Farmers, merchants, and even local magistrates, it is said, used to invest small sums in the free-traders' voyages, and one conscience-stricken MP who died in 1738 left £600 to the Crown in his will 'to compensate for the amount to which his tenants had defrauded the Customs'.

One of the most renowned smuggling gangs was led by John Carter, the 'King of Prussia'. His nickname is variously explained: some writers say that he looked uncannily like the real Prussian king Frederick the Great (1740–86), others that in order to mask his illegal activities he kept an inn called the King of Prussia, while perhaps the most humanly interesting account is that he adopted the title when he was a child, playing soldiers with his brothers.

Carter's headquarters were near Breage, in a rocky bay once known as Port Leah but since Carter's day, in tribute to his fame, called Prussia Cove, while his companions were known as the 'Cove boys'. Carter himself was celebrated as one of the great 'Gentlemen', as smugglers were sometimes called. Once when he was away from home, excise officers from Penzance visited his house and removed a cargo of contraband. On his return Carter was horrified, explaining to his comrades that he had promised to deliver the goods to his customers, and his reputation 'as an honest man' was at stake. That night, several armed men broke open the Penzance stores. In the morning, the officers discovered their loss:

> They examined the contents, and when they noted what particular things were gone, they said to one another that John Carter had been there, and they knew it, because he was an honest man who would not take anything that did not belong to him. And John Carter kept his word to his customers.

This tale comes from John Cornish's introduction to *The Autobiography of a Cornish Smuggler* by Carter's brother Harry, written in 1809 but not published until 1894. The memoirs are slanted towards repentance, since Harry had converted to Methodist Christianity and abandoned his sinful ways, but there are touches of nostalgia in his account of his early career. 'My success was rather beyond common,' he writes of his first smuggling adventures, and at one point he was 'expecting to make all our fortunes in a hurry'. Harry's ambitions were thwarted when he was taken by the French in 1777 and jailed for over two years. John Carter attempted to free him, but was himself captured, and finally the two were exchanged for French prisoners 'by the order of the Lords of the Admiralty', an unusual concession to two known criminals, suggesting that the authorities had need of their talents, since they then received commissions to go privateering against the Americans. Some while later Harry was badly wounded, and according to his own account 'afraid of the shaking of a leaf'. John took care of him at a hideout so carefully protected that the doctor had to be met a mile away and blindfolded before he was led to his patient.

The secrecy in which the smugglers had to operate explains the tantalisingly few hints about their trade contained in the memoirs. As Cornish points out, when Harry was writing in 1809, 'John Carter and the "Cove boys" were still at it, and Prussia Cove had not yet ceased to be a great centre of smugglers.'

ROCQUAINE BAY, GUERNSEY, CHANNEL ISLANDS

The Green Ray

The Green Flash or Green Ray is a name for a phenomenon occasionally seen at either sunrise or sunset, a beam of green light momentarily visible as the sun's rim appears above the horizon or vanishes beneath it. It can be seen from places with a low, distant horizon, such as the coasts of the Scilly Isles, the Isle of Man and the Channel Islands. A local resident writes that from his house at Rocquaine Bay on the west coast of Guernsey, he has seen the Green Ray at sunset about fifteen times over the last thirteen years, always in very clear and still conditions, since mist or rough seas block the sun's final dip. As he describes it, it is not a startling flash that lights up the sky, but a subtle and momentary effect: 'in my experience one needs to watch very carefully as the sun is in its final moments of setting – blink and you could miss it. As it disappears from view one can see the last remnant of it turn green and sometimes the water around it appears tinged too.'

The Green Ray is related to mirages, and depends on the density of the atmosphere. It appears when there is low pressure (since high pressure tends to produce haze), and it is therefore a sign of unsettled weather conditions. This provides a rational foundation for legends that it foretells disaster, as reported in *The Times* in August 1929 by Mona Douglas:

> In several fragments taken down by me from Manx fisherfolk, the 'flash' was seen at sunrise on the morning preceding a wreck of one or more boats, sometimes by a relative of men actually lost and in other cases by the men themselves, who took the warning and withdrew from the fated enterprise.

The Green Ray, however, known in Manx as the *soilshey-bio* or 'living light', had beneficent qualities too, and was believed to give an almost miraculous power to certain herbs if they were gathered immediately its light fell on them.

> I had this belief directly from a very old man who was, I should think, about the last survivor of the 'charmers' and who claimed to be able to cure 'all diseases of the body or mind in man, woman or child' . . . This man also told me that if any person could find what he called 'the herb

of life' at the moment when it was touched by the *soilshey-bio*, death would never touch him or anyone to whom he gave a portion of the herb to eat.

The Green Ray was also thought to confer a kind of second sight, a tradition referred to by Jules Verne in his novel *The Green Ray* (1882), where a girl wondering whom to marry hears a Highland legend that 'this ray has the virtue of making him who has seen it impossible to be deceived in matters of sentiment; at its apparition all deceit and falsehood are done away.' Ultimately, having journeyed all over Scotland in an attempt to view the Ray, she and her lover are present on top of Staffa when it appears, but fail to see it because they are looking into each other's eyes.

Eric Rohmer's 1986 film *Le Rayon Vert* adapts Verne's plot. His heroine, a hapless romantic, hears it said that anyone who sees the Green Ray will at that moment know what is in his or her own heart, and the final moment of the film depicts the real Green Ray glowing beautifully for a second – a triumph of photography – showing her that she has at last found love.

An illustration to Jules Verne's novel The Green Ray *(1882) shows a corona of dazzling beams. In reality, the phenomenon is a subtle glow. Verne's emphasis, however, is on the lovers in the foreground, gazing at each other and missing the sight for which they have travelled so far.*

LA ROCQUE, JERSEY, CHANNEL ISLANDS

A bridge from Jersey to France

There is a Jersey tradition that the channel dividing the island from France was once so narrow that it could be spanned by a bridge or a plank, and that in the sixth century the Bishop of Coutances used to cross back and forth by this means. The seventeenth-century historian Jean de Poingdestre records the legend but adds his own belief that this 'never was, unlesse it were before the Flood', and geographers agree that Jersey was already an island by Neolithic times.

Tales persist, however, that marshes between the island and the mainland were not submerged until the eighth century, when a tempest arose that lasted three whole months and completely transformed the coastline. According to legend, the storm was caused by a priest who was saying mass at a chapel at La Rocque, now on the extreme south-eastern corner of the island but then standing in the middle of the marshy plain. The service was interrupted by a crow that danced in the rafters. The priest tried to shoo it away by throwing stones, but the bird continued its antics, and eventually he lost his temper and swore at it. His profanity in a holy place, and in the very midst of the mass, brought instant retribution. The ground shook, black clouds covered the sky, and the sea swallowed the chapel and all the lands around it.

ST AGNES, SCILLY ISLES

Praying for a rich wreck

Around the end of the fifteenth century, the inhabitants of St Agnes cherished a shrine dedicated to St Warna, who was said to have sailed there from Ireland in a wickerwork coracle, and was believed to have particular power over shipwrecks – not to *prevent* them, but to *cause* them. The islanders, who made a fat profit from plundering any vessel that came to grief near their shores, used to throw crooked pins into St Warna's well and implore her to send them a rich wreck.

There was in those days no church on the island, and on one occasion when two of the residents were to be married, the entire population – consisting only of five families – set out in their boats for the nearby island of St Mary's to celebrate the wedding there. The presiding clergyman took the opportunity to rebuke the St Agnes wreckers for their misdeeds. They bitterly resented being shamed in front of the St Mary's parishioners, with whom they were generally at feud, and after some hot words they hastily set out for their own island, ignoring the signs of approaching foul weather. When they were about halfway, the storm broke with such violence that all their boats capsized, and every soul of St Agnes was drowned.

The disaster was, naturally, interpreted in folklore as a judgement on the islanders, but seems really to have happened. It is mentioned by the historian John Leland in the 1530s or 1540s as having occurred within living memory.

ST JUST, CORNWALL

'The Wrecker and the Death Ship'

Two hundred and fifty years ago, or thereabouts, a strange dark man rented a house in St Just. Nobody knew for certain where he had come from, but some said he was a pirate who had fallen out with his shipmates and been put ashore. An unusual number of shipwrecks took place near his property, and it was discovered that the stranger was in the habit of fastening a lantern to the neck of a horse, which he would then hobble by tying its head to one of its forelegs. When he drove the animal along the top of the cliffs, the bobbing motion of the lantern looked like a boat's light, leading those sailing by at night to come too close to shore, expecting plenty of sea room. Before they knew they were in danger, their vessels were ripped to pieces on the rocks. The wicked wrecker would lie in wait, and if he saw that any of the crew had escaped drowning and were trying to climb out of the water, he would cut off their hands with his axe, or swiftly knock them on the head, and then at his leisure take what he liked from their ships.

For many years he reigned as the terror of the coasts, but at last his end approached, and then the sinner became frightened. Lying on his deathbed, he cried out to those around him, 'Do save me from the devil, and the sailors, there, looking to tear me to pieces.' Clergymen who visited the dying man distinctly heard the sound of the sea roaring in his room, which remained dark all the time, although outside it was broad daylight. They saw the Devil hovering about, and tried to expel him with their prayers, but 'at last, when he took the form of a fly, and buzzed about the dying wretch, they saw it was in vain for them to try any longer.'

Meanwhile, two harvesters working in the fields heard a hollow voice calling, 'The hour is come but the man is not come.' They saw a heavy black square-rigged ship, coming in fast against the wind and tide, although not a soul could be seen on her decks. When she was near to the cliffs, black clouds gathered around her masts and advanced towards the dying man's house, then rolled back towards the 'death-ship', which at once sailed away.

> The weather immediately cleared, and nothing unusual occurred until a few men assembled to put the wrecker's ghastly remains quickly off the face of the earth; then, as the coffin was borne towards the churchyard, a large black pig came – no one knew from whence – and followed the bearers, who all declared that the coffin was too light to contain any body.

A storm blew up, and raged so violently that the coffin-bearers dropped their burden and took refuge in the church. When the tempest abated they found nothing where they had left the coffin but the handles and a few nails, the rest having been consumed by lightning.

This was the way the story of 'The Wrecker and the Death Ship' was always told, writes William Bottrell in *Traditions and Hearthside Stories of West Cornwall* (1870). It has a good deal in common with the historical reports of Cruel Coppinger at WELCOMBE, which probably helped to inspire this more supernatural version. Bottrell finds the black pig's presence in the funeral procession puzzling, but it can be assumed that this was a demon waiting to claim the soul of the dead man.

ST MARTIN'S POINT, GUERNSEY, CHANNEL ISLANDS

Guernsey donkeys and Jersey toads

In a traditional Guernsey tale, three Jersey sailors once had the brilliant idea of stealing Guernsey and adding it on to their own island. Having tied a hawser round one of the tall rocks that stand out from St Martin's Point, the mariners set sail, singing out that Guernsey was on its way. Soon they felt the rope jerk, and sailed on, full of pride at having detached their sister island from its foundations. When they reached Jersey they were much disappointed to find only a broken rope behind them, and to see that Guernsey had evidently floated back to her old place.

This story was told on Guernsey as a joke at the expense of the Jerseymen. In the Jersey counterpart, some sportive Guernseymen are said to have once tied a dead donkey to the tiller of a Jersey boat. When the Jersey sailors found why their vessel had been handling so badly, they detached the donkey and made it into pies which they then sold in Guernsey. The delicacy was much enjoyed by the buyers, who had no idea what it was they were eating, and since then, the Guernsey inhabitants have been known - at least in Jersey - as 'donkeys'.

If Guernseymen are donkeys, Jerseymen are 'crapauds' (toads), since the island was known to contain vast numbers of the animals. Philip Falle, Deputy of Jersey in the late seventeenth century, makes this a selling point:

> 'tis a received Opinion among our People, that those ugly Creatures suck in the Impurities that are in the Elements, and thereby contribute to health: which they prove by the contrary Example of *Guernezey*, which will not suffer a Toad to live in it, and yet is thought not to be so healthy as JERSEY.

Folklore accounts for the proliferation of toads or frogs with a tale that St Marculf brought two frogs with him to the island in the sixth century, and left them in St Helier marsh. When he revisited his pets, he found that frogs, like monks, were subject to temptations, and unlike monks, not trained to withstand them, so that the original two had spawned an innumerable progeny.

ST OUEN'S BAY, JERSEY, CHANNEL ISLANDS

The parish bells of Jersey

At least until the mid nineteenth century, when the fishermen of St Ouen heard a sound like bells tolling from the sea west of Jersey, they took it as an omen that a storm was coming, and stayed ashore. The ghostly sound, a peculiar booming made by the wind and tides, was explained from the late sixteenth or early seventeenth century onwards by a legend that all the church bells of the island had been sold and sent on a ship to France, a sacrilegious act punished when the vessel sank. After that, the bells could be heard to peal from beneath the waves before a gale, and when sailors were drowned between St Helier and the Minquiers, it was said that 'They have heard the parish bells.'

Historical events are supposed to underlie the tradition. In about 1558, fourteen of Jersey's church bells were pulled down by Sir Hugh Paulet, governor of the island, and sent to St Malo in France, where the ship carrying them was wrecked at the entrance to the harbour.

> Whereupon it is a by-word at this day in these parts, when any strong east-wind bloweth there, to say, 'The bells of Jersey now ring.'

This account is given in *The History and Fate of Sacrilege* (1632) by Sir Henry Spelman, who notes that all over England in the mid sixteenth century, bells were being dismantled and legends were springing up about the accidents that followed. After the death of the Catholic queen Mary in 1558 and the accession of the Protestant Elizabeth, public worship moved away from the ceremonial rites of the Roman church, and in line with the new regime, many church bells were indeed taken down. This was regarded by many as impious, and it was widely reported, Spelman continues, that the bells, shipped abroad for profit, were lost in the sea. Several English traditions of submerged bells owe their origins to this time of religious change.

SHEEPSTOR, DEVON
The pressed man

Squire Northmore, an eighteenth-century landowner of Sheepstor on the western side of Dartmoor, had a simple-witted son who was a grave disappointment to him. The thought that one day the lad would inherit all his father's property worried the squire so much that he arranged with the Plymouth press gang to kidnap his son, hoping that he would be drowned or killed in battle. The press gang's job was to enlist men into the navy, by persuasion, trickery, or force, and at the time of the wars with France they were active all around the coast, their raids leading to a spectacular battle at NORTH SHIELDS (North-East England) in 1796. In the West Country they were just as unceremonious, and having seized young Northmore while he was in the middle of haymaking, they dragged him off to sea.

When the boy learned that his own father had been responsible, he vowed revenge. He made a success of his life as a sailor, and returned years later to Plymouth in the company of a gang of seamen, who went with him up to Sheepstor. Hearing the approach of the marauders, the old squire hid himself between the rafters and the roof, and it was as well for him he did so, because they would have killed him if they had found him. As it was, they ransacked the house and carried off the title deeds, and then vanished as suddenly as they had arrived. The young man was never heard of again, but his father's fate was reported as a punishment for his callous behaviour:

> He took to drinking, and he'd ride home from the public-house that tipsy, he'd ride right into his kitchen and tumble off his hoss, and he'd get out a bottle o' gin, and sit in his settle, and make the 'oss stand there too for good company, and he'd drink to the 'oss, and when the ou'd [old] 'oss nodded his 'ead, Squire Northmore'd say, 'Same to you, sir; I drinks to 'y again.'

This story was told in the 1890s by an old woman of Sheepstor, who finished by saying that the Northmore family had to leave the property, since they had lost the title deeds, and that nothing went right for them afterwards. 'I reck'n no gude niver comes o' doin' a wrang action, does it?'

STEART, SOMERSET

'The Sea Morgan and the Conger Eels'

The tale of 'The Sea Morgan and the Conger Eels' was a Somerset tradition reported to Ruth Tongue in the 1960s. A 'morgan' was a mermaid, in Wales and around the Severn Sea, and according to the story a beautiful morgan was heard on many an autumn evening singing around Steart, so bewitchingly that any man who heard her was compelled to wade out into the sea, further and further until the quicksands swallowed him up, and then the conger eels would feast on his body. 'They always knew when the eels barked she would be about that low tide, so something was done to end her wicked ways.'

A local 'gifted woman' – a witch, or someone with magical powers – had a son who was born on a Sunday. It was sometimes said that those with this blessed birthday, the Lord's day, were proof against the charms of evil spirits, and not only was the lad fortunate in this respect, he was also deaf, so he could not hear the morgan singing. He did not think much of her looks either, what with her green hair, so when his mother sent him out to drive away the mermaid, he was not liable to be tempted. He took his mud sledge, known locally as a Steart horse, and as the morgan sang he slid over the quicksands with his eel spear, getting a fine haul of conger eels as he went along. 'When he'd speared twelve of them, she gave a skreek, and took herself off – and she never come back. All Stolford and Steart had Conger-pie that week.'

STUDLAND HEATH, DORSET

King Arthur's sword

On Tudor maps, the sea is shown stretching into Studland Heath, but the coastline has changed, and now a stretch of water known as the Little Sea is enclosed from the open waves. At certain times and tides, if a stick is dragged along its edges, a penetrating musical noise can be heard, described as the 'singing' of the sands.

Perhaps this haunting phenomenon has contributed to the idea that in the depths of the Little Sea lies King Arthur's sword Excalibur. In Thomas Malory's fifteenth-century *Morte d'Arthur*, when the king saw his end approaching he told Sir Bedivere to take the sword and 'go with it to yonder water side, and when thou comest there I charge thee throw my sword in that water'. What 'water' he means is not spelt out, and although most people have assumed that it is the same lake where he first got the sword, tales current in twentieth-century Dorset held that the sword was cast into the sea at Studland.

One of Aubrey Beardsley's 1894 illustrations to Thomas Malory's Morte d'Arthur *shows 'How Sir Bedivere cast the sword Excalibur into the water'. Dorset legend holds that the knight threw the sword into the Little Sea at Studland.*

TORBAY, DEVON

See TOTNES.

TORQUAY, DEVON

'Steer to the Nor'-West'

A sailor from Torquay, by the memorable name of Robert Bruce, was serving as first mate on a trading ship, sometime in the early 1830s. At noon one day, when the ship was in the region of Newfoundland, he and the captain went below to calculate their latitude and longitude, and Bruce, hard at work, did not notice that the captain had left the cabin again. He reported their position, and receiving no reply he looked up and saw (as he thought) the captain writing on a slate. He repeated his figures, but still got no answer, and stood up to attract the senior officer's attention. At this, the other man raised his head, and Bruce was astonished to see a complete stranger.

This, on a ship that had been six weeks at sea and was far from land, was so uncanny that Bruce rushed on deck in terror. Meeting the captain, he exclaimed that he had seen an unknown man down below, and they went down together. The cabin was empty, but there was the slate lying on the desk. The captain picked it up.

> 'Good God,' he exclaimed, 'here's something, sure enough! Is that your writing, Mr Bruce?'
>
> The mate took the slate, and there in plain, legible characters, stood the words, 'STEER TO THE NOR'-WEST.'

The writing was nothing like Bruce's, and suspecting some trick, the captain compared the words on the slate with the handwriting of every man aboard. He searched the ship for stowaways, but with no result, and deciding at last to follow the mysterious instructions, he changed the ship's course from south-west to north-west. About three o'clock that afternoon, the lookout reported a vessel trapped in ice, which proved to be a passenger ship bound for Liverpool from Quebec that had got frozen fast. The unhappy occupants had given up hope of rescue, having finished their provisions and most of their water, and now they came thankfully aboard their rescuers' ship.

As Bruce watched the new arrivals, he recognised the face he had seen in the captain's cabin. He pointed out the man, and the skipper asked the passenger to write the words 'Steer to the Nor'-West' on a slate, presenting the blank side. When he had done so, he turned it over and showed him the earlier message, in precisely the same handwriting. The passenger looked from one to the other, dumbfounded.

> 'I only wrote one of these. Who wrote the other?'
>
> 'That, sir, is more than I can tell you. My mate says you wrote it, sitting at his desk, at noon to-day.'

The passenger said he had fallen into a heavy sleep or trance a little before midday, and had dreamed of being on board a different ship. Although he had not dreamed of writing anything, he had the firm impression that the vessel in his dream was coming to their rescue, and on waking he had told his own captain that they would be relieved that very day. He had described the appearance of the ship in his dream, which exactly matched that of Bruce's ship.

Bruce later told his story to Captain Clarke of the *Julia Hallock*, who in turn told it to Robert Owen, an American minister, describing Bruce as one of the most truthful and straightforward men he had ever met. 'He always spoke of the circumstance in terms of reverence, as of an incident that seemed to bring him nearer to God and to another world than anything that had ever happened to him in his life before.' From Owen, the account reached the Reverend Bourchier Wrey Savile, who printed it in his *Apparitions* (1874).

Although Savile's (or Bruce's) is almost certainly the original narrative, a nearly identical tale was told sometime before 1930 by Robert Hughes, a Methodist minister from Holyhead, about an Anglesey ship. It was evidently a story that appealed to clergymen in search of miracles.

TOTNES, DEVON

Corineus and the giants

Ancient tradition reported that Britain was once populated by giants, offspring of the fifty daughters of Diocletian, king of Syria. Having murdered their husbands on their collective wedding night, the princesses were set adrift as punishment and came to land in south-west England. Here they found new husbands among the aboriginal demons of the island, which was known thenceforth as Albion after the eldest sister, Albina. In *The History of Britain* (1670), John Milton dismisses the tale as 'too absurd, and too unconscionably gross', but that does not stop him telling it, although he presents his account as myth rather than history. He goes on to describe how the legendary Brutus, great-grandson of the Trojan warrior Aeneas, landed at Totnes with his followers to conquer the giants, colonise the land, and retitle it Britain 'with some reference to his own name'. Totnes is some way inland, and perhaps the coastline has changed since those long-ago days, but whatever the geographical or historical facts, the town still shows the Brutus Stone marked with the hero's footprint where he first stepped ashore. (Torbay's claim to possess a stone similarly printed by William of Orange in 1688 may indicate a certain rivalry.)

One of Brutus's warriors, Corineus, wrestled the hugest of the giants, Gogmagog. As described by Michael Drayton in *Poly-Olbion* (1612–22), it was an epic battle:

> Their lusty sinewes swell like cables, as they strive:
> Their feet such trampling make, as though they forc't to drive
> A thunder out of earth; which stagger'd with the weight.

At last Corineus, by superior skill, cast his opponent into the waves with tremendous force:

> so that his violent fall,
> Strooke *Neptune* with such strength, as shouldred him withall;
> That where the monstrous waves like Mountaines late did stand,
> They leap't out of the place, and left the bared sand
> To gaze upon wide heaven: so great a blowe it gave.

For a long while, images of two men armed with clubs could be seen carved in the turf of Plymouth Hoe, and were popularly said to commemorate this event. The figures were built over in the seventeenth century, but Corineus was supposed to have had a more lasting memorial, the land exposed by the displaced sea being called Cornwall after him. The etymology, proposed in the twelfth century and often repeated, is fictional, as noted by Drayton, who remarks that the county's name is probably in fact derived from its shape, 'like a horne, which in most tongues is *Corn*, or very neere'.

TRESCO, SCILLY ISLES

A Civil War romance

At the north-east point of Tresco is a cavern known as Piper's Hole. It extends for about six hundred foot, but up to a third of this is occupied by an underground lake, so that a boat is needed to explore the full reach of the vaults.

The peculiarities of the cave are central to a love story set during the Civil War, when Tresco held out for the Crown. The island's forces were commanded by a young nobleman named William Edgcumb, who, when the Parliamentary troops proved stronger, tried in desperation to destroy his fortress, Charles's Castle, but succeeded only in blowing the roof off. Although Edgcumb's body was not found, it was assumed that he had died in the explosion.

This was not so, however, as discovered a little later – according to a nineteenth-century book of Scilly legends – by Mildred, daughter of Tresco's Roundhead conqueror Colonel Fleetwood. She was of a romantic disposition, and was not frightened to visit Piper's Hole, although it was said to be haunted:

> A Puritan sentinel, placed here on out-post duty, averred that he saw, issuing from the bosom of the earth, a grim figure clothed in white, that shook its finger at him with a menacing gesture, and so frightened him, that he fell flat upon his face, and when he regained his senses, the ghastly form was gone. Little, however, recked fair Mildred Fleetwood of these tales of horror. Perhaps they were not even displeasing to her. They gave food to her morbid appetite for novelty.

One day as she sat gazing at the sea, lost in thought, she saw a young man, his clothes torn and dirty. It was William Edgcumb. Although he was a fugitive and she a gently reared maiden, neither ran away. On the contrary, he offered to show her his hiding place, she agreed, and he conducted her through the cave to a boat, by which they rowed across the subterranean lake to a secluded chamber in which he had taken refuge. With the help of a friendly fisherman, he had survived here since his failed defence of Tresco, and had played the part of a spectre in order to discourage curiosity.

They met several times, and when at last he left the island, he asked her to go with him. She refused, for fear of breaking her father's heart, but many years later, at the Restoration, he returned in triumph and claimed his bride.

Historical sources confirm Edgcumb's and Fleetwood's roles, but say nothing of the morbid Mildred, who seems a rather literary invention.

It is probable, however, that there were authentic supernatural traditions connected with the cavern on Tresco, which was rumoured to go all the way under the sea to another Piper's Hole on St Mary's, a few miles to the south-west. The name 'Piper's Hole' indicates that these caves rank with places such as Clach-Tholl in western Argyll, Scotland, where a bag-piper was said to have entered a tunnel leading under the sea, but encountered something horrible and was never seen again. In similar tales found throughout Britain, the unfortunate musician (in England sometimes a fiddler, in Scotland always a piper) continues to play his instrument underground until the demons get him, and ghostly music is later heard to echo from beneath the earth. One more detail makes it clear that the Scilly Isles caves are of this class. In many versions, when the piper goes into the tunnel he takes along a dog, which comes running out again, howling and bald, all the hair scorched off its body by hellish fires and fumes – and a nineteenth-century account of how Piper's Hole on St Mary's connects with the one on Tresco records that local inhabitants 'tell you of dogs let in at the one entrance coming out, after a time, at the other with most of their hair off'. One guesses, therefore, at a forgotten Scillonian tale of a pipe-player who vanished in one of the caverns, survived by his dog, although later tradition only preserved the memory of a non-specific haunting.

VAZON BAY, GUERNSEY, CHANNEL ISLANDS

Invasion by the fairies

The author Victor Hugo (1802–85) lived for many years in Guernsey, and set his novel *Toilers of the Sea* there. In it he included a good deal of Guernsey folklore, some of which he might have made up or embroidered to suit his own purposes (*see* LIHOU), but much of it authentic. He recorded the local name for the fairies as *sarregousets*, and wrote that 'At night, when it thunders, if you should see men fly in the red light of the clouds and in the fluttering of air, these are the "sarregousets."'

The word is a corruption of *Les Aragousais*, 'the men of Aragon'. In the fourteenth century, a Welsh prince Ivon or Yvain, dispossessed by Edward III of England, took service with the French king Charles V, and was given command of a fleet of ships provided by Pedro IV of Castile and Aragon. Ivon's troops landed in Guernsey, made mincemeat of the local soldiers, and laid siege to Castle Cornet. After a while he was called away to other battles, but memory of the invasion lasted for centuries in the island, as recorded in 'The Ballad of Ivon de Galles', probably the oldest surviving Guernsey ballad.

Smugglers and Wreckers

As long as there have been restrictions on foreign trade, there have been smugglers. In 1215, the Magna Carta mentions wine illicitly brought into England, and more importantly, wool taken out. The Kentish Owlers (wool smugglers) made handsome profits in the thirteenth century taking duty-free fleeces to the Flanders weavers, and as time went on, their experience smuggling goods *out* of the country was turned to bringing contraband *in*. Tobacco was one lucrative line, French cognac another, and silk, lace, tea and playing cards are among the many goods that have been subject to duty at various times, all of which had their market, and their suppliers to meet demand.

Active areas around Britain were the Channel Islands, which until 1808 were exempt from customs and excise (duty on imports and exports respectively), Galloway in Scotland, and Ireland's west coast, all easily accessible from the sea and difficult to patrol inland. The richest rewards, however, were found in London, and the nearer to the metropolis the goods were shipped, the more money could be made. Devon and Cornwall were notorious smuggler territory, and in Sussex and Kent the 'free trade' was an organised industry, with capital often put up by respectable gentry, who raked in fortunes in return for their investment.

Tales of the smugglers were told from coast to coast. While there are accounts of brutal criminals, such as the gang that operated in HAWKHURST (South-East England), generally free-traders figure as brave, resourceful men (plus the occasional woman, like Bessie Catchpole of HARWICH, East Anglia), making a living against the odds, and outwitting the unpopular revenue officials. A business that relied on secrecy and darkness was often associated with hauntings, and one branch of folklore concerns fake spectres impersonated or reported by smugglers to discourage the inquisitive, although probably most were not wholly invented, but exploited pre-existing superstitions. Some crafty officers too dressed up as ghosts, as at CRIMDON DENE (North-East England), to get the better of their opponents.

Just as widespread, though never a structured affair, was wrecking, a word that can describe either intentionally misleading a ship on to the rocks (*see* for example WALNEY ISLAND, North-West England & Isle

Contraband passes between two smuggling vessels out at sea. From here, under cover of dusk, the smaller boat might make for any secluded place along the coast to unload the goods.

of Man), or the much commoner practice of plundering accidentally stranded vessels, something that still occasionally happens. In January 2007, a container ship went aground at Branscombe Bay in Devon, and goods worth thousands of pounds, ranging from motor-bikes to bottles of scent, were removed by crowds of looters.

Legally, even a 'derelict' (an abandoned ship) usually still belongs to somebody, and anything taken from it must be reported, so that its owner has a chance to reclaim their property. It has, however, often been popularly supposed that a vessel is fair game as long as there is no survivor on board, an interpretation that, at least in folklore and perhaps in fact, has led to the murder of shipwrecked sailors by greedy coast-dwellers. Some said that even animals counted as survivors, a belief that may have caused the death of an innocent monkey at HARTLEPOOL (North-East England).

According to the song, which agrees broadly with historical events as reported in early chronicles, Ivon's men arrived one morning in Vazon Bay. A shepherd gave the alarm, and the islanders assembled to make a stand on the hill above St Peter Port, where a bloody battle ensued, traditionally located near Elizabeth College on a spot known for centuries afterwards as 'La Bataille'. The men of Aragon were prevented from entering the town, but their attack on the castle was only called off when its governor paid a ransom.

In folklore, it was said that the native male islanders were all killed in the fighting save one man and boy of St Andrew's parish who concealed themselves in an oven (a detail transferred, with variation, to the tale of LE PORT DU CREUX on Sark). The dark and stocky conquerors married the Guernsey women and re-peopled the island with a short, swarthy race.

Later versions of the legend go further, identifying the invaders as supernatural beings. The story goes that a lovely girl of the island, named as Michelle de Garis, one morning found a man dressed in green lying asleep on the ground. When he awoke he told her that he was a fairy. He was very small but very handsome, and when he proposed marriage she at once accepted. The couple sailed away in his ship, leaving as a keepsake for Michelle's parents the bulb of a strangely beautiful pink and gold flower, the Guernsey lily (*Nerine sarniensis*).

Sometime later, a man walking to Vazon Bay in the dawn met a crowd of tiny men who said that they too were fairies. They had been so taken with their friend's Guernsey bride that now they all wanted one for themselves. The island men made a spirited defence of their wives and daughters, but the fairies drove them eastwards to Le Mont Arrivel, where a final hideous battle was fought. Again, only two St Andrew's men survived, the rest being slaughtered by the fairies, who then settled down happily in Guernsey. Since then, it is said, no Guernsey witch has ever needed a broomstick, since all Guernsey people are descended from the fairies and can therefore fly without help.

WATCHET, SOMERSET

The Ancient Mariner

In November 1797, Samuel Taylor Coleridge was on a walking tour of the Quantocks with William and Dorothy Wordsworth. On the walk they discussed Coleridge's idea for a sea ballad, partly inspired by a dream his neighbour John Cruikshank had had about 'a skeleton ship'. The ballad was *The Rime of the Ancient Mariner*, and Coleridge may have written its first lines at the Bell Inn in Watchet, overlooking an old harbour which matches the description of the Mariner's starting point with its lighthouse and its hilltop church.

Whether or not the Mariner starts from any actual point, he soon travels off the map and into the land of myth. The poem is a compendium of sea legend and travel lore, and although when he wrote it Coleridge had never even crossed the Channel, he had read widely about other men's voyages through frozen seas full of icebergs, and into tropical oceans where sea-snakes danced in the waves, leaving trails of phosphorescence. A whole book could be written about Coleridge's sources and inspiration – and in fact it has been, by John Livingston Lowes, whose *Road to Xanadu* (1927) is a masterpiece of literary detective work, out of print but well worth tracking down.

As in the legend of THE FLYING DUTCHMAN (p. 30), the Mariner falls under a curse condemning his ship to sail on and on without coming to land. His crime is that he has shot an albatross after it has led him and his crew through the ice, a deed for which he is reviled, and in punishment the dead bird is hung round his neck:

> And I had done an hellish thing,
> And it would work 'em woe:
> For all averred, I had killed the bird
> That made the breeze to blow.

The belief that it was most unlucky to shoot an albatross is said by many authors to have been widespread among deep-water sailors for a long time. Traditions surrounding the bird, however, have been coloured by Coleridge's poem. A passage from George Shelvocke's *Voyage Round the World* (1726), which Wordsworth had been reading just a few days before Coleridge and he discussed the new poem, tells of a 'disconsolate black Albatross' that followed the ship for several days in a continued spell of bad weather, until Shelvocke's second officer Hatley 'observed, in one of his melancholy fits, that this bird was always hovering near us, imagin'd, from his colour, that it might be some ill omen', and 'after some fruitless attempts, at length, shot the Albatross, not doubting (perhaps) that we should have a fair wind after it'. It was because the bird was *black* that it was considered ominous, not simply for its species, and it was hoped that its death would bring better weather (which it did not, but neither did anything disastrous follow).

It has been suggested that Coleridge, who had never seen the great seabirds of the far south latitudes, did not know how big the Wandering Albatross is. Doré's illustration shows the Mariner with a full-size specimen slung round his neck, in penance for the crime of killing it.

It was common practice in the eighteenth and nineteenth centuries to snare albatrosses for food, to skin their webbed feet to make tobacco pouches, and to use the long bones of their wings for pipe stems. They were also sometimes caught with baited hooks and released carrying messages. On 8 December 1847, a message was tied round an albatross's neck by the captain of the whaling ship *Euphrates*, giving details of his position. Twenty-two days later, having travelled 2,950 sea miles, the bird was shot by the captain of the *Cachelot*. Examples like this show that there was no universal taboo on killing an albatross.

One cannot, however, be certain that the superstition did not exist at all prior to the *Ancient Mariner*. Traditions that it was unlucky to kill seabirds seem to be premised on the idea that they give seafarers advance warning of rough weather, so that it was courting danger to destroy them, and also on the superstition that the souls of dead sailors became seagulls (*see* PORLOCK). As the biggest seabirds, and those that are seen further from land than any other species, albatrosses attract particular attention. They are the bird world's great gliders, soaring in seemingly motionless flight on outspread wings that can measure more than eleven foot from tip to tip. They sometimes follow ships for weeks at a time, and were once thought even to sleep on the wing, and to raise their broods on floating rafts at sea.

Some probably regarded the albatross as a seabird like any other, to be preyed on or killed for sport, while others responded more imaginatively to its grandeur. In *Moby-Dick* (1851), Herman Melville wrote of seeing his first albatross in the Antarctic, 'a regal, feathery thing of unspotted whiteness':

> At intervals, it arched forth its vast archangel wings, as if to embrace some holy ark. Wonderous flutterings and throbbings shook it. Though bodily unharmed, it uttered cries, as some king's ghost in supernatural distress. Through its inexpressible, strange eyes, methought I peeped to secrets not below the heavens.

WELCOMBE, DEVON

Cruel Coppinger

The Danish pirate and wrecker known as Cruel Coppinger lived in the West Country, probably in Welcombe, around the end of the eighteenth century, and was vividly remembered long afterwards. Later reports embellished his story to make him a legendary villain, with some touches of the supernatural. His arrival, it was said, was signalled by a hurricane from the south-west, driving before it a foreign-looking vessel that was apparently trying to make for the shore, but was beaten back by the waves and the wind. The locals gathered to watch, among them a girl called Dinah Hamlyn who had ridden down on horseback. The captain of the ship, having stripped himself almost naked, dived from the deck and swam ashore, where to everyone's astonishment he jumped up behind Dinah, seized the reins and set her horse galloping away.

The stranger announced his name as David Coppinger, and paid court to Dinah until she agreed to marry him. He made their house the headquarters of all the criminals from miles around, whom he formed into an organised gang. Smugglers' boats began to appear around the shore, led through the rocks by signal lights from Coppinger's men, and Coppinger's own schooner, the *Black Prince*, became the terror of the coast. On one occasion, Coppinger lured a revenue cutter to pursue him into a sea channel that wound through the submerged reefs, with the result that the excise boat was wrecked and every man aboard drowned, while the *Black Prince* came through unscathed.

Ordinary villagers went in fear of their lives and liberty. In 1835, a ninety-seven-year-old Welcombe man told Robert Hawker, vicar of nearby MORWENSTOW, that he had been abducted by Coppinger's crew and kept in their service for two years, until his friends ransomed him. The old man said that this was because he had seen one of the gang kill somebody, and it had been feared that he would report the murder.

At last Coppinger's money and luck ran short. A contingent of armed government vessels arrived to put down the gangsters, and the Danish pirate evidently thought the time had come for him to leave Devon. A full-rigged ship was seen waiting off shore, and when Coppinger had boarded her, according to Sabine Baring-Gould's *Vicar of Morwenstow* (1876), she vanished in a moment, like a spectre boat.

> Thunder, lighting, and hail ensued. Trees were rent up by the roots around the pirate's abode . . . and, strange to say, a meteoric stone, called in that country a storm-bolt, fell through the roof into the room at the very feet of Cruel Coppinger's vacant chair.

Coppinger's history helped inspire the more obviously legendary tale of 'The Wrecker and the Death Ship', told further west at ST JUST. Some accounts of Coppinger's own life say that he operated in Cornwall rather than Devon, so it seems that more than one place laid claim to the villain - after his reign was safely over.

WESTERN ROCKS, SCILLY ISLES

The Cursing Psalm

It was believed in western Cornwall that a person who had suffered wrong could be revenged on their enemy by reading or reciting the 109th Psalm just before death. Known as the 'Cursing Psalm', it implores the Lord to bring destruction upon the wicked and deceitful man who has slandered and opposed the speaker:

> Let his days be few; and let another take his office.
> Let his children be fatherless, and his wife a widow.
> Let his children be continually vagabonds, and beg: let them seek their bread also out of their desolate places . . .

In *Traditions and Hearthside Stories of West Cornwall* (1873), William Bottrell reports that the Cursing Psalm was traditionally held responsible for the wreck of the *Association* in 1707, when it headed on to the lethal Gilstone rock. The death toll was over two thousand, including the admiral Sir Cloudesley Shovell, and only one man (it was said) survived to tell what had happened.

The ship was returning from Toulon by way of the Scillies, and one member of crew who was a native of the islands warned Sir Cloudesley that he was on course for the Western Rocks, an area of reefs to the west of St Agnes, stretching from the Isle of Annet to the Bishop Rock. The admiral and his officers were angry at the humble sailor's interference, and when he persisted, he was summarily condemned to death for insubordination and incitement to mutiny. Just before he was hanged from the yardarm, he begged as a last favour that a psalm might be read, and when this was granted he chose the 109th. As the verses were spoken, he repeated the more threatening passages, and his last words were that Sir Cloudesley and the rest who saw him executed should never reach land alive.

His dead body was wrapped in a hammock and cast into the deep. Soon afterwards, black clouds rolled across the sky, while the gale rose to a screaming tempest.

> Then the hanged man's curse was dreaded; and lo, to the crew's consternation, they beheld his corpse – divested of its rude winding-sheet – floating near the doomed ship, which it closely followed, with its face turned towards her, – in all her varying course, through eddying currents, – until she struck on the Gilstone; when the hanged man went down with the ship and his messmates.

Sir Cloudesley's body was said to have been washed ashore stripped virtually naked by the action of waves and rocks, and could only be identified by a jewelled ring. The couple who found him buried him and sent the ring to his widow, who gave them a pension for life in gratitude, an episode Bottrell assures us is 'true history'. Much more widely repeated, however, was the rumour that the admiral came alive to land, only to be murdered for the sake of the valuable ring he wore. The Scillies generally had a terrible reputation among sailors, who believed that the inhabitants were so greedy for loot that they would kill anyone who ran aground on their shores. According to a late nineteenth-century guidebook, when a ship was once wrecked on one of the smaller islands, the distressed mariners' greatest fear was of their would-be helpers:

> When a friendly boat put out from the harbour to rescue them, they kept the brave fellows away by showers of stones till they fell to the ground with weakness from exposure and had to be carried by main force into a place of safety.

Laws on salvage stated that a vessel could not be claimed by those who found it unless it had been abandoned by all on board, so it was in the finders' interests that none of the crew should make their own case for possession (*see* SMUGGLERS AND WRECKERS, p. 56).

Many vessels have come to grief among the Western Rocks, and the souls of the drowned were believed to linger in the region. In the nineteenth century, workmen building a lighthouse on the Bishop Rock were quartered on Rosevear, an uninhabited island nearby, and were convinced that it was haunted. A blacksmith who spent a solitary night there reported hearing unearthly music from the sea, and after that none of the labourers would stay on the island without company. Even today, it is said that few locals willingly take their boats among these rocks alone.

WESTON-SUPER-MARE, SOMERSET

The man who didn't believe in luck

A contributor to *Notes and Queries* in the mid nineteenth century wrote that on a hill above Weston-super-Mare was a heap of stones, on which every fisherman setting out to his day's work would add another stone for luck. The practice was still followed in the 1960s, according to the Somerset folklorist Ruth Tongue, who goes on to tell a warning tale of a 'clever Dick of a fellow' who said he didn't believe in luck, and refused to lay a stone. Her story is a composite one, made up from different versions she heard from seafaring people of the district, and it includes several widespread superstitions (which she italicises).

> First go off, *he went to sea on a Friday*, yes, he did. And he'd hardly gone a quarter mile when blest if his old woman didn't come *running after him* in her white apron *calling his name*.

Taking this in reverse order, calling somebody's name aloud was ill-omened, because you never knew who, or what, might be listening, and it was bad luck to see a woman on your way to the sea. Friday was generally regarded as an unfortunate day to start a voyage – and when a man of GREAT YARMOUTH (East Anglia) once defied the taboo, he found that no good came of it.

On the man went, though, and his wife watched him out of sight – another thing she shouldn't have done, and she might have expected never to see him come back. He put no stone on the cairn, so it served him right to come face to face with the parson, even worse for a sailor than to meet a woman.

> But he still wouldn't go home. Down by the harbour what should run across the road but a *black pig*, and you should have heard the fool swear, but still he went on down to the boats, and as he sailed out he heard a voice from the cliffs shouting, '*Good dog, after that rabbit now.*' But even then he wouldn't turn back.

Seeing a pig was bad news, whatever its colour (*see* ST MONANS, Scottish Lowlands), though black is always ominous. The rabbit too was an unlucky animal for mariners (*see* RABBIT ISLANDS, Scottish Highlands & Islands), and Tongue records a piece of conversation she overheard on the Somerset coast in 1948: 'I can't go on the water, you bin talking of rabbits.'

It was no surprise, after all this, that the 'clever fellow' was never seen again.

WEYMOUTH, DORSET

Admiral Tryon's ghost

On 22 June 1893, Lieutenant Reginald Bacon of the Royal Navy visited the Whitehead Torpedo Works at Weymouth, and afterwards had lunch with his fellow officers. Suddenly the stem of a wine glass snapped. 'That should mean a big naval disaster,' said one of the company. In popular tradition, the ringing of a wine glass meant the death of a sailor, and to break it was worse.

At that very moment, 3.44 p.m., HMS *Victoria* was wrecked off Beirut.

Her captain, Vice-Admiral Sir George Tryon, had recklessly ordered the leading ships of his line to turn towards each other, leaving too little room for the manoeuvre. His own ship was rammed by the *Camperdown*, sinking almost at once, and among the many dead was Sir George himself, who refused a lifebelt and was heard to admit that the collision was his fault.

The same afternoon, the admiral's wife was giving a party at her home in Eaton Square, London. Several guests, who knew that Sir George was at sea, were surprised to see him in the house, and some even claimed he spoke to them. Lady Tryon herself saw nothing, and confirmed at the time that she was not expecting her husband. It was said that when she entered the library, where he was supposed to have been seen, she found a wet footprint on the floor behind the admiral's desk, and a wet fingerprint on a globe of the world, marking the place in the Mediterranean where the *Victoria* had gone down, while the clock had stopped at 3.44.

The history of this ghost is an interesting one. Although the apparition was described by many writers, none were first-hand witnesses, nor did they quote anyone who was. So how did the rumour get going? The most likely culprit is the magazine *Review of Reviews*, which reported the wreck of the *Victoria* and the *Camperdown* in August 1893, and a few months earlier, in its Christmas issue for 1892, printed a story called 'A Ghost in a Ballroom' which dealt with a party guest who spoke to nobody, and was later realised to have drowned, miles away, at that very time. It seems, therefore, that a reader got the two items confused, or perhaps was inspired by them to manufacture the tale of Sir George's wraith.

WIMBORNE MINSTER, DORSET

Isaac Gulliver the gentle smuggler

Isaac Gulliver (1745–1822) was described in an official report of 1788 as 'the greatest and most notorious smuggler of the West Country'. His territory stretched the length of the Dorset coast, where he was held in respect and even affection, and he was called the 'gentle smuggler' because his gang was forbidden to use firearms.

Gulliver first went to sea at the age of twelve, but in his later career he preferred to leave the hands-on transport of spirits and tea to others, concentrating his own formidable powers on organisation. Importantly, he had a rich patron in the person of John Fryer, a banker with considerable property in land and ships. Plenty of respectable men invested in the 'free trade', as smuggling was often called, which was seen by many as semi-legitimate and was certainly profitable (see SMUGGLERS AND WRECKERS, p. 56). Gulliver put his money to excellent use, acquiring houses and farms which he used to store his goods and stable his horses, and even buying Eggardon Hill as a 'signalling station' for the benefit of his vessels out at sea.

The fifty or so men he employed usually dressed in smocks and powdered their hair, and were known as the 'white wigs'. The mayor of Lyme Regis later described how the 'white wigs' used to wait in a cave at the mouth of the river Lyme, less than a hundred yards from the custom house, until their services were required to unload a boat. It seems that a blind eye was generally turned, but sometimes official notice had to be taken, as on one occasion when the authorities were tipped off that Gulliver would be at Wimborne market and turned up to intercept him, but paid no attention to a grey-haired old shepherd who had come in early with his flock. The shepherd was, of course, Gulliver himself.

On another occasion, the smuggler had a narrow escape when customs officers pursued him to his very door but had no warrant to enter. They kept watch for several days until their search warrant finally arrived, and felt confident that they had him at last. When they knocked, however, it was to be greeted by a tearful Mrs Gulliver, who showed them a pallid corpse lying in a coffin. Embarrassed, the men muttered a few condolences and left, upon which Gulliver hopped out of the coffin and filled it with contraband ready for the hearse to take away.

For much of Gulliver's career, England was at war with France, and he may have taken a covert role, spying for his country. It was not unknown for smugglers to involve themselves in espionage, since they moved freely across the Channel and had access to all sorts of informants. Many were traitors, working on the French side, according to Napoleon himself. Imprisoned on St Helena, the defeated emperor boasted that he had had more than five hundred English smugglers in his pay:

> I had every information I wanted through them. They brought over newspapers and despatches from the spies we had in London. They took over spies from France, landed and kept them in their houses, then dispersed them over the country and brought them back when wanted. They came over in boats not broader than their bath.

Gulliver, however, was a patriot. One of his descendants reported that 'Isaac Gulliver was certainly a much more successful British agent than the ordinary spy. He always knew exactly where the French fleet lay; and often much more than just that.' Among his family, the story went that he, or one of his gang in France, heard of a plot to assassinate George III, at that time staying in Weymouth. Isaac hastened to the king and warned him, and received a royal pardon in gratitude.

After his retirement from crime, Gulliver settled down to respectability and even became a churchwarden, but died considerably richer than most upright citizens, leaving a fortune of £60,000. His memorial can be seen in Wimborne Minster.

ZENNOR, CORNWALL

A mermaid in church

Hundreds of years ago, a beautiful and richly dressed lady would occasionally put in an appearance at Zennor church. Nobody knew her name, nor where she lived, nor anything about her except that she sometimes attended a service, and had done so for many years – so many, in fact, that it was wonderful how young and pretty she still looked. They also remarked on what a sweet voice she had.

The best singer in the parish, a young man called Mathew Trewella, once followed her when she left the church. He was never seen again afterwards, and neither was she, and that might have been the end of things but for a curious accident.

One morning a ship cast anchor near Pendour Cove, a little to the north-west of Zennor, and soon afterwards a mermaid came alongside and hailed the captain, telling him either that she wanted to go to church, or that she had just come back (accounts varied), but in any case that his anchor was blocking her doorway. The sailors displayed wonderful energy pulling up the anchor and setting the sails, since none of them wanted to hang around the mermaid's home any longer than necessary. When the villagers of Zennor heard that a mermaid lived near Pendour, and that she had talked of going to church, they concluded at once that it was this lady who had been part of their congregation, and lured Trewella off to live with her.

A mermaid carved on a bench in Zennor church is supposed to commemorate these events, although the sequence was probably the other way round, the fifteenth-century carving inspiring the story. Mermaids quite often appeared in churches – or at least their images did. Generally they were symbols of lust and vanity, warnings against temptations of the flesh, but a more positive spin was given by a Zennor gardener in the 1950s, who quoted an old Cornish passion play to the effect that 'Half a mermaid is human; a woman from the head to the stomach. Like that is the Jesu.' The dual nature of the mermaid could thus be understood as representing the sacred synthesis of God and man.

SOUTH-EAST ENGLAND

Hampshire, Kent, London, Sussex, Isle of Wight

ALTON, HAMPSHIRE

Sweet Fanny Adams

In 1867, a young girl of Alton was murdered and dismembered by Frederick Baker, a solicitor's clerk. His diary entry for that day read: '24th August, Saturday. Killed a young girl. It was fine and hot.'

Baker was hanged at Winchester in December, but the case was not forgotten. At around the same time, the Royal Navy had had its traditional fare of salt tack (pork or beef salted down in barrels) changed to cheap tinned meat, 'preserved mutton' as it was fairly euphemistically called, and a gruesome joke started going the rounds suggesting that there was a direct link to the murder – that, in fact, what the sailors were being fed was the chopped-up body of the girl, some of whose remains had been found at Deptford Dockyard. Some say the rumour began when a sailor discovered a button in his dinner. Accordingly, the name of the unfortunate girl, Fanny Adams, was applied to the new navy rations, and as her initials were the same as those for a crude phrase meaning 'nothing at all', the phrase 'sweet Fanny Adams' was substituted for 'sweet f— all'.

The new tins may not have been very much worse than the old salt tack rations, which the navy kept in circulation for longer than you would have thought possible. Salt beef, phased out from the 1860s, was finally withdrawn in 1904, from which date no new barrels were salted down, but existing supplies reportedly lasted until 1913, and those of salt pork until 1926.

ARUNDEL, WEST SUSSEX

Halcyon days

The folklore collector Edward Lovett wrote in 1928 of seeing a dead kingfisher hung up as a sort of barometer:

> It was at Arundel when I saw my first example of this interesting thing. The bird was a stuffed specimen and it was suspended by a piece of common string about two feet long, from the beam of a room in a small old-fashioned house; the string being fixed to the back of the bird so that it hung horizontally.
>
> The owner assured me that the bird always pointed in one direction during fine weather, but turned and pointed the other way 'when it rained,' or when it was going to rain.

This practice was common in France along the Loire valley, where a local name for the kingfisher is *vire-vent*, 'turn-in-the-wind'. It began among fishermen, who used to hang the birds in the rigging of their boats.

The custom of using kingfishers, also called halcyons, as weathervanes was known in England from at least the sixteenth century. In *King Lear* (1608), Act II scene 2, Shakespeare has Lear's loyal servant Kent rail against sycophants who 'turn their halcyon beaks / With every gale and vary of their masters', and it continued to be a received opinion in the mid seventeenth century that 'a Kingfisher hanged by the bill, sheweth what quarter the winde is', as noted by Sir Thomas Browne, who experimented by hanging up two kingfishers, but found that they gave contradictory results.

Although we tend to think of kingfishers as river birds, in winter they congregate around the coast, and are occasionally seen hovering near ships, when they are taken as a fortunate omen, as in a seventeenth-century poem by Robert Wild:

> The peaceful Kingfishers are met together
> About the Decks and prophesie calm weather.

In Greek legend, the association of kingfishers with pleasant weather was explained by the story of Queen Halcyone. Having thrown herself into the sea in grief when her husband was drowned, she was changed into a kingfisher, and for her sake, it was said, the gods stilled the waves for a while every winter. Since kingfishers were thought by ancient naturalists to make their nests on the open sea, this period of calm (a few fine days which

Hertfordshire
Oxfordshire
Buckinghamshire
Stepney
London
Bermondsey
Isle of Dogs
Berkshire
ENGLAND
Surrey
Hampshire
Alton
West Sussex
Devil's Dyke
Arundel
Bosham
Chichester
Hove
Portsmouth
Shoreham-by-Sea
Selsey
Cowes
Newtown
Isle of Wight
Brading
Dunnose Point
0 5 10 15 20 mi
0 10 20 30 40 km

Essex
Thames Estuary
The Nore
Isle of Sheppey
Chatham
Whitstable
Margate
Ramsgate
Sandwich
Flats
Goodwin
Sands
Kent
Dover
Hythe
Folkestone
Hawkhurst
East Sussex
Hastings
Newhaven
Cuckmere
Haven
The English Channel

generally do in fact occur in the Mediterranean around the winter solstice) was supposed necessary in order for the birds to hatch their eggs in safety. Later observation proved that kingfishers actually nest in riverbanks, but the phrase 'halcyon days' continued to signify sunny, peaceful times.

BERMONDSEY, LONDON

A haunted ship

A curious statement was drawn up in October 1870 at Bermondsey, by Captain John Blacklock of the *Robert Lowe*, and signed by seven of his crew. It bore witness to the appearance of a whole family of wraiths on board the vessel, a steamship employed in repairing telegraph cables.

The statement began by relating how W. H. Pierce, an engineer, had come down with typhus, and died on 4 October. The stoker, David Brown, described as 'a strong healthy man . . . not likely to be led astray by imagination', had looked after him until the day before his death. At 3 p.m. on that day, Pierce had been delirious and wanted to get out of bed, but Brown had stopped him. Brown said:

> I was standing on one side of the bunk, and while trying to prevent Pierce from rising, I saw on the other side of the bunk *the wife, two children, and the mother of the dying man, all of whom I knew very well*, and they are all still living. They appeared to be very sorrowful, but in all other respects were the same as other ordinary human beings. I could not see through them: they were not at all transparent. They had on their ordinary clothes, and perhaps looked rather paler than usual. The mother said to me, in a clearly audible voice, '*He will be buried on Thursday, at twelve o'clock, in about 1400 fathoms of water.*'

The apparitions then vanished. Pierce had seen nothing. Brown ran from the cabin and told the captain that he would not stay with the invalid any longer. Not wanting to frighten the rest of the crew, he did not immediately say why, but about an hour later he told Captain Blacklock and Andrew Dunbar, the chief engineer, what had happened. He had never, he said, seen anything like it before, and insisted that at the time he had been 'perfectly cool and collected'.

Blacklock confirmed that Brown had come to see him and declared that he would no longer attend Pierce. Blacklock had tried to persuade him, but Brown refused, left in tears, and was so upset that he became quite ill.

The statement by the captain and crew was made at general request, since rumours of the haunting had got about and caused gossip. The men concerned declared that what they had said was truc 'so far as cach of thc circumstances came under our personal notice', but committed themselves to no opinion as to the cause of the phenomenon.

Reporting this case in *Apparitions: A Narrative of Facts* (1874), the Reverend Bourchier Wrey Savile adds that Pierce's widow had had a premonition of her husband's death. While staying with her mother-in-law, she was woken one night by loud knocking which sounded as if it came from the front door, but when she looked out no one was there. 'It came into her mind that she would soon receive bad news. Afterwards she dreamed that she was a widow, and that her children were dressed in deep mourning.' This happened on 28 September, a few days before the actual death.

BOSHAM, WEST SUSSEX

Vikings steal a bell

Today one of West Sussex's prettiest villages, in former centuries Bosham was an important town with royal associations. Some accounts state that its harbour was where King Canute commanded the tide to go out, and although that claim is disputed by places including WALLASEY (North-West England & Isle of Man), the eleventh-century Danish king may well have stayed in Bosham.

Bosham as it was in the late nineteenth century. Low tide exposes the mudflats, but somewhere in deeper water, according to the legend of the town, lies a bell stolen from the church by Vikings.

Long before that, however, the Danes had already made their presence felt here. According to legend, Vikings once attacked the church of Bosham monastery and carried off the great tenor bell. In triumph they took it to their longship and set sail, but as soon as the pirates were gone, the monks ran back to their church and rang a peal on the remaining bells. The tenor bell answered with one booming note, and then broke through the planks of the Vikings' boat to sink forever beneath the waves at Bosham Deep, once known by reference to the legend as Bell Hole. The Vikings themselves were either drowned, or, in an alternative version, immediately converted to Christianity when their shattered vessel miraculously mended.

Like other 'drowned bells', those of SELSEY for example, the lost bell is said to sound underwater, in this case in answer to its former companions in Bosham church. A local rhyme records that it cries:

> Ye bells of Bosham, ring for me,
> For as ye ring, I ring wi' ye.

Much later, the people of Bosham tried to recover their bell, and under instruction from a wise man they harnessed a team of pure white oxen to pull it up. When it had nearly broken the surface of the waves, the rope snapped and it sank once more, because one of the cattle had a single black hair in its tail. According to a local harbourmaster, however, it was a female voice that caused the disaster. The instruction was that no woman must speak while the bell was being raised, but one could not restrain herself, and cried 'Oop she comes!' – at which, of course, the bell vanished forever. This twentieth-century version of the tale reflects the common sailors' belief that women brought bad luck at sea, a superstition that inspired folk tales such as that told at DOVERCOURT (East Anglia).

BRADING, ISLE OF WIGHT

The Druids' well

Until the late nineteenth century, boats entering the estuary of the river Yar could sail two and a half miles upstream to Brading, at that time the main eastern port of the Isle of Wight. At low tide, the banks were an expanse of mudflats criss-crossed with salt creeks, and at the flood the whole area was a brackish lake, known as Brading Haven.

In the 1870s, the land was successfully drained and reclaimed, in order to build a railway line from Brading to Bembridge. Several earlier attempts had failed when the sea breached the dykes, but in the course of one such enterprise in about 1622, an ancient stone well was found near the middle of the haven, showing that the area had once been solid ground.

From this discovery sprang a legend accounting for the drowned land. It was said that the harbour had once held a great forest of oaks, and, in a rhymed version of the tale composed by W. H. Davenport Adams in 1856:

> Midway in this wood, – so the tale they tell, –
> There yawn'd a deep chasm, a haunted well.

Here a Druid or magician kept a water-spirit captive, and put a charm on the well to bring disaster on anyone who uncovered it. Ages later, in the days of William the Conqueror, a Norman knight wanted to hunt in the forest and had the undergrowth cleared away, revealing the structure:

> And he vowed, in defiance of charm or spell,
> To pluck every stone from the Druid's well.

When he put the spell to the test, a deluge followed that drowned both the knight and the land.

This account is influenced by older traditions of overflowing wells in places such as CARDIGAN BAY (Wales). An alternative account, purportedly taken from an old manuscript and given in Abraham Elder's *Tales and Legends of the Isle of Wight* (1839), is more sensational. According to Elder, the Druids of old had several holy places on the island, of which the most mystic and terrible was 'the Hexel ground' – presumably from 'hex', a witch or a curse – at Brading Haven:

> Here was their most sacred grove of oaks, and once in every year they built a tall pile of dry wood, on the top of which it was said that they bound a living man, confining him in a vessel of wicker-work, and they sacrificed him with fire.

The stone well was the spot at which they made their human offerings. A time came, however, when three Christian pilgrims arrived to convert the islanders. The Druids took the missionaries prisoner, put them in their wicker cage, and had just lit the pyre when thunder rolled, lightning struck, and with a rushing sound a fountain of water fifty foot high rose from the well. 'Then came a dreadful roaring, like all the storms of the ocean roaring together. The sea came rushing in upon them in one mighty wave. Trees and men were hurled along to one wild ruin.'

The waves put out the fire, and the wicker vessel floated the three Christians to safety. 'But the sea never returned again to its deep; and the Hexel ground of Yar . . . still lies beneath a wide lake' – which in Elder's days was true.

CHATHAM, KENT

Ghosts of the dockyard

Ghosts have been reported at Chatham Dockyard in Medway since at least the seventeenth century. While visiting Sir William Batten, Surveyor of the Navy, in April 1661, Samuel Pepys stayed the night at Hill House and wrote in his diary that 'Sir William telling me that old Edgeborrow, his predecessor, did die and walk in my chamber, did make me somewhat afraid, but not so much as, for mirth sake, I did seem.'

Waking at three in the morning, 'by the light of the moon I saw my pillow (which overnight I flung from me) stand upright, but, not bethinking myself what it might be, I was a little afraid . . .' He soon went back to sleep, however (possibly helped by the previous evening's 'very merry' supper), and was not disturbed again.

Although Pepys did not see Edgeborrow's ghost, several people are supposed to have seen a more famous phantom, that of Horatio Nelson, who joined his first ship *Raisonnable* at Chatham in 1770. He is said to appear not as a one-armed veteran, but as a young man – very young, perhaps, since he was only twelve years old when he arrived there.

CHICHESTER, WEST SUSSEX

The headless frogman

In 1957, the gruesome discovery was made in Chichester harbour of a headless and handless corpse dressed in frogman's gear (a rubber suit with flippers and oxygen cylinder). The body was identified as that of Lieutenant Commander Lionel 'Buster' Crabb, a veteran diver and underwater demolitions expert. He had gone missing more than a year earlier, while diving in Stokes Bay, Gosport, near the Soviet warships that had brought the Russian statesmen Khrushchev and Bulganin to Britain, and the mystery of his fate had been a nine days' wonder in the papers. Various theories had been put forward, including a Russian kidnap, a Russian murder, and a publicity stunt gone wrong, but the authorities had remained tight-lipped about the whole affair, making only a brief announcement that

Crabb was 'presumed dead as a result of trials with certain underwater apparatus'.

The fishing-up of the mutilated body provoked another outburst of interest, but little more information was forthcoming, and no answers were given as to why Crabb's head and hands had been removed, nor why the Portsmouth hotel where he had supposedly stayed the night before his disappearance had the relevant page torn out of its register. People suspected a cover-up, and wondered whether the dead man was really Crabb. In 1960 the matter exploded into the papers yet again, with headlines shouting CRABB ALIVE, and CRABB WORKS FOR RUSSIA. It was claimed that he had defected, been smuggled back to the Soviet Union on one of the Russian ships, and had been helping them ever since with trials of underwater weapons. The Russians themselves denied all such rumours – but then they would, wouldn't they?

Cold War paranoia was at its height, and several books were written about the affair, all full of tantalising theories, none conclusive. It was noted that the body washed up at Chichester had been identified neither by Crabb's estranged wife nor by his girlfriend Patricia Rose, but by a naval officer, who might have been following an official script. Rose, with whom Crabb had a rather complex relationship – they had been engaged, split up and each married somebody else, but had gone on seeing each other afterwards – was convinced that he was alive, and insisted that she had had messages from him promising that he would come back to the West. She also reported that he had said to her before he disappeared, 'I love you but my Queen and my country come first,' implying that his mission to Russia was as a double agent.

The last episode in the saga (so far) was in 1987, when the government announced that rather than releasing the relevant papers under the thirty-year rule, they were going to keep the file closed for a further seventy years – a clear hint that there are secrets here. However, we will not know what they are until 2057.

COWES, ISLE OF WIGHT

Launching a ship

The launch of a ship is usually also the occasion of her naming, a ceremony of great symbolic importance. Toasts or libations of wine or spirits seem to have been involved for a long time in the christening of boats, and are sometimes said to be a survival of ancient pagan blood-sacrifice. Up until the late seventeenth century, the ship's health would be drunk from a silver cup that was then thrown into the sea, an extravagant practice that was replaced with the smashing of a bottle on the bows, ideally champagne, but often something much inferior, another economy measure. In the late nineteenth century, a shipyard office boy revealed that his employers used 'christening wine', very cheap and probably rather nasty, since it was never intended to be drunk.

If anything goes wrong with the naming, it is held to be a bad omen for the vessel's future. One early occasion was apparently a disaster, when a Hanoverian princess 'threw the bottle with more energy than accuracy and it struck one of the spectators, causing severe injuries'. Since then, according to Commander A. B. Campbell's *Customs and Traditions of the Royal Navy* (1956), the bottle has always been attached to a cord, but the performance is still prone to error. In about 1860, at the launch of the tea clipper *Chaa-sze*, the thrower did not use enough force, and the bottle swung back unbroken. A carpenter leapt on to a log and flung the bottle again at the ship's side, shattering it, but the bungled ritual gave the ship an unfortunate reputation that hung about her for the rest of her career.

A more recent example of an ill-omened launch was in 1973, when the *Morning Cloud*, the third racing yacht of that name owned by Edward Heath (then the British prime minister), entered the water at Cowes. A crowd had gathered on the wall along the slipway to watch, and just as the newly christened boat passed, the wife of one of the crew stumbled and fell on to the concrete below, and had to be rushed to hospital with severe concussion. 'This will be an unlucky boat,' Heath heard somebody mutter, and the prediction proved true. The third *Morning Cloud* had been designed specifically to win the Fastnet race from Cowes to Plymouth, but in the event it proved to be 'the most frustrating ocean race in which I have so far taken part', wrote Heath, since the yacht was constantly becalmed, every breeze dying away as soon as they caught it.

Much worse was to come. In 1974, the yacht was due to compete in the Admiral's Cup, and a crew including Heath's godson Christopher Chadd set off to bring her from Burnham in Essex round to Cowes. A Force 9 gale drove up the Channel, throwing the ship on her side and washing two men overboard. One was rescued, but the other, Nigel Cumming, was lost, his lifeline broken. A second wave hit the boat, and Chadd, who was wearing his life jacket but had not clipped on his lifeline, was swept from the deck and drowned. The remaining five members of crew abandoned the badly damaged yacht and crammed themselves into a four-man life raft, which eventually came ashore at Brighton.

On the same day, the tempest, later known as the '*Morning Cloud* storm', smashed the original *Morning Cloud* against the rocks of Jersey, where she was moored, though without any loss of life. That boat's beginnings too had been unpropitious, since at her christening the wine bottle had to be struck three times before it broke.

CUCKMERE HAVEN, EAST SUSSEX

A Sussex moonraker

> Still as a mouse if you chance to sit
> By a deep, deep lane in Sussex
> Where the trees grow arching over it,
> You may hear the chink of a ghostly bit,
> And watch the men and horses flit
> Down the hollow ways of Sussex.
>
> On a pitch black night, in a thick sea mist
> They ran ashore in Sussex:
> Yet you may know from the double twist
> Of bloody rag round the Captain's wrist
> That the revenue sharks were keeping tryst
> In the lonely ways of Sussex.

The 'hollow ways' of Bernard Darwin's poem still wind through the Sussex woods, sunken paths along which men and horses could pass hidden even in daylight, taking contraband from the coast. Cuckmere valley, where the lonely shore gave easy access to hills and covered lanes ideally suited to secret journeys, was one of the free-traders' favourite routes. A traditional anecdote tells how a Sussex shepherd, sitting on a hill above Seaford, saw a boat come in to Cuckmere Haven, from which two barrels were carried up and sunk in a dewpond. He waited until nightfall, and then by the light of the moon went to explore the pond with his crook, but was interrupted by two excisemen, who asked him what he was doing. The quick-witted shepherd put on his most exaggerated country accent, and replied that he was trying to catch the moon, which seemed to have fallen into the pool. The revenue officers rode away, laughing at the foolish peasant, who got his brandy after a little more fishing.

In Wiltshire, a similar story was told of 'simple rustics' fooling the authorities, and gave rise to the nickname 'moonrakers' for the county locals.

DEVIL'S DYKE, WEST SUSSEX

St Dunstan and the Devil

Looking around Sussex one day, according to a popular tale, the Devil was furious to see so many churches and to hear the sound of prayer and bells. He decided that he had had enough of all these holy people, but in his gentlemanly way he offered St Dunstan, the Archbishop of Canterbury, the choice of how Sussex was to be destroyed. The saint considered, and after a while he said that drowning in sea water might be best, but he doubted whether the Devil could cut through the Downs to let the sea in. Conceitedly, the Devil declared that he could do it in a single night. Done, said Dunstan, and if you haven't managed it by dawn, then you'll have to give up.

The Devil dug and dug, flinging great lumps of chalk about, and by three o'clock in the morning only a slender barrier remained between the sea and the low-lying land north of the Downs. St Dunstan saw that he would have to step in, and set himself to pray. In answer, every cock crowed all over the Weald. The Devil stopped work and looked up, amazed that dawn

Satan was prevented from flooding Sussex by St Dunstan, Archbishop of Canterbury, who then pulled his infernal adversary's nose with a pair of tongs. Here the Devil is portrayed in reptile shape, making the image a spoof on dramatic pictures of St George slaying the dragon.

could have come so soon. Seeing the sky still dark, he realised that he had been tricked, but he had lost precious minutes, for now the sun really was beginning to rise, and his chance was gone.

Fuming, the Devil flew off to find St Dunstan, who invited him cordially into his bishop's palace at Mayfield. Here the saint had a workshop with a small forge, and while the Devil explained how cross he was, and how unfairly he had been treated, Dunstan occupied himself heating up a pair of tongs. 'Yes! Yes! Proceed!' he said. 'But I have an argument against all this!' At last the Devil asked him what his argument was, and in answer the saint pulled the Devil's nose with the red-hot tongs until he flew away.

You can see the very tongs he is said to have used, at the Old Palace in Mayfield, and as for the Devil's digging, the evidence for that is the big cleft in the hills north of Hove, running south-west from Poynings towards the sea.

The narrative composed by Hilaire Belloc in 1912 neatly combines one famous Sussex legend, the origin of the Devil's Dyke, with another, how St Dunstan pulled the Devil's nose. Other versions of the former often attribute the Devil's defeat not to the archbishop, but to a woman who shone a candle through a sieve so that it looked like the sun rising, tricking the cocks into crowing and the Devil into suspending his labours. A variation on the Devil's Dyke tale, recorded from Poynings in the mid twentieth century, records that when his Satanic majesty flew off in chagrin, he had a big lump of Sussex clay stuck to his heel, which fell off in the Channel and became the Isle of Wight.

DOVER, KENT

Giant ships

In a children's story of 1884, *Her Majesty's Bear* by Elizabeth Harcourt Mitchell, a French girl exclaims at her first sight of the Dover cliffs, 'Oh, how white and how high! Why are they so white, so much whiter than the other side?' Her companion, a boy of Dover, tells her the reason he has heard:

> 'Once upon a time, a great ship from Norway came down into these seas, and she was so big that she could not get through the straits, but stuck quite fast. The captain, I cannot tell whether the tale be true, then said to the crew, "Soap her sides, my men," and they soaped and soaped until she could slip through quite easily, but she left the soap upon the cliffs, and ever since they have been as white as snow.'

This was a widespread fable. The great ship had a name, the *Merry Dun of Dover*, and to give an idea of her impossible dimensions, it was said that the captain had to ride about the decks on horseback to give orders, and sailors who climbed the rigging when young came down again with grey beards. An early nineteenth-century ballad says that the *Merry Dun* could take on board coal from Sutherland at one end, while at the same time unloading a cargo at London, and that while tacking to leave the English Channel,

> Like some mountain, the dread of the people,
> Her flying-gib-boom then knock'd down Calais steeple!

One basis for tales of enormous ships is Noah's Ark, capable of holding at least a pair of every animal (strictly speaking it held a pair of every 'unclean' beast, those that could not be eaten, and seven of every 'clean' animal fit for food); another may be the medieval Irish legend of the *Roth Ramhach*, a magic land-and-water ship that sailed as well on land as on sea, and had a thousand beds each containing a thousand men (*see* MALAHIDE, Southern Eire). Norse mythology has *Skidbladnir*, constructed by the dwarves and given to the god Frey, a ship so capacious that when it sailed it could contain all the gods with their weapons and armour, but contrived so ingeniously that it could be folded up and put into a pocket.

In the nineteenth century, when ever-larger craft were being built, tales of giant ships developed in many nations. To Frisian sailors, the stupendous vessel is the *Mannigfual*, and in America she is the *Courser*, captained by Old Stormalong, hero of a sea chanty and of many tall tales. The French equivalent is called *Le Grand Chasse Foudre* ('Lightning-chaser'), although the nineteenth-century historian Auguste Jal heard a coarser version which he gives only as *Le Grand Chasse F . . .* , omitting the obscenity. Swedish yarns of the *Refanu* describe her as so large that it takes three weeks to get from one end of her to the other, and say that a schooner once got lost in her soup pot.

All these huge ships are portrayed as comfortable places, with plenty to eat and a happy crew – and, in the case of the *Refanu*, a pub in every pulley. Their virtually infinite size suggests that they are an image of a mariner's heaven, although other fables set the sailor's preferred afterlife firmly on land (*see* FIDDLER'S GREEN & DAVY JONES'S LOCKER, p. 376).

DUNNOSE POINT, ISLE OF WIGHT

Visions of a ghost ship

While in charge of a submarine in the 1930s, Commander F. Lipscombe was convinced that he had nearly run into a fully rigged nineteenth-century frigate off the Isle of Wight. Puzzled at seeing such an old-fashioned ship, he made enquiries on the island, where everyone told him that he must have encountered the phantom of the *Eurydice*.

Built in the 1840s, the *Eurydice* was converted into a training ship in 1877 and sent off to the Caribbean with a crew of young sailors under instruction. She returned in March 1878, and approached Dunnose Point in mid afternoon. Several onlookers reported that she was carrying too much sail, so that if a strong wind blew, she would be at risk. They could see from the cliffs, as apparently those on deck could not, that ominous clouds were blowing up from the north-west, and as the ship rounded the point, a blizzard caught her, throwing her to starboard. Her portholes were open, letting water pour through, and before any action could be taken, she had capsized. Only two men survived of the three hundred aboard.

Several reports were later made of premonitions connected with the disaster. The sister of one of the sailors said that at the time of the wreck she had had a feeling of terrible anxiety, relieved for a second when she thought she heard her brother's footsteps approaching. When she opened the door and saw nobody there, she was at once convinced that he was dead. More extraordinarily, at a tea party in Windsor, Sir John MacNeill (many of whose family were said to have second sight) exclaimed out of the blue, 'Good Heavens! Why don't they close the portholes and reef the topsails!' One of his companions asked what he was talking about, and MacNeill said that he had suddenly had a clear vision of a ship racing up the Channel under full sail, her portholes gaping and a black storm gathering.

Not long after the tragedy, it began to be said that the *Eurydice* was haunting the scene of her wreck. Fishermen said that a frigate would appear, speeding towards them, but vanished when it got near, and so often was this apparition seen that when Commander Lipscombe had his encounter, it was at once assumed that this was another of the *Eurydice*'s appearances.

Figureheads

From early times, carvings or pictures of deities, animals, or human beings have adorned ships, a practice serving several purposes. A symbol of a god or saint protects a vessel, while giving the ship a face helps identify her in a crowded harbour, and also lets her 'see her own way', a function sometimes fulfilled simply by painting an eye on either side of her bows.

The Vikings made serpents or dragons of their longboats, and animals, particularly lions, continued for centuries to be popular emblems. In the mid eighteenth century, fantastically complex images were carried by some British ships, such as the *Victory* of 1765, which displayed on its bows among other things a bust of the king, four cherubs, figures of Britannia and Victory, and a five-headed hydra, but towards the end of the century, naval cutbacks curbed excesses like this. From around this time, half-length human statues became common, and began to be known as 'figureheads'.

Considered to personify their ships, figureheads could be treated with delicate consideration. In 1778, a hammock was tied over the eyes of the *Royal George*'s figurehead to save it from seeing a humiliating retreat, and when the *Brunswick*'s figurehead had its hat shot off in battle in 1794, the captain reportedly substituted his own hat.

While it was shameful for a male image to be bare-headed in face of a foe, it was not unusual for a female one to go bare-breasted. The widespread portrayal of half-naked women as figureheads is sometimes linked to pre-Christian dedications of ships to goddesses, or to a superstition that a naked woman could calm a storm at sea, although a display of female flesh in an all-male environment does not necessarily call for in-depth explanation.

Occasionally, figureheads could exert a malign influence. In the mid nineteenth century, a charming female figurehead from a captured French warship was built into an American vessel, which sank on its first voyage. The image was rescued and transferred to a new ship that also went down soon afterwards. Yet again the figure was salvaged, and when her next ship, the *Maritana*, ran on the rocks near Boston, Massachusetts, in 1861, she was found floating among the bodies from the wreck. This time she was kept on land, first at

A tenth-century Anglo-Saxon ship carries an impressive figurehead of a horse.

Lincoln Wharf and then in the Old State House – both of which were devastated by fire during her residence, leaving the figurehead, however, intact. Accidents continued to occur around this sinister lady, and it was said in the 1880s that she appeared as a phantom, strolling along the quays by night.

This is perhaps the only ghostly figurehead on record, although there is a story that a nineteenth-century curate at Morwenstow in Cornwall, on passing the churchyard at twilight and seeing a naked woman apparently rising from a grave, hurried to report to the vicar that his dead parishioners were on the move. It was in fact a figurehead, erected in the West Country tradition as a memorial to a shipwrecked crew.

FOLKESTONE, KENT

The Herring Spear and the Seven Whistlers

The sound of birds passing overhead on their annual migration across the Channel was held to be a sign of good fishing, but was nonetheless feared by nineteenth-century Folkestone sailors. They called the flocks the 'Herring Spear' or 'Herring Piece'.

In his *Curiosities of Natural History* (1860), Francis Buckland writes that 'Strange things are seen and heard by the fishermen when drifting silently along, far out at sea, through the murky darkness of midnight.' An old Folkestone man named Smith told Buckland that he liked to hear the Herring Spear, since 'we always catches more fish when it is about,' but when Buckland's friend Roberts heard the noise himself out in a boat near Dover, the men he was with were afraid of it:

> All of a sudden, a curious rushing and a rustling sound was heard over and about the boat, accompanied by a low musical twittering; this phenomenon was not continuous, it passed away, and in a few minutes was again repeated; meanwhile all was darkness and mystery. He asked the men what strange noise that was; they shook their heads in ignorance; they did not know what it was, whether earthly or unearthly; they evidently regarded it with somewhat more than awe, and were unwilling to converse about it; their dread of it was great; all they knew was that it was the 'Herring Piece.'

Buckland identifies the birds concerned as redwings, who migrate around the herring-fishing time. The nineteenth-century folklore collector Fletcher Bassett says they are curlews, and reports that fishermen in the Channel thought they caused the east wind, but he conflates the Herring Spear with the Seven Whistlers, another mysterious night-calling flock of birds mentioned by Smith, who had told Buckland about the Herring Spear. The Whistlers were a worse omen:

> 'I never thinks any good of them,' said old Smith; 'there's always an accident when they comes. I heard 'em once one dark night last winter. They come over our heads all of a sudden, singing "ewe-ewe," and the men in the boat wanted to go back. It came on to rain and blow soon afterwards, and was an awful night, sir; and, sure enough, before morning, a boat was upset, and seven poor fellows drowned.'

Smith said that the noise was made by curlews, but knowing what bird made the noise is not the same as ascribing the event itself to natural causes. Allowance must be made for popular belief not only in the direct intervention of Providence in human affairs, but also in the existence of soul-birds (*see* PORLOCK, South-West England & Channel Islands). Some said that the Seven Whistlers were the spirits of the dead, and that their cries foretold more death to come.

GOODWIN SANDS, KENT

The ship-swallower

In *A Perambulation of Kent* (1570), William Lambarde describes the Goodwin Sands as 'either sea, or land, or neither of both'. Towards the end of the eleventh century, he writes, due to 'a sudden and mightie inundation of the sea', the place was 'violently overwhelmed with a light sand, wherewith it not onely remaineth covered ever since, but is become withall . . . a most dreadfull gulfe, and ship swalower', a description as true today as it was then.

The Goodwins are extensive shoals lying off the coast of Kent near the entrance to the English Channel from the North Sea. Partially exposed at low tide, they present an innocent face to the sky, and a surface firm enough to walk on or even for a game of cricket, but when the sea rises, the land quakes, becoming an unstable bog. Any ship unlucky enough to go aground here can be sucked down completely – swallowed, as Lambarde suggests – or shattered by the currents that swirl the sand away from underneath and leave the vessel unsupported, so that it breaks apart under its own waterlogged weight. Repeated attempts to build a lighthouse warning sailors of their peril have always failed, due to the shifting ground, and the Goodwins are now marked by lightships.

The very real dangers of the Goodwin Sands are accompanied by a considerable body of legend. It was believed that they were the last remnant of the 'Island of Lomea', an estate belonging to Edward the Confessor's adviser Earl Godwin, and the local phrase for stormy weather was 'Earl Godwin and his court are hungry', suggesting that a spectral earl lurked beneath the sands in wait for prey. As with other 'drowned lands' such as those at BLACKPOOL (North-West England & Isle of Man) and CARDIGAN BAY (Wales), bells of lost churches are said sometimes to be heard from beneath the waves. In the case of the Goodwins, however, there seems no evidence that this was ever solid ground: investigation has revealed marine matter and sand all the way down to the chalk base.

This picture shows an early attempt to set up a warning beacon on Goodwin Sands.

Rumours that the land was encroached on by the sea may have started after drainage of the area was abandoned in the fifteenth century, when money intended for the project was diverted to build a church at Tenterden. From this abortive enterprise came a proverb current in the late eighteenth century, 'Tenterden's steeple's the cause of Godwin's sands', apparently a statement made during an inquiry into the silting-up of nearby Sandwich harbour. Because the examining official did not wait to hear the rest of the story, he thought it was nonsense, and thus the phrase became common parlance for any absurd explanation.

The notorious hazards of the sandbanks are mentioned in *The Merchant of Venice* (1598), Act III scene 1, where Antonio's ship is reported to have foundered 'on the narrow seas; the Goodwins I think they call the place, a very dangerous flat, and fatal, where the carcases of many a tall ship lie buried'. Shakespeare was writing not long after a Spanish galleon, part of the 1588 Armada expedition, had headed on to the Goodwins in flames when her powder magazines caught fire.

This ship is said to have reappeared in 1703, when a fleet of British men-o'-war were driven on to the sands. Several were lost, including *Northumberland* and *Mary*, and survivors from the latter reported seeing 'a great warship of Drake's day, her sails tattered, burning from fore to aft and her guns firing', which 'sailed right through our ship and finally disappeared before our eyes into the depth of the Sands'. Fifty years later, another phantom appeared. The captain of an East India clipper recorded in his log that while his vessel lay at anchor off the Goodwins, 'an armed frigate came driving down on my ship, her masts gone, her decks and hull in fearful shape'. As it came nearer, the crew realised that it was the long-vanished *Northumberland*. 'It was a spectacle far too terrible to dwell upon, to see this ghost of what was once a fine warship going to her doom a second time.'

In the eighteenth century, the most famous wreck occurred, that of the schooner *Lady Lovibond*. A romantic tragedy is sometimes said to lie behind her destruction. The captain had just married, and his bride was aboard, but so too was a rejected rival for her love, and out of jealousy he set the ship directly on course for the Goodwins. Another version lays the blame, again, on a phantom, holding that just before the disaster the helmsman had been seen transfixed in horror, watching a ghostly ship on a collision course with his own.

These reports from twentieth-century ghost-hunters, neither well sourced nor consistent as to dates, are cited as supporting evidence for modern appearances of the phantom ship, now reported as that of the *Lady Lovibond* rather than the sixteenth-century Spanish galleon or the eighteenth-century *Northumberland*. The *Lady Lovibond* is supposed to have appeared every fifty years since 1784 (or, according to some accounts, 1748), but missed her most recent appointment. Her last recorded appearance was in 1934, when a lifeboat was sent to a ship on fire near the Goodwins, but found no ship and no trace of wreckage when they got to the place.

Whatever the truth of the ghost ships, the Goodwins have an eerie atmosphere. In *Sea Phantoms* (1963), Warren Armstrong writes of riding at anchor in a fog just off the coast:

> A bell buoy tolled funereally nearby; it was as if we were in a tomb that narrowed with each passing hour. It was an uncanny feeling; more so when I heard a strange sound, like something infinitely evil crawling aboard along our anchor chain. But the captain of the supply tender grinned at me.
>
> 'You'd get used to it if you had much to do with the Goodwins,' he explained. 'That sound comes from big hermit crabs climbing the links of the anchor chain of any ship in these parts!'

HASTINGS, EAST SUSSEX

See HOVE.

HAWKHURST, KENT

The Hawkhurst Gang

Smugglers were not necessarily popular members of their community. While folklore tends to paint a picture of wit, bravery, and resourcefulness, history has instances of very different behaviour, and some of the larger gangs tyrannised over their neighbours like a local mafia. In the early eighteenth century, the Hawkhurst Gang grew from modest beginnings to become a huge and brutally well-organised outfit, involving a network of associates from Folkestone as far as Poole in Dorset. Their leader, Arthur Gray, became so rich from his illegal trade that he was able to build a mansion (later known as Gray's Folly) at Seacox Heath near Hawkhurst, and he boasted that within an hour he could raise a force of five hundred armed and mounted men. Gray's successor Thomas Kingsmill, an even tougher proposition, increasingly abandoned traditional smuggling in favour of extortion and robbery.

Contemporary reports may have exaggerated some of the gang's exploits, being written, like modern tabloid scares, more for sensation than accuracy. One story tells how some of the mob met an aged woman

A drawing, probably made soon after the event, of the Hawkhurst Gang breaking open the Poole custom house.

one day and asked how old she was. 'Eighty-one,' she replied, at which one of the men struck her on the head with his pistol, exclaiming, 'You have lived long enough!' An identical tale was told of the Ransley Gang, who operated further east around Hythe; whether it was true of either is debatable.

Certainly, however, the gang terrorised the south Kent villages, to such an extent that the inhabitants of Goudhurst contemplated wholesale evacuation. A local soldier, George Sturt, decided to make a stand, and organised his neighbours into 'The Goudhurst Band of Militia', who fought a pitched battle with their oppressors in April 1747. The militia won the day, but the Hawkhurst Gang was not done yet. Not long afterwards, a cargo of contraband tea was intercepted by the revenue officers at Poole. Sixty of Kingsmill's men attacked the custom house and seized the goods by force, but one of them was recognised by an onlooker, a shoemaker named Daniel Chater, who informed on him. Escorted by an old customs official, William Galley, Chater set out to give a formal statement, but the two men were overtaken and murdered by the smugglers - strapped on horseback and whipped to death in Galley's case, while Chater survived his beating for a few days and was then thrown in a well with stones pitched in on top of him.

Many people knew who was responsible but nobody dared speak out, and it was several months before one of the culprits was named in an anonymous letter. When arrested, he identified some of his confederates, and the conspiracy of silence was over. Thirteen of the ringleaders were executed, and that, finally, was the end of the Hawkhurst Gang. Or was it? At least one of the members, George Beesley, is rumoured to make appearances still. He was captured in Staplehurst, a few miles north of Hawkhurst, by dragoons who broke into his home just in time to stop him escaping up the chimney, and it is said that his ghost has been seen running through the house, only to disappear as it reaches the fireplace.

HOVE, EAST SUSSEX

Smugglers and vicars

Vicious thugs like the HAWKHURST Gang may have given Kentish smugglers a bad name, but the Sussex free-traders were regarded on the whole with affection. Their business was, indeed, so widespread at one time that it was virtually impossible to draw a line between those who were and were not involved. Even children were warned to turn a blind eye to any unusual night-time comings and goings.

> If you wake at midnight, and hear a horse's feet,
> Don't go drawing back the blind, or looking in the street,
> Them that asks no questions isn't told a lie.
> Watch the wall, my darling, while the Gentlemen go by!

Rudyard Kipling's 'Smugglers' Song' (1906) is based in reality. In the mid nineteenth century, a woman from Burwash, where Kipling lived for some time, reported to the village rector that as a child she had often been put to bed with the instruction, 'Now, mind, if the "gentlemen" come along, don't you look out o' window.' Peeping was punished severely, as it could lead to identification.

Burwash is a long way from the sea, but was one of the staging posts for the transport of contraband from the coast towards London. The sailors who brought the goods over in their boats were reliant on their inland counterparts, who received the tubs and cases on shore and conveyed them to hiding places from which they could be taken to the capital. Sometimes these caches were in private houses with big cellars or secret recesses, but in emergencies a church could be used. Belfries and churchyards were unlikely to be searched, particularly if a rumour of a ghost was set going to explain any odd noises and discourage the inquisitive.

Some clergymen were at least passive participants. A story is told of a discreet vicar who feigned illness one Sunday in order not to have to see inside his church, since he knew that the smugglers, hard-pressed by pursuing excisemen, had lodged their cargo in his pews. Others discovered their own involvement after the fact. Webster Whistler, vicar of Hastings, was woken up one night to take delivery of a barrel of brandy, his share of a consignment that had been concealed, unknown to him, in his church tower.

More innocent yet was the vicar of Hove and Preston, who (in the days before Hove had been devoured by Brighton) used to alternate services between the two villages. Turning up one 'Hove Sunday', he found the bell silent and the door locked. The sexton told him he'd got the week wrong, and should be in Preston. The vicar, however, was certain that this was not a 'Preston Sunday', and insisted that he be let into his church. In the end the sexton told him, 'It's no use, sir. You can't preach today.' 'Why not?' demanded the vicar. 'Because the church is full of tubs, and the pulpit's full of tea.'

HYTHE, KENT

St Nicholas

St Nicholas, patron of sailors, was revered in all coastal areas before the Reformation, and in many places his cult continued among mariners long afterwards. In England alone, around four hundred churches and chapels were dedicated to him, and were visited before and after voyages by sailors who presented propitiatory tributes to the saint beforehand, and thank-offerings on their safe return. One such place was the chapel of St Nicholas near Hythe, mentioned in Lambarde's *Perambulation of Kent* (1570). 'This is one of the places,' he says:

> Where such as had escapte the Sea
> Were woont to leave their guifts.

Fishermen who had survived a storm would come here to give thanks, and leave one or two of their best fish in gratitude. Similar rites were followed in Liverpool at the fourteenth-century chapel of St Nicholas, where a statue (now vanished) of the saint was courted with gifts, and at a chapel in Norwich which issued lead tokens to its votaries as a kind of receipt.

Lambarde writes with Protestant disapproval that St Nicholas held among Catholics 'the same Empire that Neptune had in Paganisme', being able with no more than a gesture to 'appease the rage and wallowing waves of the Sea, and also preserve from wrecke and drowning so many as called upon his name'. It is not entirely clear why such powers were attributed to St Nicholas. One of his most memorable miracles was raising to life some children who had been butchered and thrown into a tub of brine, and it has been suggested that pictures of this event were misinterpreted - the tub being taken for a boat, and the children for sailors. Later accounts of Nicholas's divine career state that he rescued three sailors from a tempest, an episode that may stem from the brine-tub images.

See also HAWKHURST.

ISLE OF DOGS, LONDON

The jinx on the Great Eastern

The *Great Eastern* was by far the largest of the mid nineteenth century's revolutionary iron ships. Designed by Isambard Kingdom Brunel, and built in John Scott Russell's shipyard on the Isle of Dogs, she had a tonnage (capacity) nearly four times that of any other ship afloat, and her vast size made for unprecedented problems. She was built on a sloping platform parallel to the river so that when she was finished she could be moved down sideways, but she kept getting stuck on the mud embankments, so that several attempts had to be made before she was finally waterborne. An abortive launch was considered an ominous portent for any vessel (*see* COWES), and throughout her short career the *Great Eastern* was considered a jinxed ship.

By the time she set off on her maiden voyage in 1859, she had already bankrupted Russell, her builder, and caused Brunel's death at the age of fifty-three, from the stress of overseeing her construction. Hardly had she left the Thames estuary when an explosion shattered the jacket on one of her funnels, filling the ship's saloon with steam, broken glass and iron fragments. Luckily the room was almost deserted at the time, but six firemen who had been working on the funnel were scalded to death. A stopcock that should have been open turned out to have been inexplicably wedged shut. In 1862, the ship struck a reef off Montauk, New York, tearing a hole in her outer hull, and bad luck continued to pursue her wherever she went. She claimed the lives of over thirty of her own crew, and caused at least ten accidents to other vessels, costing her owners over £130,000.

Suggestions that she was haunted seem to have begun during her construction, caused by knocking that came from inside the vessel when nobody was at work. Similar sounds were said to have preceded the explosion in 1859, interpreted (with hindsight) as a warning, and riveters repairing the damage from the 1862 accident went on strike for a while, complaining that they heard hammering outside their workroom. This turned out to be caused by a piece of metal hitting the ship's side as she rose and fell in the waves, and doubtless all the other noises could have been simply explained, but by this time the ship's uncanny reputation was thoroughly established. Travelling on the *Great Eastern* in 1868, the novelist Jules Verne was told several tales about the ship's ghost, one man claiming that it was a passenger who had lost his way in the hold and died there, another saying that it was an engineer boiled alive in the steam box.

Verne did not hear what came to be the most popular legend, that when the ship was built, a riveter (and his young assistant, in some versions) had been walled up alive in the ship's double hull. This became so firmly believed that when the *Great Eastern* was taken out of service in the 1880s and broken up at Birkenhead, it was widely anticipated that bones would be found, although in the event nothing of that nature was discovered.

MARGATE, KENT

The Lion's Mane

> If the bather or shore wanderer should happen to see, either tossing upon the waves or thrown upon the beach, a loose, roundish mass of tawny membranes and fibres, something like a very large handful of lion's mane and silver paper, let him beware.

This passage from J. G. Wood's *Out of Doors* (1874) may ring a bell with readers of Sherlock Holmes. In 'The Lion's Mane' (1927), Conan Doyle quotes from the nineteenth-century naturalist, who had an almost fatal encounter with *Cyanea capillata*, a poisonous jellyfish. One morning while swimming off Margate, Wood saw something like a patch of sand in the midst of the waves. Curious, since he knew that there was ordinarily no sandbank there, he swam nearer. As he did so, he began to notice a slight tingling in the toes of his left foot, which he took to be a touch of cramp. The sensation became stronger, like a nettle sting, and began to extend up his leg. 'Suddenly the truth flashed across me, and I made for shore as fast as I could.'

He found that dozens of slender transparent threads were dangling from his arm, still attached to the jellyfish, although he was now forty or fifty yards away from it. Although delicate as spiderweb, each filament was 'armed with a myriad poisoned darts that worked their way into the tissues, and affected the nervous system like the stings of wasps'. By the time he reached shore he was in severe pain, and when he left the water it was absolute torture. Wherever one of the threads had touched his skin, it had left a scarlet line of minute dots or pustules, and it felt as if each dot was a red-hot needle.

> Severe, however, as was this pain, it was the least part of the torture inflicted by these apparently insignificant weapons. Both the respiration and the action of the heart became affected, while at short intervals sharp pangs shot through the chest, as if a bullet had passed through heart and lungs, causing me to fall as if struck by a leaden missile.

His heart stopped beating 'for a time that seemed an age', before starting again with several convulsive leaps. His lungs refused to work, 'and I stood gasping in vain for breath, as if the arm of a garotter were round my neck'. The chest pains, the erratic action of his heart and the stoppage of his breathing continued as he staggered the two miles to his lodgings, where he caught a glimpse of himself in

The Headless Sailor

Sir Robert Pye and his wife Lady Anne lived at Faringdon in Berkshire in the seventeenth century. Their eldest son, Hampden Pye (his mother's maiden name was Hampden), married beneath him, to the fury of his family and particularly his mother, and in order to escape the unpleasantness he went to sea, joining Sir George Rooke's expedition to destroy Spanish treasure ships at Vigo, northern Spain, in 1702. In the battle he was decapitated by a cannon shot, and it was rumoured that this had been no accident, Lady Anne having arranged for her son to be placed in the position of greatest danger.

Following her crime she knew no peace. According to Faringdon gossip, 'whenever she went out in her carriage her son stood at the door with his head under his arm, handed her in, and took his seat opposite her', and after she died, his ghost became troublesome to

That Hairy-faced man is sallow and wan,
And his great thick pigtail is wither'd and gone;
And he cries, 'Take away that lubberly chap
That sits there and grins with his head in his lap!'

the whole neighbourhood. He was finally 'laid' in a pond near Faringdon House by a clergyman who bound the restless spirit there for a hundred years, and when the time was up in the 1830s, the locals became nervous of passing the pool in case Hampden should pop out of it, brandishing his head.

The tale of Pye's misalliance may or may not be true, but his death at Vigo is certainly fabrication. He must have already met his fate, whatever it was, by 1701, when Sir Robert died and was succeeded by his second son Edmund, and since Lady Anne followed her husband within months to the grave, she could not have been haunted by any apparition in 1702. Nevertheless, the scandalous story gained wide credence from the verses of Richard Barham, who made it the subject of one of his *Ingoldsby Legends*. Barham's talent for embellishment has contributed to other traditions, such as that of Sir Robert de Shurland at the ISLE OF SHEPPEY (South-East England), and his ballad of 'Hamilton Tighe', first printed in 1837, ensured that the tale of the murder and the ghost entered legend. He rechristened his hero in order, he said, to prevent 'all unpleasant feelings', but a footnote in later editions throws discretion to the winds, stating explicitly that the real name was Hampden Pye.

Barham took liberties with the original story, making the wicked lady a stepmother and so supplying her with a better motive, since she wants her own son to inherit the Pye (or Tighe) estates. He also invented a villainous gunner, 'Hairy-faced Dick', who aims the fatal cannon on instructions from his captain, later an admiral (possibly Rooke himself), and the final verses describe how the three conspirators are haunted by their guilt in visible form, a man who holds his head on his lap and follows them wherever they go:

> The Admiral, Lady, and Hairy-faced man
> May say what they please, and may do what they can;
> But one thing seems remarkably clear, –
> They may die to-morrow, or live till next year, –
> But wherever they live, or whenever they die,
> They'll never get quit of young Hamilton Tighe!

the mirror and hardly knew the white, shrivelled face he saw. He anointed himself with salad oil, and gulped brandy as if it had been water, remedies that helped to some extent, but it was days before he could walk, and three months later he was still getting shooting pains in his chest.

Wood counted himself lucky the creature's filaments had only touched his knee and arm, since had they fastened on his chest or stomach, he thought, he would probably not have been able to get to shore, and could easily have drowned. He earnestly advises his readers to look out carefully for the creature, and never under any circumstances to swim across its track, no matter how distant it may be, since it can send out its filaments to an 'almost interminable length', and they can sting even when detached. If you see one, he writes, you should allow at least a hundred yards, and do your best 'to keep out of the way of a foe who cares nothing for assaults, who may be cut into a thousand pieces without losing one jot of his offensive powers, and who never can be met on equal terms'.

Conan Doyle made this account central to one of his great detective's last achievements. Unusually, Holmes himself is the narrator, having retired to the South Downs, where Watson only visits him from time to time, taking no part in this adventure. One morning, Holmes finds one of his neighbours on the point of death, having apparently been terribly flogged. Before dying, the man gasps a few unintelligible words and a final shriek, 'the Lion's Mane'. Eventually Holmes finds the culprit lurking in a pellucid green pool under the cliff:

> 'Cyanea!' I cried. 'Cyanea! Behold the Lion's Mane!'
>
> The strange object at which I pointed did indeed look like a tangled mass torn from the mane of a lion. It lay upon a rocky shelf some three feet under the water, a curious waving, vibrating, hairy creature with streaks of silver among its yellow tresses. It pulsated with a slow, heavy dilation and contraction.

'It don't belong to Sussex,' says the local police inspector, to which Holmes replies, 'Just as well for Sussex.' The Lion's Mane, however, is no exotic foreigner. Although it rarely comes as far south as Kent or Sussex, it is native to the cold sea around the north coast of Britain. It is commonly less lethal than Wood suggests (he was sensitive to wasp stings and may have been unusually susceptible to *Cyanea*'s venom), but his warning to steer clear should be heeded.

NEWHAVEN, EAST SUSSEX

A sea-serpent

Charles Dawson, the self-proclaimed discoverer of 'Piltdown Man', once saw – or said he saw – 'the most conventional old sea-serpent you could imagine', in the Channel between Newhaven and Dieppe.

In April 1906, Charles Dawson was crossing from Newhaven to Dieppe on the steamer *Manche*. In the early afternoon, he went on deck with a pair of field glasses, and caught sight of 'what seemed a large, cable-like object struggling about'. It was about two miles away from the ship. A couple of other passengers saw it too, and one said, 'Hallo! what's that coming, the sea-serpent or what is it?'

When he first saw it, the thing, or creature, had been coming straight towards the ship, but now it turned away, offering its side view:

> I could not see any head or tail but a series of very rounded arched loops like the most conventional old sea-serpent you could imagine, and the progressive motion was very smart and serpentine for when one loop was up its neighbor was down.

The loops, Dawson said, were fully eight foot high from the water, and the length of the whole beast was sixty or seventy foot 'at the smallest computation'. He watched it as long as it was in sight, and took several snaps with a small Kodak camera – none of which, when developed, showed any detail further than a few yards. He discussed the matter, however, with his fellow passengers, and exchanged cards with several, all of whom agreed that it was 'too bold and too large to be anything like a school of porpoises sporting after one another'.

Dawson described all this in a letter to Arthur Woodward, keeper of the Natural History Museum's geology department and an expert in marine palaeontology, in October 1907. This was about eighteen months after the 'sighting', and he did not name any of the other people who saw the serpent. That, perhaps, would not be enough to cast doubt on his account, but Charles Dawson was the man later responsible for the notorious Piltdown hoax. In 1912, he pretended to discover a fossilised skull that puzzled anthropologists for forty years – having himself artfully concocted the relic from a human cranium and an orang-utan's jaw. His report of the sea-serpent may therefore, as suggested by John Evangelist Walsh in his book on the Piltdown fraud, have marked 'that moment in Dawson's life when he came to an overriding belief, no doubt to him quite thrilling, in his unlimited power to deceive'. On the other hand, he may have genuinely believed he saw the serpent. Sightings of similar beasts, including those at THORPENESS (East Anglia) and many others, have been documented by people with no history of deceit, and some are far more detailed than Dawson's.

NEWTOWN, ISLE OF WIGHT

The Pied Piper

Long ago, around Newtown harbour, there was a flourishing town called Franchville, but by the early nineteenth century little trace of it remained. Only one cottage stood there, and although the names Gold Street and Silver Street were still in use, they referred to green country lanes, not urban roads. Memory of the town survived, however, in a popular tradition, according to Abraham Elder's *Tales and Legends of the Isle of Wight* (1839). The story goes that in the fourteenth century, the town was terribly infested with rats. Large rewards were offered to anyone who could get rid of the vermin, and eventually a stranger appeared in the district, dressed in fantastically coloured clothes, and playing on his pipe such enchanting music that when he went down the street, all the rats followed him. He got into a boat and sailed away across the sea, still playing, and the rats, still following, were drowned. The piper returned for his reward, and when this was refused, he took a terrible revenge. Playing a new tune, he led all the children off through the town and into the countryside. 'What became of them nobody ever knew: for neither the Pied Piper nor the children were ever heard of again.' Further disaster fell on Franchville some years afterwards, for when it was attacked by the French in 1377, the enemy found it inhabited only by old people, whom they were easily able to overcome, and the town was destroyed.

This story, which Elder attributes to an almost certainly fictional antiquary, 'Mr Winterblossom', will be familiar to most people from Robert Browning's 'Pied Piper of Hamelin', first published in 1842. The poem's probable source is a considerably earlier work, A *Restitution of Decayed Intelligence* (1605) by Richard Verstegan, who dates his tale in 1376, sets it in Hamelin, Brunswick, and ends it – as does Browning, though not Elder – with the lame boy who cannot follow the other children, and therefore survives to tell people what has happened.

Whether Elder had heard a genuine Isle of Wight tradition one cannot now be sure. A town called Franchville or Francheville did once stand on the site of Newtown, and was indeed destroyed by the French. One reason why it was short of defenders when the French attacked was that the Black Death had decimated the Isle of Wight's population in the fourteenth century, and since the plague was understood to be spread by rats, it is quite possible that a rat-related legend was adopted to account for Franchville's disappearance.

THE NORE, KENT

Mutiny at the Nore

The last mutinies in the British navy occurred in 1797, during the war with revolutionary France. The men's grievances included low pay, poor food and inadequate shore leave, all fully justified complaints. A sailor could be at sea for years at a time, and when he finally returned he was not immune from being immediately forced into service again by the press gangs (*see* NORTH SHIELDS, North-East England), while his wife and children might be in dire need of money. Since 1653, an able seaman's pay had stuck at twenty-four shillings a month, nowhere near enough to keep a family. A petition was sent to the Admiralty but was ignored for two months, and when the fleet at Spithead was signalled to sea on 15 April, the men refused to raise their anchors until their demands had been dealt with.

Pay was duly increased and the other demands met, or at least promises were made, and the mutineers were pardoned. Some sailors, however, did not believe the government would keep faith with them, and a riot at Portsmouth, in May, was followed by a second mutiny at the Nore, a sandbank at the mouth of the Thames used as a naval anchorage.

A piece of naval folklore explains an anomaly in time-keeping at sea by reference to the mutiny. The seaman's day is divided into periods called watches, and every half-hour the bell on the quarterdeck is struck a certain number of times, beginning with one stroke at the beginning of the watch, two after half an hour, and so on. According to the sequence, 6.30 p.m. should be five bells, but in practice the bell sounds only once. This is said to be because the Nore mutiny was timed to start at half past six in the evening, but the officers got wind of this and arranged for one stroke of the bell instead of five, so that potential rebels would fail to take part.

If this was really their plan, it was ineffectual. Most of the North Sea fleet joined the mutiny, and the 'Floating Republic', as it was called, blockaded the Thames under the leadership of Richard Parker (generally an ill-fated name for a sailor – *see* SURVIVAL CANNIBALISM, p. 176). This Parker was an able seaman who gave himself the title of admiral for the occasion, and hoisted the revolutionary flag on his ship, HMS *Sandwich*. The authorities, having already granted concessions, refused to give way any further, and brought in warships to restore order. One by one the rebellious crews abandoned the struggle, and at last the ringleader was isolated even on his own vessel, which was surrendered. Together with twenty-four men loyal to him, 'Admiral Parker' was hanged at the yardarm on 30 June.

A ballad, 'The Death of Parker', was composed almost immediately, and became very popular. It is sung in the person of Parker's bereaved wife:

> Ye gods above, protect a widow!
> And with pity look on me.
> Help, O help me out of trouble;
> Out of sad calamity!
> It was, by the death of Parker,
> Fortune prov'd to me unkind;
> And though hung for mutiny,
> Worse than he were left behind.

She tells how she asked three times to be given his body, but was three times refused. Later, according to the song, she and two friends crept secretly to the gallows and removed the corpse, which they took to London for burial.

In the mid nineteenth century, a woman whose sailor father had been present at the mutiny (though not himself a mutineer) used to sing a version of the ballad to children she looked after, and 'in low tones that made one's blood run cold' would add: 'Men have been hung at the yard-arm for singing that song. It was condemned throughout the Fleet.' Although not explicitly supporting the mutiny, the words 'Worse than he were left behind' were evidently thought contentious, if not treasonous.

PORTSMOUTH, HAMPSHIRE

Jack the Painter

If you find yourself in Portsmouth on a windy night, down by the harbour at midnight you may hear a metallic rattling, or even see a phantom figure dragging chains. This is the ghost of 'Jack the Painter', who continues to haunt the place he tried to destroy.

Jack the Painter in the dock, after his arson attempt.

Jack – whose real name, according to various accounts, was Aitken, or Hill, or Hinde – was born in Edinburgh, but became an enthusiastic supporter of the American independence movement. Having spent some time in the colonies, he returned to England and joined the military, his career a sequence of desertions and re-enlistments under a new alias every time, his purpose sabotage. He was hired by an American agent who provided him with a French passport and instructions on how to make an incendiary device, and having constructed one, on 6 December 1776 he tried to set light to the Rope Walk in Portsmouth, where cables were made. His first attempt was foiled by damp matches, but the next day he was more successful, starting a blaze that caused £60,000 worth of damage.

He nearly got away with it, since at first the fire was thought to be accidental, but then Jack's home-made tinderbox was discovered in the wreckage, and a witness reported seeing him leaving the scene. His lodgings were searched, the fake passport and instructions from the American spy were discovered, and Jack was arrested. Condemned to death for treason, he was hanged from the mainmast of the *Arethusa*, and his body was afterwards displayed in chains at Blockhouse Point, at the mouth of the harbour, where it stayed for several years.

Some drunken sailors once stole the skeleton and left it in a pub in Gosport, supposedly in payment for beer, giving rise to the lines:

> Whose corpse by pondrous irons wrung
> High up on Blockhouse Beach was hung,
> And long to every tempest swung?
> Why, truly, Jack the Painter.
> Whose bones some years since taken down
> Were brought in curious way to town,
> And left in pledge for half a crown?
> Why truly, Jack the Painter.

RAMSGATE, KENT

The fata Morgana

The fata Morgana, meaning the 'fairy Morgan', is the name given originally to mirages seen in the Strait of Messina between Sicily and the southern tip of Italy, said to be the work of the sorceress Morgan le Fay of Arthurian legend. Although best known in English and French tradition, tales of her enchantments were associated with Sicily in several medieval texts, perhaps brought to the region by the Normans, who ruled the island from the eleventh century. The mirages attributed to her were described in lyrical terms, as in one account from 1643 of this 'most wonderful, delectable vision'. The observer was on the Italian side of the strait, and from his viewpoint the sea nearer Sicily seemed to rise like a ten-mile chain of dark mountains, while the waters in the foreground were smooth as a polished mirror:

> On this glass was depicted, in *chiaro scuro*, a string of several thousands of pilasters, all equal in altitude, distance, and degree of light and shade. In a moment they lost half their height, and bent into arcades, like Roman aqueducts. A long cornice was next formed on the top, and above it rose castles innumerable, all perfectly alike. These soon split into towers, which were shortly after lost in colonnades, then windows, and at last ended in pines, cypresses, and other trees, even and similar. This is the *Fata Morgana*, which, for twenty-six years, I had thought a mere fable.

Commenting on this in *Travels in the Two Sicilies* (1783–6), Henry Swinburne notes that certain conditions are essential to see such a 'pleasing deception'. The weather must be calm, the tide at the flood, and the spectator must stand in a high place, facing west, so that the mountains of Messina form a dark backdrop. As soon as the sun rises high enough to make a forty-five-degree angle on the water, 'every object existing or moving at Reggio will be repeated a thousand fold upon this marine looking-glass, which, by its tremulous motion, is, as it were, cut into facets.'

'Fata Morgana' came to be applied as a name to similar spectacles elsewhere. The phenomenon, which is scientifically explained as reflections from water vapour, has been seen from places including YOUGHAL (Southern Eire) and Ramsgate, as in an account sent to the *Gentleman's Magazine* in 1789. Sitting on a bench with a wide view across the Kentish downs and the sea, the correspondent found the French coast 'very plain to be seen', and, in the air above it, 'exactly the same coast, with the white cliffs and land above it, reflected as in a mirror, but in a fainter degree, and on the summit of the reflection a faint resemblance of towers, &c.'. Amazed and delighted, he gazed at the sight before him:

> After some time, the reflected cliffs disappeared, and another object presented itself; the small towers . . . assumed a magnified and magnificent appearance, rising, in some parts, to a great height above the land, insomuch that I could scarcely believe but that my eyes deceived me, so much was I lost in surprise and admiration.

The phantasmal city remained in view for about half an hour, after which clouds gathered at the tops of the imaginary buildings and soon swallowed them in mist.

SANDWICH FLATS, KENT

The Dunkirk Armada

In May 1940, when the German blitzkrieg threatened to cut off Allied lines of supply and retreat in France, around a quarter of a million men had to be evacuated, at short notice and high speed, from the beaches of Dunkirk. An unprecedented hybrid of a fleet was assembled for the task, comprising many hundreds of vessels ranging from troopships manned by the navy to fishing boats and coal barges sailed by volunteers. A naval officer who went with the first convoy describes it as consisting of everything from smart motor yachts to converted lifeboats and Thames river launches.

The gallantry of the untrained civilians quickly became a national legend. A typical passage can be found in Helen Foley's novel *A Handful of Time*, written in 1961 when memories of the war were still quite fresh:

> Against a lamp-post where the shrimp-man normally stood with his basket was a blackboard with a roughly chalked message: 'No shrimps today – gone to fetch our boys.'

When the vessels of the 'Dunkirk Armada' reached the Normandy coast on 27 May, they operated a relay service, the boats with the shallowest draught going close in to the shore to fetch off the soldiers, while larger craft waited in deeper water to receive them, all the while being machine-gunned and bombed by the Luftwaffe. Over the next few days they managed to rescue almost all the British Expeditionary Force, and many French soldiers and civilians too.

Tales of individual courage proliferated, not just of the men, but of the boats themselves. After losing her crew to a bomb at Dunkirk, a Thames barge dragged her own anchor free and floated unmanned to Sandwich Flats, where she arrived slightly damaged but still seaworthy, and even contrived to anchor herself again. The barge's identity is a little mysterious – she appears in different sources as the *Ena* and the *Singing Swan* – but her adventure could have happened.

SELSEY, WEST SUSSEX

St Wilfrid converts Sussex

Returning from Rome with a party of monks, Wilfrid, the seventh-century bishop of Ripon and York, was caught in a south-easterly gale and ran ashore on the Selsey peninsula. The tide went out and the party was stranded, at the mercy of the local 'heathen', who advanced in force to capture the Christians, take their money, seize their ship and murder anyone who resisted. According to Wilfrid's friend and biographer Eddi, the bishop offered a ransom for their release, and spoke calmly to the Saxons, wishing to convert them:

> But they with stern and cruel hearts, like Pharaoh would not let the people of the Lord go, saying proudly that all that the sea threw upon the land became as much theirs as their own property. And the idolatrous chief strove like Balaam to curse the people of God, and to bind their hands by his magic arts. Then one of the Bishop's companions hurled, like David, a stone blessed by all the people of God, which struck the cursing magician in the forehead and pierced his brain, whom an unexpected death surprised as it did Goliath, and he fell back a corpse.

Enraged by the death of their leader, the army attacked, and the monks had to fight back. Wilfrid prayed to God, and in answer to his entreaties the tide rose and floated the Christians' ship to safety on the open sea.

Not long afterwards, the plucky bishop elected to return to Sussex as a missionary, and was granted land at Selsey, the very place of his battle. Having performed many miracles, including teaching the locals how to fish properly, Wilfrid became popular, and so did his religion. From being England's last stronghold of paganism, Sussex went over wholesale to Christianity, so much so that the epithet 'saelig' (holy, or innocent) became attached to the county. Its neighbours in later centuries often chose to misunderstand the nickname as 'silly Sussex', but

When St Wilfrid asked for heaven's help, God 'immediately ordered the sea to return a full hour before its wont', taking the saint and his companions out of danger.

although the root of the word is the same, no insult was originally intended.

St Wilfrid's cathedral, built in honour of Sussex's first bishop, later fell prey to the marauding sea. Its site is about a mile out from the present shore, near the Owers Light that flashes to warn sailors of the shallow sandbanks beneath, and fishermen have reported hearing its bells ring from beneath the water.

ISLE OF SHEPPEY, KENT

The tale of Grey Dolphin

Sir Robert de Shurland, Baron of Sheppey in the early fourteenth century, became famous for the manner of his death. As told by the historian Edward Hasted in 1782, the story went that Sir Robert, having fallen out with a priest and gone so far as to bury the man alive, was banished from the court of Edward I. Eager to obtain pardon, he swam on his horse two miles out to sea to where the king's ship lay at anchor. Forgiveness was granted him, but when he returned to shore he was told that his horse had accomplished the feat by magic. As a precaution against sorcery, Sir Robert cut off the horse's head, but about a year later, when he was hunting, his new horse stumbled, throwing the rider. Just at the place where he fell lay the skull of his former mount, and Sir Robert bruised himself so badly on it that he died.

A version of this tale became well known through Richard Barham's *Ingoldsby Legends* (1840), where the horse is named as Grey Dolphin and the narrative is slanted to make it a fatal prophecy. The story begins when Sir Robert is indignant because no prayers are being said at the burial of a drowned sailor. When the friar refuses to perform a short service, the baron kicks the obstinate monk so hard that he flies into the open grave and is killed. Accused of murder and sacrilege, Sir Robert decides to throw himself on the mercy of the king, who has arrived on

Gerald du Maurier's image of Sir Robert de Shurland and his noble steed Grey Dolphin, meeting the witch who foretells that the horse will cause the baron's death.

the Kent coast recruiting for the French wars. The baron calls for his dapple-grey steed, and swims out into the Thames nearly to the Nore: 'the stream was broad and the current strong, but Sir Ralph and his steed were almost as broad, and a good deal stronger.' He asks pardon, which is granted, and his gallant horse bears him back to dry land, where an ugly old woman cries to him that Grey Dolphin has saved his life this time, but is destined to be the means of his death. To forestall his fate, Sir Robert cuts off Grey Dolphin's head with his sword, but three years later he is riding across the same beach when he sees an apparition of the old crone sitting on something like a large white rock. The witch vanishes, and the stone turns out to be the skull of a horse. Sir Robert, banishing his grief for Grey Dolphin, kicks the skull into the water, where it laughs and sinks, and the baron finds that one of its teeth has stuck in his toe. The wound becomes inflamed and festers, and Sir Robert dies of gangrene.

The horse legend was originally inspired by Sir Robert's tomb in Minster church, which bears the effigy of a recumbent warrior with a horse's head behind one leg, apparently rising from the waves. The emblem is explained by some historians as showing Robert de Shurland's right to 'wreck of the sea', a privilege which, as noted by Thomas Philipot in 1659, 'is evermore esteemed to reach as far into the Water, upon a low ebb, as a man can ride in, and touch any thing with the point of his Launce'. A 'nag's head' forms part of the heraldic crest of many families, often those whose names suggest the equine (Horsley or Chevall, for example), and has contributed to folklore in other places, notably at LAND'S END (South-West England & Channel Islands), where a knight's steed was said to have galloped faster than the rising waters and saved his master.

SHOREHAM-BY-SEA, WEST SUSSEX

A postponed vow and a phantom ship

Every 17 May, it is reported, a phantom ship glides along the south coast, and comes to grief on a rock near Shoreham harbour, a vision said to go back many hundreds of years to a tragedy arising from a neglected vow.

Long ago, Earl de Warenne, lord of Lewes, was engaged in a feud with the lord of Pevensey, and matters came to a head one May, when a pitched battle was fought between the opposing sides. Watching from the battlements of Lewes Castle, Lady de Warenne saw Lord Pevensey gallop towards her husband and raise his battleaxe above the earl's head. Holding up her baby son, she vowed to St Nicholas, patron saint of sailors, that if de Warenne's life were spared, the boy should not marry until he had made a pilgrimage to Byzantium. Her prayer was answered. The earl dodged the axe, Pevensey lost his balance, and as he swayed in the saddle, de Warenne ran him through with his sword. Seeing their lord fallen, Pevensey's troops fled, and de Warenne returned in triumph to his castle and his wife.

Twenty years passed, and the earl's son Lord Manfred had not yet gone to Byzantium. Each anniversary of the battle was marked with a feast, and this May the event was particularly splendid, since it also celebrated Manfred's betrothal to the beautiful Edona, daughter of Lord Bramber. Dancing had begun, when all at once a cold wind blew through the hall. The torches went out, the din of war was heard, and across the walls in lightning flashes played out the battle between Pevensey and de Warenne. At the moment when Pevensey raised his axe to strike, the vision abruptly ceased, and all was darkness and silence.

The terrified guests, in no mood to continue the party, withdrew, and the following day Earl de Warenne, reminded of his duty, commissioned a ship to take his son to Byzantium. As soon as it was finished, Manfred set sail, promising that the moment he returned he would marry Edona. Having performed his sacred task, he set out on his return voyage, and on the next 17 May his sail was seen off Worthing Point. The earl's family and retinue assembled to see the ship come in, with Edona first among the watchers. All was joy, until the dreadful moment when the ship, passing Shoreham, ran full on to the hidden rock, listed to one side and sank.

> The Earl and every soul around him stood motionless; not a word broke the silence of that sad scene. To move was useless. One sad last long drawn sigh burst from Edona, and she fell never more to rise. The Earl passed his hands over his eyes, dropped his head on his bosom, no smile ever rested on that face again.

One sailor survived, but all the rest, including the young nobleman, were drowned. The earl built the church of St Nicholas on the hill where he had stood when he saw the vessel sink, as an everlasting reminder that vows must be fulfilled. Edona was buried where she had fallen dead of grief, and a cross was raised at the entrance of the church to mark her grave, while the ghostly ship continued to hold its melancholy anniversary, sailing every year to its doom.

This ornate piece of legendry appears in a history of St Nicholas church, Brighton, written by its nineteenth-century verger T. W. Hemsley. The church dates from the mid fourteenth century, which would fit with the death of Earl John de Warenne, last of his line, in 1347, but that is about as far as history supports the tale.

STEPNEY, LONDON

Customs of the sea

Births rarely occur at sea, but they do happen. In the nineteenth century, according to Fletcher Bassett's *Legends and Superstitions of the Sea* (1885), it was held by sailors (English sailors, it is to be assumed) that any baby born on board ship belonged to the parish of Stepney.

Another tradition, reported by Admiral Sir Herbert King-Hall from his father's experience, laid down that the child had to be given the name of the ship on which it was born. In 1848, King-Hall senior was in command of HMS *Growler*, homeward bound from Bermuda with some troops and their wives. He was woken one night by the boatswain, who reported that one of the women had been safely delivered of a little girl, and the next day he read an entry in the ship's log giving the child's name as Growler. When he asked why:

> 'Custom of the sea, sir,' answered the boatswain. 'Any child on board ship is given the ship's name.'
>
> 'But,' protested my father, 'this poor child cannot go through life named Growler.'
>
> 'Custom of the sea, sir; always 'as been,' was all that the boatswain would say.

The captain, though a respecter of naval tradition, was a merciful man. Declaring that on this occasion the custom would not be observed, he ordered the boatswain to erase the name.

WHITSTABLE, KENT

Sharks' teeth

Visiting Whitstable in the 1920s, Edward Lovett noticed a tray full of fossilised sharks' teeth in a shop window. He was investigating the topic of modern magic, and asked the shopkeeper if he had anything for sale in the way of charms or mascots.

> His reply was an emphatic 'No'; adding that the people of Whitstable were far too well educated to believe in such rubbish. I then asked him what the fossil teeth were for, and he told me that they were Cramp Stones and if anyone suffered from cramp it would be at once cured by carrying one of these cramp stones in the pocket.

Lovett ventured to suggest that this was a form of superstition, but the shopkeeper denied it. He knew scores of people, he said, who had had their pain relieved by carrying a shark's tooth. Lovett bought a few of the teeth, and showed them to an old sailor who had lived in Whitstable all his life, telling him what the shopkeeper had said. 'I carry one myself,' the man replied, taking a tooth out of his own pocket, and he added that everyone he knew had a charm like it.

Other fossils often carried to cure or prevent cramp were ammonites or 'snakestones' (*see* KEYNSHAM, South-West England & Channel Islands, and WHITBY, North-East England). The shark's tooth, although evidently a well-known remedy for cramp, was also often used against snakebite, as suggested by its folk names 'adder-tongue' and 'tongue-stone', both of which come from its remarkably tongue-like appearance.

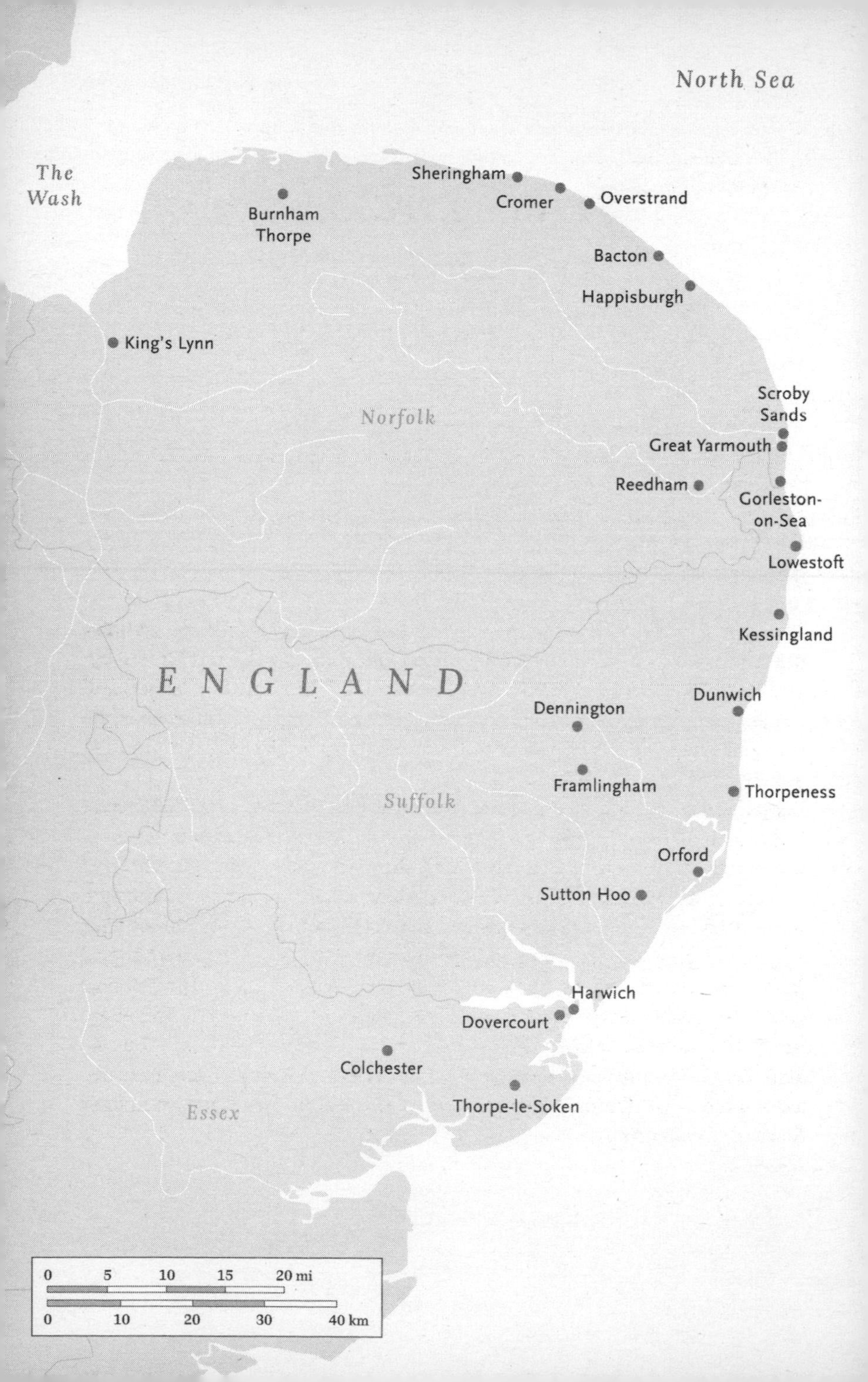

North Sea

The Wash

Burnham Thorpe

Sheringham

Cromer

Overstrand

Bacton

Happisburgh

King's Lynn

Norfolk

Scroby Sands

Great Yarmouth

Reedham

Gorleston-on-Sea

Lowestoft

Kessingland

ENGLAND

Dunwich

Dennington

Framlingham

Thorpeness

Suffolk

Orford

Sutton Hoo

Harwich

Dovercourt

Colchester

Thorpe-le-Soken

Essex

0 5 10 15 20 mi

0 10 20 30 40 km

EAST ANGLIA

Essex, Norfolk, Suffolk

BACTON, NORFOLK

The Long Coastguardsman

An eerie apparition used to be seen in the nineteenth century, walking nightly towards Mundesley from Bacton (now the site of the North Sea Gas on-shore terminal), just as midnight struck. He left no footprints, and could not be seen when the moon was shining. Loving a high wind, he shouted and sang at the top of his voice during storms, and in moments of calm could be heard laughing loudly.

His history was unknown, and Ernest Suffling reports in his *History and Legends of the Broad District* (c. 1891) that nobody he asked had actually seen the phantom themselves, though often they knew someone else who had – a common problem in ghost-hunting. They had all heard of him, though, and called him 'the Long Coastguardsman'. His behaviour resembles that of some elementals – supernatural beings associated with the weather and particularly with storms, such as Shuck (*see* OVERSTRAND), the *Dooiney Hoie* (*see* LAXEY, North-West England & Isle of Man), and Jan Tregeagle of GWENVOR SANDS (South-West England & Channel Islands) – but he also used to tempt people into danger by crying for help, a nasty practice associated with demonic beings such as the Kelpie (*see* SOLWAY FIRTH, Scottish Lowlands).

BURNHAM THORPE, NORFOLK

Horatio Lord Nelson

To Tennyson, as to most of nineteenth-century England, Horatio Nelson was 'The greatest sailor since our world began'. Even before his death he was a hero, and after it he became a mythic figure, literally identified with that other champion of the sea, Sir Francis Drake, by some who said that Drake had been reincarnated in the person of Nelson to save his country in her hour of need (*see* SCAPA FLOW, Scottish Highlands & Islands).

Nelson was born in Burnham Thorpe in 1758, and Norfolk regards him as particularly its own. Friends of his childhood were keen to repeat anecdotes of his precocious valour and nobility. From John Glyde's *Norfolk Garland* (1872) comes a tale of his schooldays, when all the boys longed for the ripe pears in the headmaster's garden but nobody dared to steal them. Horatio, then about ten years old, volunteered for the task, and was lowered down one night from his bedroom window by a rope of sheets. Collecting the pears, he distributed them among his fellows without keeping one for himself. 'I only took them,' he said, 'because every other boy was afraid.'

In later life he lost the sight of an eye at the capture of Corsica in 1797,

The death of Admiral Lord Nelson, engraved from a painting by Ernest Slingeneyer (1820–94).

and in 1798 an arm, in an attempt to take a Spanish treasure ship near Tenerife. A one-armed, one-eyed warrior sounds like a joke, and was sometimes treated as such by the man himself. At the battle of Copenhagen, he ignored a misguided signal to break off the action, remarking that he had a blind eye, and was sometimes entitled to use it. That story may well be true, but certainly apocryphal is an old Norfolk churchman's tale of how Nelson was given the freedom of Great Yarmouth. Asked to put his right hand on the Bible to take his oath, he put his left hand on the book.

> 'Your right hand, my lord,' repeated the clerk. 'That is at Teneriffe,' was the quiet reply.

Nelson's dying words at the battle of Trafalgar, 21 October 1805, were reported throughout the land: 'Take care of my dear Lady Hamilton, Hardy, take care of poor Lady Hamilton. Kiss me, Hardy.' The image of the great man expiring in Admiral Hardy's arms became the theme of countless mementoes, including ballads, broadsides, pottery and pictures. At Portisham in Dorset, Hardy's home town, a short episode was inserted into the traditional Christmas mumming play, and 'for half a century at least', writes Hardy's biographer A. M. Broadley, 'never failed to provoke roars of applause':

> NELSON: Hardy, I be wownded.
> HARDY: Not mortually I hopes, my Lord.
> NELSON: Mortually, I be afeard. Kiss me, Hardy, thank God I've done my duty.

It is often said that the three rows of white tape on sailors' blue collars commemorate Nelson's victories at the Nile, Copenhagen and Trafalgar, although the decoration was introduced only in 1857. Similarly, the black silk handkerchief worn by sailors, in fact used long before 1805, was supposed to be a sign of mourning for the admiral's death.

COLCHESTER, ESSEX

Oysters and oyster shells

Colchester has had exclusive rights over the oyster fisheries of the Colne since the twelfth century. The town's ancient privileges are celebrated at the annual Oyster Feast, held in October, and also at the beginning of the oyster-dredging season in early September, in a ceremony said to go back to 1540. The mayor and other officials, dressed in their full regalia (often with wellies underneath), proceed in a fleet of small boats from Brightlingsea into Pyefleet Creek, and a proclamation is read out, beginning: 'Be it known, The Several Fishery of the River Colne and waters thereof hath, from time beyond which memory runneth not to the contrary, belonged and appertained to the Borough of Colchester.' A toast is then drunk in gin, and gingerbread is eaten – both gin and ginger being thought to counteract seasickness – and the mayor trawls in person for the first oysters of the season.

The mystique of oysters comes mainly from their association with pearls, although the Colchester oysters are not pearl-bearing (unlike British mussels, which can produce small pearls, as they do for example at RAVENGLASS, North-West England & Isle of Man). As a food, oysters were traditionally eaten by the poor, being plentiful and easily gathered, and it is only since the mid nineteenth century, when stocks began to decline, that they have become a dish for the elite. Before then, oyster feasts were held in many places, and the great quantities of empty shells were traditionally made into miniature caves or grottoes, a custom that lasted into the mid twentieth century in some places. A correspondent to *The Times* in November 1957 reported that in one London suburb, 'Grotter Day' (local pronunciation) was a big event:

> In the streets surrounding a large factory at Mitcham, Surrey, home-going workers are besieged once a year in July with requests to 'Please remember the Grotter' and find themselves picking their way carefully over structures of stones, shells, and flowers.

The grottoes were hollow, with a candle lit inside (perhaps their decline is linked to the rise of the Hallowe'en pumpkin), and children used to beg for 'a penny for the grotto' much as they once asked for 'a penny for the guy'. The custom was a relic from the days of pilgrimage, when it was an act of grace to visit the shrine of St James in Compostela, and those unfit for the long journey to northern Spain could pray before a home-made shrine instead. St James's symbol was the cockleshell, in England translated into the more easily available oyster. Naturally, however, the Mitcham children did not know the history of their grottoes, any more than most November guy-makers could tell you much about the Gunpowder Plot.

CROMER, NORFOLK

St Elmo's fire

> Wednesday evening last, about half-past nine o'clock, many of the inhabitants of this place were surprised at seeing several lights on the top of the church steeple, and many others on the chimneys of some of the houses situate on the cliff. On attentively viewing them, it was discovered that they were the kind of lights sometimes seen at sea on ships' masts and yards, called 'Corpus Sant.'

This newspaper cutting, included in W. B. Gerish's unpublished collection of Norfolk folklore (1916–18) held in Great Yarmouth public library, is headed 'Cromer, April 6' but is otherwise undated. The article reports that the mysterious lights were said to be caused by a large insect (some sort of firefly, perhaps) or were 'some substance raised from the surface of the German Ocean by the wind', but they must in fact have been atmospheric electricity, which in stormy weather tends to gather around exposed points such as church spires and the tips of ships' masts, producing a crackling glow.

St Elmo's fire, or the corposant. Here there are three lights, rather than the one or two mentioned by Pliny and other authorities, leaving it debatable whether they are to be regarded as a good or bad omen.

The phenomenon, most often observed at sea, has given rise to a variety of supernatural explanations. The ancient Greeks and Romans identified the lights as the earthly sign of the Heavenly Twins or Dioskuroi, Castor and Polydeuces (in Latin, Castor and Pollux), semi-divine figures who were the special protectors of sailors. The constellation Gemini, 'the Twins', was said to be the brothers' immortal manifestation in the skies, and early sailors, accustomed to navigate by the stars, must have looked at the flickering sparks that sometimes visited their mastheads and thought that these were stars come down to man's level. They were fortunate omens if two appeared at once, but one solitary light was considered dangerous. In his *Natural History* (77 CE), Pliny records that:

> When they occur singly they are mischievous, so as even to sink the vessels, and if they strike on the lower part of the keel, setting them on fire. When there are two of them they are considered auspicious, and are thought to predict a prosperous voyage . . . their efficacy is ascribed to Castor and Pollux, and they are invoked as gods.

Later, Christian legend continued to associate the lights with holy figures, including Jesus, St Peter Gonzalez (a thirteenth-century monk revered by Spanish fishermen), and particularly St Elmo, a patron saint of sailors. According to one story, the saint died at sea during a storm, having promised the crew that he would return and show himself to them if they were destined to survive. Soon after his death, a glowing light appeared at the masthead, taken to be either the saint himself or his sign. A variant version told in Brittany is that St Elmo was once in danger of drowning when he was saved by a Breton captain, and in gratitude sent the glimmer as a warning of storms to all mariners. Common names for the lights include 'St Elmo's fire', 'St Peter's fire', and 'corposant' or 'corpus sant', as in the Cromer article, from Old Spanish or Portuguese *corpo santo*, 'holy body'. More unusual local variants are 'ampizant' and 'Corbie's Aunt' (both corruptions of 'corposant', the latter with a Scottish twist, 'corbie' being the Scots word for a raven), and 'Jack-a-lantern', by association with the land phenomenon the will-o'-the-wisp or jack-o'-lantern, caused by the marsh gas methane.

The opinion mentioned by Pliny, that two lights were propitious but one was ominous, persisted until at least the seventeenth century. When the *Sea Venture*, commanded by Sir George Somers of LYME REGIS (South-West England & Channel Islands), was caught in a storm in 1609, according to an eyewitness account a 'sparkling blaze' hovered around the masts, which 'the Grecians were wont in the Mediterranean to call Castor and Pollux, of which if only one appeared without the other they took it for an evil sign of great tempest'.

Some seamen, however, believed that it was not so much the number of lights that mattered, but where the corposant appeared and how it moved. In *Two Years Before the Mast* (1840), the American author Richard Henry Dana tells how he was sent aloft with an English boy to take in sail, and when they came back down they found the rest of the crew staring upwards. Right above where they had been standing was 'a ball of light, which the sailors name a corposant'.

> They were all watching it carefully, for sailors have a notion that if the corposant rises in the rigging, it is a sign of fair weather, but if it comes lower down, there will be a storm. Unfortunately, as an omen, it came down, and showed itself on the top-gallant yard-arm.

Dana's English companion was uneasy at having been so near the corposant, since it was held a fatal sign to have its pale light cast upon one's face. On this occasion, however, no evil consequences followed.

DENNINGTON, SUFFOLK

Shadowfeet

Ancient travellers reported many marvels from the far corners of the earth, recorded in medieval treatises and finding their way on to maps and into ecclesiastical imagery. Many British churches display carvings of monstrous, gigantic or exotic creatures said to inhabit distant lands or oceans, including the mermaid, the basilisk, the whale, and the unicorn. Not only strange animals, but unfamiliar races of men were said to live in remote lands, and these too were shown in churches. The Blemmyae, men with no heads but with mouths and eyes in their chests, were mentioned in Pliny's *Natural History* (77 CE) as living in Africa, and an image of one such grotesque is in Norwich Cathedral. The Sciapods (Greek for 'Shadowfeet') were so called because they used their feet as parasols. An engaging Shadowfoot, wearing a hat as extra protection although his foot shields his entire body, appears in Dennington church, and gets a mention in M. R. James's *Suffolk and Norfolk* (1930):

> A carving of a Sciapus on a bench end does not escape notice; he lies on his back and holds his huge foot (or feet, but there should be but one)

> over him to shade him from the sun. Such men were to be found, if not in Africa, then somewhere else.

Again, one source is Pliny, his information attributed to the fourth-century BCE physician Ctesias, who writes of 'men, who are known as Monocoli, who have only one leg, but are able to leap with surprising agility. The same people are also called Sciapodae ['making a shadow with their feet'], because they are in the habit of lying on their backs during the time of the extreme heat, and protect themselves from the sun by the shade of their feet.'

Descriptions like this were still being offered as 'natural history' in the fourteenth century, when Sir John Mandeville's *Travels* appeared, gaining an eager readership with its accounts of wonders seen in the far quarters of the world. Mandeville locates the Shadowfeet in Ethiopia: 'In that land, too, there are people of different shapes. There are some who have only one foot, and yet they run so fast on that one foot that it is a marvel to see them. That foot is so big that it will cover and shade all the body from the sun.'

Some commentators suggest that the Shadowfeet originated in early travellers' attempts to describe people afflicted with elephantiasis, a tropical disease that can greatly enlarge parts of the body. This rationalising explanation, however, gives scant acknowledgement to the great vigour and antiquity of the 'sailor's yarn' or tall tale, as embodied in the tradition of WONDER VOYAGES AND LOST LANDS (p. 458), from Odysseus and Jason to Maelduin and St Brendan (*see* BRANDON CREEK, Southern Eire).

Imaginative use was made of the Sciapod tradition by C. S. Lewis in the third volume of his 'Chronicles of Narnia', *The Voyage of the* Dawn Treader (1952), which imitates a medieval wonder journey. Here the one-legged race are known as Monopods, or (their own preferred name) Dufflepuds.

> Each body had a single thick leg right under it (not to one side like the leg of a one-legged man) and at the end of it, a single enormous foot – a broad-toed foot with the toes curling up a little so that it looked rather like a small canoe.

They eventually learn to use their feet as boats, and are delighted with their new means of transport.

The Picts, early inhabitants of Britain displaced by Celtic invaders, were often seen as equivalent to the fairies or pixies, and were sometimes described as having feet so large that they could shelter from the rain by lying down and raising their legs over themselves, imagery derived from that of the Sciapods. Similarly, the Fomhóire, sea-pirates of Irish myth (*see* ROCKABILL, Southern Eire), were sometimes said to have one arm and one leg apiece.

DOVERCOURT, ESSEX

'The Four-Eyed Cat'

It was, and perhaps still is, a general belief among fishermen that women were unlucky at sea. The superstition is found in the navy and merchant service too, in spite of the fact that up to the mid nineteenth century women were frequently employed on large ships as cooks and nurses, and it was common practice for girls to visit ships in port for the benefit of sailors without shore leave.

In smaller boats, women were always considered ill-omened passengers, a belief connected with the dread of witchcraft. Both elements are clear in the story of 'The Four-Eyed Cat', told to the folklore collector Ruth Tongue in 1955 by the twelve-year-old daughter of a lightship sailor from Dovercourt near Harwich.

There was once a beautiful lady, the girl began, who was bad at heart, and was supposed to know 'more than a Christian should'. The locals wanted to 'swim' her – throw her into the water and see whether she floated, proving her a witch, or sank, in which case she would be innocent, though much good it would have done her – but because her father was rich and powerful, nobody dared try it. The enchantress put a spell on a poor fisherman, and although he was due to be married in a week, he deserted his sweetheart and followed his new love wherever she went. The couple decided to run away to sea, and they went to the fishing grounds, although they were careful to keep it a secret from the other mariners.

> A storm blew up and the whole fishing fleet were lost to a man, for they had on board a woman with them at sea, though none knew of it but her lover. It was she that had whistled up the storm that had drowned her own lover, for she hated everyone. She was turned into a four-eyed cat, and ever after she haunted the fishing fleet.

That was why, the storyteller said, fishermen (including her own uncles) would never cast their nets before cockcrow, and why they always threw some fish back into the sea, 'for the cat'.

DUNWICH, SUFFOLK

The drowned city

Tales of 'drowned lands' often have a factual basis. All along the East Anglian coast, the sea has eaten at the shoreline, consuming whole towns and even the regional capital. In the eighteenth century, the historian Thomas Gardner was told by old inhabitants that

> Dunwich (in antient Time) was a city, surrounded with a Stone-Wall, and brazen Gates; had fifty-two Churches, Chapels, Religious Houses, and Hospitals, a King's Palace, a Bishop's Seat, a Mayor's Mansion, and a Mint.

This was an exaggeration, but not an outrageous one. In its glory days in the mid twelfth century, Dunwich had at least eight churches, and was a bustling port until tempests in the fourteenth century first choked its harbour and then devoured a great part of the town, destroying more than four hundred houses and leaving it less than a quarter of its original size. Erosion has continued ever since. By the early nineteenth century Dunwich had a population of fewer than two hundred and fifty, of whom only twelve were entitled to vote, although it continued to return two Members of Parliament – most of whose constituency was underwater – and it was rumoured that voters had to take to the sea to cast their ballot, going out in boats to where their town hall used to be, or riding horses into the waves. In 1832, it was disenfranchised under the Reform Act.

An engraving from 1831 shows Dunwich just before it was disenfranchised. The accompanying article comments that 'It was once an important, opulent, and commercial city, but is now a mean village . . . we learn that its ruin is owing chiefly to the encroachments of the sea.'

Lord Leighton's 1857 painting of a fisherman and an amorous siren imagines the mermaid's tail as long and snakelike enough to wrap around her victim's legs. He puts up no resistance, and will probably be dragged underwater and drowned.

This picture from a fifteenth-century German Bible shows the waves around Noah's Ark populated not only by mermaids but by a little mer-dog, a reflection of the idea that the ocean contained versions of all creatures found on land.

St Nicholas was credited with powers over the waves, and this fifteenth-century painting shows him calming a tempest in answer to the prayers of desperate sailors. Perhaps the storm was raised by the mermaid on the bottom left who seems to be fleeing the saint's anger.

A 'mermaid' or strange fish exhibited in London in 1822, sketched by George Cruikshank. The creature was 'a hideous combination of a dried monkey's head and body, and the tail of a fish', but people crowded to see it simply because it was so ugly.

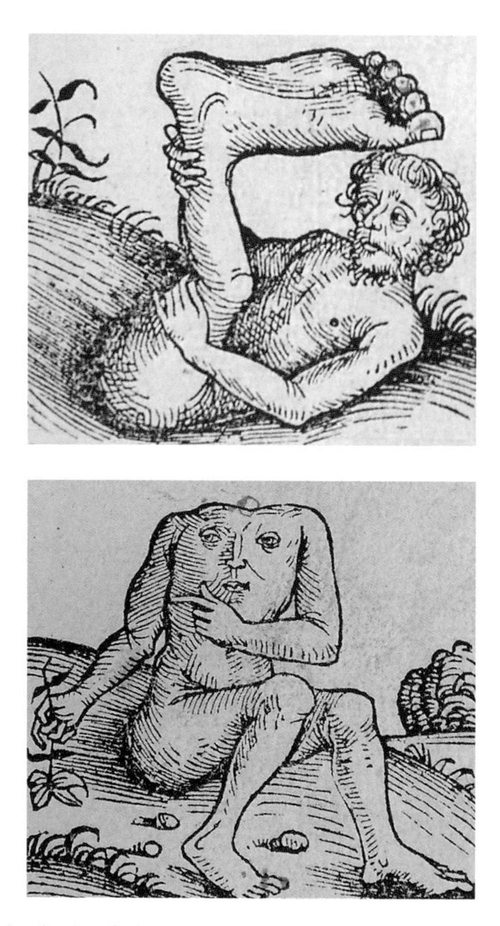

Sciapods ('Shadowfeet') and Blemmyae (whose faces were in their torsos) were among the foreign wonders reported by sailors to those back home. These fifteenth-century illustrations make a good guess at what such extraordinary beings might look like, and carvings can be found in churches including Dennington (East Anglia).

A fifteenth-century church carving from Zennor (South-West England) gave rise to a famous legend of a mermaid who came to church there, and enticed away the town's star singer.

After his three days and nights in the whale's belly, the prophet Jonah is cast up onto dry land, in a painting after Vernet.

The marine monster Leviathan, from Gustav Doré's illustrations to Milton's Paradise Lost. *Most majestic among God's creatures, the biblical beast was as huge as a whale, and had dragon like characteristics that helped to inspire tales of the sea serpent.*

A detail from a map by Abraham Ortelius (1527–98) includes a variety of legendary elements. The mythical island of Brasil (Hy Brazil) is located west of Ireland, flanked on the other side by a merman with bird's feet.

FRAMLINGHAM, SUFFOLK

Watery fiends

In days of popular belief in witchcraft and demons, 'experts' wrote of the various species of evil spirit that existed, as well as about techniques of witchcraft. In *A Discourse of the Polity and Kingdom of Darkness* (1681), Henry Hallywell described six different kinds of fiend, of which one was 'Aquatick or watry . . . drowning men often, raising storms at Sea and sinking Ships'.

Reports like this fed into European witchcraft trials of the seventeenth century. East Anglia's obsessive witch-hunter Matthew Hopkins tried the eighty-year-old parson of Framlingham, Reverend Lowis, keeping him awake and making him walk with Hopkins up and down his cell for days and nights, until he was ready to admit to his 'crimes'. The Nonconformist minister Richard Baxter talked to 'many understanding, pious, and credible persons' who had visited the prisons and heard the statements of the accused, and wrote that the old parson 'confessed that he had two imps, and that one of them was always putting him on doing mischief, and (he being near the sea) as he saw a ship under sail, it moved him to send him to sink the ship, and he consented, and saw the ship sink before him'. Lowis was hanged.

That witches could and did cause shipwrecks was commonly believed, particularly after the notorious case of the coven in NORTH BERWICK (Scottish Lowlands) said to have attempted the murder of King James in 1590.

GORLESTON-ON-SEA, NORFOLK

A smuggler outwits the Customs

The Reverend Forbes Phillips, vicar of Gorleston, wrote *The Romance of Smuggling* (1909) under the pen-name Athol Forbes. Whenever people found out that he lived by the coast, they always asked whether his parishioners smuggled, to which his reply was, 'Yes, whenever they get the chance.' Forbes's own house had once been a centre of the smuggling trade, and not in the distant past:

> Beneath my feet as I write are large and roomy cellars, once used for the storage of imported goods, and until a few years ago a subterranean passage connected these with a landing-stage by the waterside; and, let the full truth be told, the designer of all was the vicar of the parish, and this house was, and is still, the vicarage.

The White Horse Inn at Gorleston, says Forbes, was a favourite meeting place for the smugglers. Built in the reign of Edward III, by Forbes's day it was a ruin, and said to be haunted – 'not without reason', he adds. Neighbours reported that sometimes they were woken in the night by drunken shouts, but when they opened their windows to protest, they saw nobody there. Excavations had taken place from time to time, sometimes unearthing a skeleton or a skull.

In the days when the White Horse was still a working pub, a bet was once laid there between Leggatt, a well-known smuggler, and a customs official named Hacon, who had been sent to the district to put down the rampant trade in contraband. He was fresh from successes in Cornwall, and had little doubt that he would triumph in Norfolk. When Hacon boasted of his intentions, Leggatt was at no pains to hide his own involvement with the free-traders. On the contrary, he wagered fifty guineas that within three weeks he would bring a hundredweight of tobacco over the South Town Bridge between Gorleston and Yarmouth, and deliver it at the officer's own front door.

Hacon saw no difficulty in accepting the wager. As Leggatt had been reckless enough to name the very route that he would take, all that was necessary, surely, was to keep a strict watch on the bridge, and accordingly he set two men to watch the Gorleston end, two more at the Yarmouth end, and himself patrolled between. On several occasions men attempted to cross the bridge bearing suspicious-looking loads, but when the bundles were searched they contained nothing but the most innocent goods. This was clearly a ploy of Leggatt's to provoke the authorities, and Hacon doubled his guards. Throughout the district, everyone was agog to see how the drama played out. Crowds lined both approaches to the bridge, and neglected their proper work: 'fishing ceased, bloaters were scarce, herrings at a premium.'

The day before the three weeks were up, a wretched little schooner put into Gorleston, flying a foreign flag upside down as a sign that there was illness aboard. A doctor went out to the vessel, and found a man who seemed to be in a high fever, raving, singing and cursing in a variety of languages. The doctor prescribed something he thought might calm the patient, but the next day the boat was flying its flag at half-mast, showing that the man had died. Members of the crew duly came ashore that evening at Gorleston with the coffin. The deceased, they said, had been a native of Yarmouth, and had wanted to lie in his home town. Permission was accordingly given for them to cross the bridge, thronged as it was with locals eager to see the outcome of Leggatt's wager. The procession passed, and so did the time. Hacon declared that the deadline had been reached.

Unperturbed, Leggatt shook his hand. 'When will it be convenient to pay me my fifty guineas?' he asked. 'The boot's on the other leg,' Hacon protested, but the tobacco was at that moment on his doorstep. 'You go home and have a look at it, and keep the coffin as a small present from yours truly,' said Leggatt.

GREAT YARMOUTH, NORFOLK

Unlucky Friday

A widespread tradition, often told by sailors and fishermen as an unlocated cautionary tale, but attached specifically to Great Yarmouth in Ernest Suffling's *History and Legends of the Broad District* (c. 1891), deals with the common superstition that Friday was ill-starred for all new enterprises, and particularly for starting a sea voyage. A Yarmouth man, who began with bad prospects since his name was Friday, decided to make a thorough job of defying the fates. He laid the keel of his new boat on a Friday, finished her on a Friday, christened her *The Friday*, and launched her on Friday. 'Then, when all was ready, she was taken out of harbour on a Friday by her skipper, whose name was Friday (a relative of the owner), and everything of any importance to her occurred on a Friday.'

At first things seemed to go well for her. Her first voyage, of course, took her a week, so that she returned on a Friday morning, with a record haul of herring. Again she set out on a Friday, and not only made a good catch, but picked up two large anchors with their cables, and found a derelict coal ship, for which good salvage money was paid. Now, however, her luck ran out. She embarked once more on a Friday, and this time never returned.

That Friday is the unluckiest day of the week is a general principle drawn from the idea that Good Friday, the day of Christ's crucifixion, is a day of sorrow. There were also practical reasons not to begin a journey on that day, since the sailors would miss their weekend ashore. In recent times the taboo has been less regarded, and in 1976 the Cunard liner *Countess* was launched on Friday 13 August, a date that would probably have been strictly avoided in more superstitious times. Several twentieth-century writers, however, report a continuing prejudice. Commander Campbell's *Customs and Traditions of the Royal Navy* (1956) tells us that his sailors always preferred not to sail on a Friday, and in *Folklore and the Sea* (1973), Horace Beck records a conversation he had with an old captain:

> 'Do you think Friday is a bad day to start a voyage?' 'I certainly do.' 'Did you ever sail on Friday?' 'Sail on Friday? Hell no. Who'd think of sailing on Friday?'

Herring

A mainstay of the British diet and economy until recent times, herring have plenty of associated folklore. In Sussex, where the herring shoals generally arrive off the coast in early November, it is said that they come to watch the Guy Fawkes bonfires, while in the MORAY FIRTH (Scottish Lowlands), the abrupt disappearance of the fish in the eighteenth century was explained as a supernatural punishment.

Proverbially, 'to throw a sprat to catch a herring' means to bait your hook with a little fish, in other words forgoing a small advantage in the hope of something better, and 'herring-gutted' is a colloquial expression still used in East Anglia of a tall, thin person (like a herring that has been prepared for smoking by having its guts removed). A 'red herring' (literally one that has been dried, smoked and heavily salted) is metaphorically something interesting but irrelevant, and its longer form, 'to draw a red herring across the path', means to distract from a subject by raising a side issue, an image allegedly derived from fox-hunting. According to *Brewer's Dictionary of Phrase and Fable*, 'A red herring . . . drawn across a fox's path destroys the scent and sets the hound at fault.' This would probably work – red herrings smell strong – but who would want to do it and why is not stated.

From the twelfth century onwards, the centre of the English trade was Great Yarmouth in Norfolk. Even before the town was built, fishermen and buyers congregated in the area every autumn, a gathering that became the Yarmouth Free Fair lasting from Michaelmas to Martinmas (29 September to 11 November). As described by Daniel Defoe in 1738:

> During the Fishing-fair, as they call it, one sees the Land cover'd with People, and the River with Barks and Boats, busy Day and Night, landing and carrying off the Herrings, which they catch here in such prodigious Quantities, that it is incredible.

In Yarmouth and Lowestoft alone, Defoe heard, four million herring were taken in one season, and other sources confirm the vast numbers caught and eaten throughout Britain, particularly before

Norfolk herring fishermen at work, in the mid nineteenth century.

the Reformation when everyone was expected to have fish rather than meat on Wednesday, Friday, and Saturday, and during Church fasts. In Lent 1265, the Countess of Leicester's household consumed up to a thousand herring daily, and in the year 1307–8, Durham Priory bought 242,000. Cheap and plentiful, herring remained a staple food in Britain until 1951, when the catch dropped so catastrophically that a ban was imposed on North Sea herring fishing.

A more unexpected traditional use was in marriage divination. In nineteenth-century Galloway, girls used to throw herring fat against the wall, and if it stuck upright, they would get an upstanding man, but if it was bent, they could expect a crooked husband. On Tyneside, a young man who wanted to dream of his future bride could do so on St Agnes Eve (20 January) if he ate a raw herring, bones and all, a supper that could certainly be expected to stimulate dreams, though not necessarily pleasant ones.

HAPPISBURGH, NORFOLK

The legless spectre

In Whimpwell Street, Happisburgh, there used to be a pump, and before that a well. When houses got their own water pipes in the twentieth century, the pump fell into disuse, but for some time it was not removed, because people dreaded a return of the groaning ghost that used to haunt the spot before the pump was put in place. In the early nineteenth century, the spectre was visible as well as audible, and a horrible sight it was – a legless figure that glided along with dreadful speed, its head hanging down its back between the shoulders as if its neck were almost cut through. It wore sailor's clothes with a pistol in the belt, and had a seaman's long pigtail that, because its head dangled so, almost touched the ground. In its arms it carried a long bundle.

In *History and Legends of the Broad District* (c. 1891), Ernest Suffling tells us that two brave farmers decided to follow the apparition and see what it was up to. For several nights they watched in vain, but at last it appeared.

> Although its head was reversed and turned away from the direction in which it was going, it still kept a straight course along the middle of the road until it came to the well; here it paused, and balancing its burden in its arms, dropped it endways down the mouth, and after gliding aimlessly around for a minute or so, quietly disappeared down the well also.

It was decided to search the well, and a young man named Harmer volunteered himself to be lowered down on a rope, taking a lantern. For forty foot he descended, seeing nothing, but on his way back up he glimpsed a piece of dark blue cloth caught on a brick, which convinced him that something might be further down. Back he went, and fished around with a pot-hook tied to a clothes line until he caught a heavy, sodden sack. This was hauled up, and found to contain a pair of boots, with the legs of their one-time owner still inside.

Harmer was too scared to try again, so a fisherman went in his stead, and at last retrieved the decomposing body of a man whose head was nearly severed, attached only by a small flap of skin at the back of the neck. Just like the ghost, he wore a sailor's frock, a broad leather belt with a big brass buckle, and a pistol.

The search was widened, and in a deserted cowshed near Cart Gap a patch of blood was found, together with some fragments of empty bottles, three or four gold coins, and a pistol, the pair to that found in the well. It seemed clear that there had been a smugglers' quarrel, and that one of the men had been killed, his legs hacked off to make carrying the body easier, and all the remains thrown in the well. Why the culprits had carried the corpse nearly a mile inland, instead of burying it in the sand, was never explained.

When men often wore tall boots, the notion of finding a pair of them with the legs still inside was an urban myth of its time, and is mentioned in several local horror stories.

HARWICH, ESSEX

Bessie Catchpole the smuggler

Female smugglers were rare but not unknown. In the early nineteenth century, when Bessie Catchpole's husband was killed trying to bring in a cargo of tobacco and spirits, she came aboard his yawl the *Sally* wearing men's clothes, a pipe between her teeth and a cutlass at her side, and announced that she was the new captain. Most of the men cheered, but one laughed. She knocked him to the deck with her fist, and after that there was no more trouble.

For more than two years she ran the business successfully, respected by her crew and also by the law-abiding citizens of Harwich in Essex and Ipswich in Suffolk, between which she divided her time. Her cunning became famous. On her way with contraband from Dunkirk, the Channel full of shipping, she steered directly for a naval gunship – which promptly sailed in the other direction, seeing that she was flying a yellow signal, sign of sickness aboard.

Bessie was careful of her reputation and always observed the Sabbath as a day of rest, insisting that her men did likewise, and she could play the part of an innocent woman with aplomb. Pursued by a government cutter one evening, she ran her ship on until dark fell, then sank her barrels of brandy, tied together in a long line, and anchored the *Sally*, showing no lights. At dawn the revenue officer spotted the boat and came aboard to search. Of course he found nothing untoward, and demanded why Bessie had evaded him the night before. 'Had I known you were on board and coming to see me, I would have waited,' she replied. 'But some of you King's men have very naughty characters, and I wished to protect myself.'

Although Bessie had to prove herself in a man's world, there seems to have been no feeling that it was risky to sail with her just because she was female.

Generally, women were considered unchancy company at sea, at least in smaller craft (*see* for example DOVERCOURT), but smugglers may have been more cosmopolitan than fishermen, or perhaps it was simply that as the owner and captain of her boat, Bessie Catchpole had privileged status.

KESSINGLAND, SUFFOLK

See THORPENESS.

KING'S LYNN, NORFOLK

The bookfish

In June 1626, a cod caught at King's Lynn was found to have something quite out of the ordinary in its stomach. When it was sold at Cambridge and slit for gutting, inside was a package wrapped in sailcloth, which proved to contain century-old religious treatises. The case of the 'bookfish' became notorious. Some thought it was a hoax, but Mr Mead, a Fellow of Christ Church, Oxford, wrote that he had not only seen but smelt the evidence – 'He that had his nose as near as I yester morning, would have been persuaded there was no imposture.' A verse was made by some student wits:

> If Fishes thus do bring us Books, then we
> May hope to equal Bodlyes Library.

There are many stories about odd things found inside fish. The oldest example, told by the Greek historian Herodotus in the fifth century BCE, is the tale of Polycrates, who was advised that his unbroken prosperity would make the gods jealous. He chose his most valued possession, a gold ring, and cast it into the sea, but soon afterwards he was presented with a beautiful fish, which proved to have swallowed the ring. This boded no good to Polycrates, whose fortune soon deserted him.

The tale has been often recycled, most famously as a miracle of St Mungo of Glasgow, said to have found a ring belonging to the queen inside a salmon, and it also forms part of a popular ballad set in YORK (North-East England). Unlike the original tale of Polycrates, whose finding of his ring signalled disaster, most later legends make it an entirely propitious omen.

Similar episodes have been told as fact. In 1558 or 1559, a Mr Anderson was said to have discovered, inside a salmon, a ring that he had earlier dropped into the Tyne, and several examples can be found in the

This picture of the 'bookfish' from 1866 shows a very small volume in a very large cod.

twentieth-century press. *The Times* of 24 June 1995 had an account of a fisherman catching a five-pound bass in Southampton Water and finding a silver ring in the stomach (although it was not his *own* ring, the detail that makes a legend), and a more bizarre discovery was reported in the *Eastern Daily Press* on 2 March 1993, under the title 'Tony finds what makes a cod tick'. A young man named Anthony Grey had caught a cod at Aldeburgh and found inside it a wristwatch, 'still working and accurate to the second' – a good advertisement, if they had mentioned the watchmaker.

LOWESTOFT, SUFFOLK

A demon dog

A swarthy foreigner, said to be Italian, once stayed in Lowestoft at some unspecified time in the past. His arrival coincided with the appearance of a large black dog in the neighbourhood, and it was thought that the animal must belong to the Italian, although man and beast were never seen together.

The stranger, who spoke good English, became friendly with a young fisher-boy, and when he had to leave Lowestoft he invited the lad to go with him. The boy refused, however, and so the foreigner asked him instead to look after his dog.

After the Italian had gone, the dog and his new master became inseparable. They used to swim together in the sea, and one day the lad went out further than usual, the dog paddling along behind him. When they were a good distance from shore, the boy turned to make for land, but the dog growled at him. The boy persisted, the dog snapped, and in the end both of them swam further and further out, the dog chasing the boy onward with horrible snarls. Hearing the sound of panting almost at his ear, the boy looked round and was appalled when, instead of the dog's shaggy head, he saw the face of the Italian stranger, who grinned evilly, then resumed canine shape and launched himself savagely at the boy's neck. Just as the lad was about to sink, a sailing ship came within call, and he was taken on board, his neck fearfully torn by the dog's teeth. The dog, meanwhile, 'dived like a whale' and vanished.

This tale appears in a late nineteenth-century work by Morley Adams, who sets it no more precisely than 'in a small seaside hamlet'. The East Anglian folklorist Enid Porter, however, locates it in Lowestoft. She and Adams both identify the sinister dog as the well-known canine phantom Shuck, who haunted many places including OVERSTRAND. His usual behaviour, however, is quite unlike that of the Italian's demonic alter ego. Shuck mostly appeared as a big black dog, and occasionally as other animals, but never as a human, and although he is sometimes seen as a menacing apparition, he is unlikely to make a direct attack. The Lowestoft shape-shifter was probably a sorcerer who had learned his trade in the Black School of Padua, famed as an academy of the occult. To say that someone had 'studied in Italy', or even that they looked Italian, was often (to the British, at least) equivalent to calling them a magician.

ORFORD, SUFFOLK

A wild man of the sea

Orford Ness, a shingle spit stretching along the East Anglian coast, is a difficult obstacle for sailors to negotiate in heavy weather, and its dangers have been said to include regular appearances of a phantom boat. In the 1960s, Captain Bob Roberts told the folklore collector Horace Beck that a small sailing vessel was often seen during storms, making for Orford, but if any crew tried to follow, they would find their craft wrecked on the bar across the mouth of the bay, while the decoy boat disappeared.

The shingle barrier itself was attributed to supernatural causes, according to Roberts, the channel to Orford harbour having been blocked by a mermaid in revenge for ill-treatment she had received. Offended mermaids have been blamed for the silting up of other harbours, such as PADSTOW (South-West England & Channel Islands), but in the case of Orford there is an unusually definite and probably historical incident behind the legend. In the twelfth century, a strange bald-headed long-bearded creature was caught by some local fishermen, and kept for several months at Orford Castle. Described in John Stow's *Chronicles of England* (1580) as 'a Fish having the shape of a man in all poyntes', he ate anything he was given, and was particularly fond of raw fish. The soldiers at the castle inflicted various torments to persuade him to speak, without success, and took him often to church, 'where he shewed no tokens of adoration'. He was sometimes indulged with a swim, and on one occasion managed to break through a triple barrier of nets into the open sea, but oddly enough returned to captivity for a while before escaping again, this time for good.

His description is quite unlike that of closely observed and described animals such as that seen at YELL (Scottish Highlands & Islands), or indeed of legendary mer-people, since he had no tail, spoke no words, and accomplished no magical feats. Well into the Middle Ages and even later, people were sometimes found in remote places living a savage existence and apparently untouched by civilisation, and it is likely that the Orford captive was an unusually aquatic 'wild man' of this kind. Local traditions of the 'monster' are said to have persisted into the late nineteenth century, and it seems that in later legend the fish-man was transmuted into a more traditional mermaid, who maliciously or vengefully obstructed the sailors' passage to the sea.

OVERSTRAND, NORFOLK

Black Shuck

In Britain, the creatures most likely to be noticed roaming alone in populated areas are dogs. Farm animals are generally kept within bounds, and for a long time very few large beasts have lived wild in our islands. Domestic cats of course may be seen, but rarely inspire surprise or fear. A big dog, on the other hand, is potentially dangerous, and so will be remembered and mentioned, giving a rational basis for the fact that probably the most commonly reported animal apparitions are 'phantom dogs'. Usually they are black, an indication of their mysterious nature and also because they are often seen at night when anything dark-coloured looks black. Otherwise their characteristics vary: they may be menacing, warning or even protective, acting as guardians of lonely travellers. In coastal areas they are often thought to function as weather omens, appearing before a storm.

In East Norfolk, there are many accounts of a canine spirit. John Glyde writes in 1872:

> A headless dog with saucer eyes is said to pass nightly over Coltishall Bridge, while another, 'Old Shuck' by name, travels between Beeston and Overstrand, the terror of the neighbourhood. A lane in the latter parish is called after him 'Shuck's Lane'.

Overstrand was believed to mark the south-eastern limit of his territory, but he was known to appear further west in Sheringham, where a black dog named Shock used to emerge from the sea and run along 'Shock's Lane' towards the hills. 'Shuck' and 'Shock' mean ragged-haired or shaggy, as in 'shock-headed', and there may be a connection with Old English *scucca*, a demon.

Mentioned without comment by Glyde, but noted as an impossible combination by other nineteenth-century folklorists, are the dog's headlessness and his 'saucer eyes'. Shuck's big luminous eyes almost always figure in his description, and he is often described as headless into the bargain. Both are traditional marks of ghosts, though generally not attributed to the same apparition.

See also THORPENESS.

REEDHAM, NORFOLK

The Viking invasion

In the year 866 or thereabouts, a Dane named Ragnar Lothbrok (the second part of his name means 'Hairy-breeches') went out one day in a small boat to catch seabirds with his hawk. An unexpected storm carried him far out to sea, where he spent several days and nights until at last he was driven on to the English coast at Reedham. Edmund, king of East Anglia, welcomed the stranger to his court, showing him so much favour that the royal huntsman, Berne, became uncontrollably jealous, and killed Ragnar in the woods. For days the crime went unsuspected, but at last a greyhound that had grown fond of Ragnar led the king to where the body lay buried.

Convicted of murder, Berne was sentenced to be cast adrift in Ragnar's own boat, without sails or oars. By chance, or fate, he drifted to the coast of Denmark, where the vessel was recognised, and Berne was tortured to make him reveal what had become of its owner. Treacherously, he declared that Ragnar had been executed on the orders of King Edmund. The Vikings assembled a force of twenty thousand men, led by Ragnar's sons Hinguar

A Viking ship, one of the vessels that crossed the North Sea in the ninth century bringing invaders to the English coast.

and Hubba, and with Berne as their guide they set sail to take vengeance for Ragnar's death.

This story, which accounts for the great Viking invasion of the ninth century as a drama of personal revenge, was current in England from at least the thirteenth century. There is no reason to suppose that it contains any truth beyond the fact of the Danes' attack. Reedham, for one thing, was never a royal court, the East Anglian kings ruling from Rendlesham in Suffolk, not far from the royal cemetery of SUTTON HOO. The Danish conquest, however, was so momentous that legends flourished around it, and the semi-mythical figure of Ragnar was at the centre of tales in both England and his native Denmark. The twelfth-century Danish historian Saxo Grammaticus relates that Ragnar arrived not in East Anglia but in Northumbria, whose king threw him into a snake pit. In true Viking style Ragnar laughed at his fate and sang a dying song, and was revenged by his sons, who punished the Northumbrian king by carving on his back the 'blood-eagle', a possibly fictional Viking retribution that involved tearing the ribs from the spine and pulling out the lungs like wings.

SCROBY SANDS, NORFOLK

An appearing island

One of the most prevalent sea legends is that of an island that magically comes and goes. EYNHALLOW (Scottish Highlands & Islands) and INISHBOFIN (Northern Eire & Northern Ireland) are just two of the many examples. One origin of such tales is the fact that real islands do appear and disappear, created or destroyed by volcanic action, for instance, or by tidal accretion and erosion.

Opposite Great Yarmouth is a sandbar known as Scroby Sands, which is generally underwater but occasionally appears above sea level. In 1578, it emerged from the waves and remained stable for several years. The mainland citizens were charmed by their new island, on which they held fairs and picnics, and after a while they declared it their own, naming it 'Yarmouth Island'. Their claim was disputed by Sir Edward Clere, local lord of the manor, but the burgesses would not give up their pleasure ground. Clere prepared a lawsuit against them, and proceedings were about to begin in 1582 when the sea intervened. A storm washed the sandbank down to its former level, and the island was never seen again.

SHERINGHAM, NORFOLK

Omens of storm and drowning

Upper and Lower Sheringham were once distinct communities, the former well-off farmers, the latter rough, impoverished, and superstitious fisherfolk. A contributor to *Notes and Queries*, basing his information on talks with an old sailor, wrote in 1864 that the Lower Sheringham inhabitants were constantly haunted by fear of wreck at sea, since 'drowning is almost the commonest shape in which death visits the village'. A variety of storm warnings and other apparitions were reported:

> A little way out to sea there is a spot, they say, just opposite a particular cliff, where the captain of some old ship was drowned, and there more than once fishermen have heard sounds like a human voice coming up from the water: whichever way they pull, the voice is in the other direction, till at last, on a sudden, it changes, and comes just beneath their boat like the last wild cry of a man sinking hopelessly.

If they were wise, they would pull for shore, to escape the gale that was sure to follow. Slightly later accounts speak of a whole shipful of sailors, drowned because the Sheringham fishermen would not go out to save them, whose spirits made weird noises before a gale of wind, and were known as the 'Yow-yows'.

Another ill-omened sound was reported at a gap in the cliffs on the boundary of the parish, where a hundred years earlier twelve drowned sailors had been washed ashore, their bodies thrown into a ditch without burial rites, and covered in a heap of stones. If anyone ventured there in bad weather, they would distinctly hear a sound 'which my old friend illustrated by taking a handful of shingle, and dropping them slowly one by one on to a big stone'.

Surprisingly, given its long coastline, Norfolk is short on mermaids, but Sheringham has one (and she is shown in duplicate on the Upper Sheringham town sign). The story goes that a little mermaid once came to a church service. 'Git yew arn owt,' cried the horrified beadle ('Get you on out') – 'We carn't hev noo marmeards in hare!' and slammed the door in her face. When no one was looking, she crept back in and sat on the end of the pew nearest the door – where she remains to this day, as a carving.

See also OVERSTRAND.

SUTTON HOO, SUFFOLK

Ship-burials

In one of British archaeology's most dramatic episodes, a barrow on a private estate in south-east Suffolk was opened in 1939, revealing a magnificent royal tomb dating from the end of the seventh century. A huge rowing boat, nearly seventy feet long, had been buried here, containing splendid armour, gold coins and jewelled ornaments, silver vessels, and ceremonial articles including a harp. Some of the treasure can be seen in the British Museum in London, and some in Sutton Hoo's own exhibition hall.

The mound presents certain riddles. It was probably raised to the memory of Raedwald, king of East Anglia, but whether it was his grave or his monument has not been conclusively determined. If it never contained a body, that may have been because of conflict between Christian and pre-Christian influence. Although some Christian symbols have been identified in the tomb, including crosses and a fish symbol, the burial itself is in the pagan tradition, containing lavish supplies of grave-goods to accompany the deceased into the afterworld, and sited in a cemetery that had certainly been used for cremated remains (cremation at that period being a pagan rite). It has been proposed that the king's body was buried in consecrated ground, while this monument in the pagan manner was raised in a pagan cemetery, to satisfy old-fashioned public taste.

Before the discovery, it had been generally assumed that Anglo-Saxon life, after the Romans left, was squalid and impoverished. Contemporary descriptions of courtly life in *Beowulf* and other works, with their images of high-roofed halls in which warriors drank mead while harpists played and minstrels sang, were taken as poetic licence. Now, however, it seemed that a heroic and privileged lifestyle was still possible in seventh-century East Anglia.

Nor was this the only respect in which the excavation threw light on ancient poetry. *Beowulf*, written in the eighth century but referring to events some two hundred years earlier, gives details of two burials. The hero Beowulf himself is cremated and a mound raised over his remains on a high promontory, 'widely seen by sea-farers', a description that could easily apply to the raised cairns of the Sutton Hoo plateau overlooking the mouth of the river Deben. At the beginning of the poem, the warrior Scyld's funeral is described. As he has requested before his death, his friends carry him to the edge of the sea, where a ship lies at anchor, and set his body down by the mast:

> There were brought many treasures, ornaments from far-off lands. Never have I known a keel more fairly fitted out with war-weapons and battle-trappings, swords and coats of mail. Upon his breast lay many treasures, which were to travel far with him into the power of the flood.

Unlike the Sutton Hoo vessel, Scyld's ship is actually launched to find its destiny amid the waves, but the symbolism is the same. The notion of death as a journey, accomplished in a ship as the vessel that could go fastest and furthest, was common to many cultures influenced by the seafaring ancient Greeks. The widely travelled Norsemen in particular were predisposed to believe in death as a sea voyage, and the same idea inspired Celtic myths of islands of the dead such as THE BULL (Southern Eire).

THORPE-LE-SOKEN, ESSEX

The body in the boat

In the summer of 1752, a scandal unfolded in Essex involving a clergyman and an aristocrat, bigamy and a beautiful corpse. A boat put in on the Colne estuary just below Colchester, and a large chest was handed on shore, events that were quickly reported to the revenue officers, who hurried to the scene. When they opened the chest, however, instead of contraband they found the embalmed body of a woman.

They demanded an explanation, only to find that the young man in charge of the boat spoke no English. This made things even more suspicious, and having deposited the corpse in Hythe church until a better place could be found, the officers took the foreigner into custody. An interpreter was fetched, who discovered that in fact this was Lord Dalmeny, son of the Earl of Rosebery. He had been born in Florence, and had never been in England before, which was why he did not know the language. A few years earlier, in Verona, he had married a lady slightly older than himself; she had recently died of an illness, leaving a written statement that had dumbfounded her husband. Under the name Kitty or Catherine Cannom, she confessed, she had already been married to Alexander Gough, rector of Thorpe, so that her subsequent marriage to the young nobleman had been illegal. Her dying wish was to be buried in her former home, and this was why he had had her body preserved and brought it to Essex.

The Reverend Gough was sent for, and at first wanted to run the young man through with his sword. On consideration, however, he realised that his rival had acted innocently, and the lady's two husbands attended her funeral together.

THORPENESS, SUFFOLK

Sea-serpents of East Anglia

In December 1933, *The Times* printed a short account of a sea-serpent seen at Thorpeness a couple of years earlier by Sybil Armstrong. She said that on a calm, clear evening, she had observed two black, rounded humps, with a series of smaller humps in between, extending to a length four or five times greater than that of an ordinary rowing boat - about fifty or sixty foot in all, perhaps. The animal sped past a sandbank and then turned towards the open sea, churning up the water with what looked like a pair of large grey fins.

Thirty years later, Mrs Armstrong expanded on her report in a letter to Tim Dinsdale, printed in *The Leviathans* (1966). Although so long ago, events seemed fresh in her memory. Together with her governess and cook, she wrote, she was in the sitting room of a bungalow at the edge of a low cliff with nothing obstructing her view of the sea. It was about eight in the evening, but there was still plenty of light. Through the french windows, she saw something she thought at first was a man swimming, and commented that he was out late, but as she said this, she realised that it was too big to be a man's head, and was approaching at too great a speed. Going out for a better view, she saw a serpentine length behind the head, and exclaimed in surprise, bringing the other two women out to see what the matter was.

A quarter of a mile or so out to sea was the sandbank mentioned in the *Times* report, and when the creature came level with the bank it crawled up on top of it, then down the other side, where it 'beat on the sea with enormous oyster coloured *fins*', making a towering spray of water, five or six times the height of a man, she thought. She was reminded of a swan

Strange Fish and Jenny Hanivers

Monstrous sea-creatures, or 'strange fish' as they were generally known, were popular attractions from the sixteenth century onwards. The public appetite for marvels was stimulated by accounts of extraordinary creatures seen in distant lands – men who had one foot large enough to shelter them from the sun (*see* DENNINGTON, East Anglia), mermaids and other anomalies – and nobody at home could say which were true and which false. The 'sea-bishop' and 'sea-monk' were said to have human features with appropriate clerical head-dresses, and were thought to raise storms at sea. It has been suggested that the legend originated in the sight of a skate or ray from underneath, since the lower side of these flatfish can look like a grotesque face, but certainly the delightful sea-bishop drawn by the French zoologist Rondeletius (1507–66) bears little resemblance to any known species. The 'straunge Fish' seen by Martin Frobisher's crew in the 1570s, 'who had a Boane in his Head like an Unicorne', sounds just as fabulous, but he was probably a narwhal.

A 'sea-bishop', from a drawing in Gulielmus Rondeletius' Libri de Piscibus Marinis *('Book of Sea Fishes'), first printed in 1554.*

Fairs and sideshows exhibited such beasts, some genuine rarities, but many of them doctored specimens. Live people could be shown, more or less disguised. The Spanish writer Cervantes (1547–1616) told of a deformed sailor rescued from a desert island, who was made to stay 'naked, besmeared and tattoo'd' in a tub, and shown to curious observers, and Shakespeare's Trinculo has a similar fate in mind for the wild man Caliban, when he exclaims in *The Tempest* (1611), Act II scene 2:

A strange fish! Were I in England now . . . and had but this fish painted, not a holiday fool there but would give a piece of silver.

Exotic provenance was half the battle. Jasper Mayne's comedy *The Citye Match* (1639), Act III scene 2, features a young cockney who is dressed up and put on show at a shilling a view. Someone protests that they saw another 'strange fish' for only fourpence. 'Gentlewoman, that was but an Irish Sturgeon,' she is told. 'This came from the Indies.' She pays up.

Displays continued into the nineteenth century. In the 1820s, a picture of a beautiful siren at Bartholomew's Fair in London enticed visitors inside the booth to be shown something very different, a dried monkey's head and body attached to a fish's tail. Although the spectacle was soon known to be a hideous one, hundreds of people queued daily to see it. In *Curiosities of Natural History* (1860), Francis Buckland reports seeing a sort of mermaid in Folkestone, which was in fact a large shark ray 'tortured into something like a human shape'. Fakes like this, skates or rays distorted and dried to look like mythical beasts, are known as 'Jenny Hanivers', a name overlooked by dictionaries, although it has been recorded since the 1920s. It could derive from Jenny (denoting a female, as in Jenny Wren) d'Anvers, 'of Antwerp', a port where they may have been manufactured. Alternatively, a Jenny Haniver might be analogous to a Hanover Jack, a nineteenth-century term for a counterfeit sovereign.

beating the water with its wings. As the animal pulled the length of its body over the sandbank, she could see how 'incredibly *vast*' it was, later estimating that it was about five times as long as a fishing boat. 'I do promise you,' she assures Dinsdale, 'it is exactly what I saw.'

The East Anglian coast seems to have been rich in monsters in the late nineteenth and early twentieth centuries. In July 1891, the vicar of Overstrand saw something like a large snake with a brown shaggy mane and a spiny back, travelling very fast through the sea, and another speedy creature was reported in 1912 by several observers, starting with Lilias Rider Haggard, who wrote to her father Sir Henry on 20 July from Kessingland Grange, near Lowestoft:

Like this creature depicted in Robert Hamilton's Natural History of the Amphibious Carnivora (1839)*, the sea-serpent observed at Thorpeness by Sybil Armstrong in the 1930s had a mighty pair of fins, and raised a spray of water several yards high.*

> We had a great excitement here this evening, and are convinced we have seen a sea serpent! I happened to look up when I was sitting on the lawn, and saw what looked like a thin dark line, with a blob at one end, shooting through the water at such a terrific speed it hardly seemed possible anything alive could go such a pace . . . I tore into the morning-room, and got the glasses, and though it had in that moment nearly vanished in the distance we could make out it had a sort of head at one end, and then a series of about 30 pointed blobs, which dwindled in size as they neared the tail. As it went along it seemed to get more and more submerged, and then vanished. You can't imagine the pace it was going. I suppose it was about 60 feet long.

Sir Henry sent her letter to the *Eastern Daily Press* in the hope of finding out what this might have been. The editor suggested porpoises, and contributed a sarcastic comment that Rider Haggard, 'in his capacity as a master of fiction' (he was the author of *King Solomon's Mines* and *She*), 'has discovered many extraordinary and wonderful beings in his time; and a sea serpent more or less must be quite an ordinary incident for him'. Many readers, however, turned out to have seen the same thing, and wrote to say so. One of them had had a good look through binoculars, and thought it was a line of birds flying low over the water, an explanation the editor was happy to adopt.

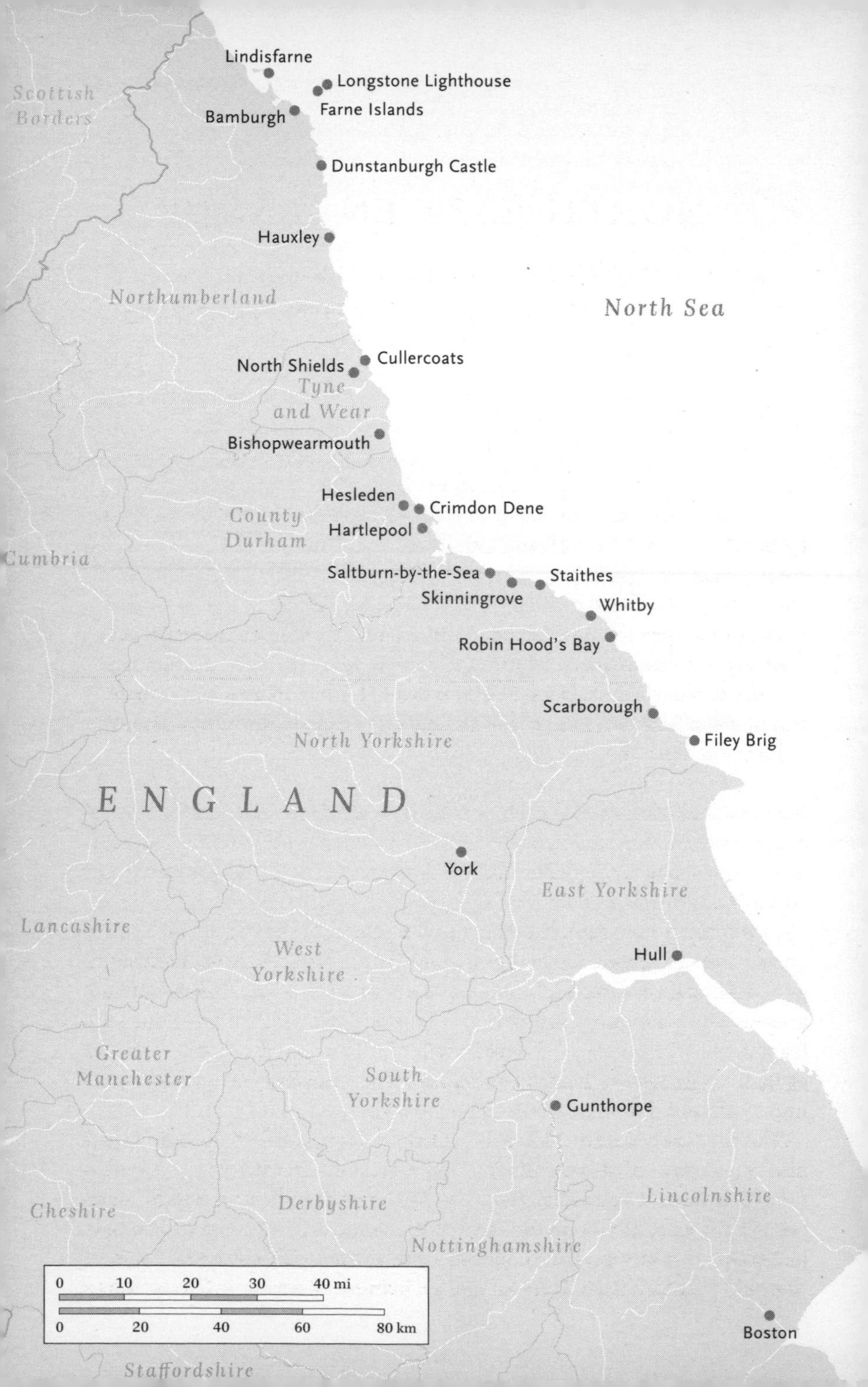

Lindisfarne
Longstone Lighthouse
Farne Islands
Bamburgh
Scottish Borders
Dunstanburgh Castle
Hauxley
Northumberland
North Sea
Cullercoats
North Shields
Tyne and Wear
Bishopwearmouth
Hesleden
Crimdon Dene
Hartlepool
County Durham
Cumbria
Saltburn-by-the-Sea
Staithes
Skinningrove
Whitby
Robin Hood's Bay
Scarborough
Filey Brig
North Yorkshire
ENGLAND
York
East Yorkshire
Lancashire
West Yorkshire
Hull
Greater Manchester
South Yorkshire
Gunthorpe
Cheshire
Derbyshire
Lincolnshire
Nottinghamshire
0 10 20 30 40 mi
0 20 40 60 80 km
Boston
Staffordshire

NORTH-EAST ENGLAND

County Durham, Lincolnshire, Northumberland, Tyne & Wear, Yorkshire

BAMBURGH, NORTHUMBERLAND

The magic ship and the loathly worm

> Once on a time – a time long ago – Bamburgh Castle was the residence of a witch stepmother, who, from hatred and jealousy, banished her lord's son beyond the seas, and changed his fair daughter into a toad, and this loathsome shape she was to endure until her brother could return and dissolve the enchantment.

Many times the brother tried to return, but in vain, for the queen had set charms around the coast so that whenever his ship neared the shore, it was driven back by invisible forces, or came to pieces as the very nails sprang from the beams.

At last the prince countered magic with magic. He built himself a vessel entirely out of rowan, a wood famed for its power against enchantment, and as an extra safeguard he bound the sails and ropes with red thread, another time-honoured protection from evil spells. This marvellous ship leapt across the waves as if it had a will of its own, and sailed to harbour at Bamburgh, where the prince was able to free his sister from her horrid imprisonment and restore her to her own shape.

The folklore collector Michael Denham, who was told this story in the mid nineteenth century, calls it a 'pure version' of what had become a famous ballad, 'The Laidley [Loathly] Worm of Spindleston Heugh', composed by Robert Lambe in the eighteenth century. Lambe claimed to have had access to a very ancient manuscript, and although his supposed source, 'the old Mountain Bard, Duncan Frasier, living on Cheviot, A.D.1270', was a

The enchanted ship sails towards Bamburgh Castle, bringing the prince to rescue his sister from the spells of the wicked queen.

fiction, he seems to have used authentic local tradition, and perhaps some fragments of a genuinely old song, as inspiration for his own verses. In this version Margaret the princess became a more terrible monster than a toad: a loathly worm, or in other words a giant serpent or dragon. The creature lurked in the Spindleston Hills west of Bamburgh, and so poisonous was its breath that for miles around nothing would grow. People feared that the whole of the north country would soon be a desolate waste.

The wicked stepmother rejoiced at the destruction she had caused. When she looked out of her window and saw a gallant ship sailing towards the castle, she knew that the princess's brother was on his way, but thought that she still had the upper hand.

> When she beheld the silken sails,
> Full glancing in the sun,
> To sink the ship she sent away
> Her witch-wives every one.

As in Denham's tale, however, the prince was protected by the magic rowan:

> Their spells were vain. The hags returned
> To the Queen in sorrowful mood,
> Crying, That witches have no power,
> Where there is rown-tree wood.

Next the queen sent armed soldiers, and finally the serpent itself rose up to fend the vessel off with its tail, but the prince evaded all these perils and came ashore on Budle sand. He rode to Spindleston, flung his horse's reins over a tall stone pinnacle later known as the Bridle Rock, and drew his sword, swearing that he would kill the creature if it did him harm. The serpent cried out to him to put down his weapon and give her three kisses, for if the spell was not lifted before sunset, it would be too late. Then he kissed the worm, and it crept into its cave and emerged again as Margaret, his own sister.

The prince and princess went up to Bamburgh Castle hand in hand, and the king was glad, but the queen trembled. 'Woe be to thee, thou wicked witch,' said the prince, and told her that as she had done to his sister, so should be done to her. She became 'a loathsome toad', and to this day she sits on the sand spitting poison at every maiden, and will never resume her own shape until the end of the world.

Robert Lambe calls his hero 'Childy Wynd' – 'childy' or 'childe' being an archaic word for a noble youth. In one oral version of the ballad he became 'Child o Wane', a name for a champion that remained in colloquial use into the early nineteenth century, when, according to the song collector George Kinloch, Bamburgh schoolgirls still applied the title 'to any boy who protects them from the assaults of their school-fellows'.

BISHOPWEARMOUTH, TYNE AND WEAR

Unlucky houses

Wrecks at sea might be attributed to several supernatural factors, including witchcraft, sacrilege, and ignoring a bad omen or taboo. A shipmaster's wife at Bishopwearmouth, now part of Sunderland but in the nineteenth century still a separate coastal community, gave a novel reason for her husband's misfortunes, reported in William Brockie's *Legends and Superstitions of the County of Durham* (1886). She blamed his misadventures on 'unlucky houses', and as the couple prepared to move house yet again, she lamented:

> My hoosband was three times shipwrecked in our first house. Then we removed; but we had ney sooner gettin' into the next house than he was browt on shore in the life-boat. Aw's shour I hope we'll hev better luck in the house we're gan tey.

Having survived no fewer than four wrecks, this was surely not an unlucky but a very *lucky* man.

BOSTON, LINCOLNSHIRE

Cormorants

An unusual portent was observed at Boston church on Sunday 29 September 1860, when a cormorant alighted on the steeple, 'much to the alarm of the superstitious among the townspeople'. Although it flew away for a couple of hours, it returned and stayed there all night, and next morning the church caretaker shot the bird. 'The fears of the credulous were singularly confirmed', according to the account in *Fenland Notes & Queries* (1889–91), when news came that on the very day the bird had first been seen, the *Lady Elgin* had sunk with three hundred passengers on board, including the MP for Boston and his son.

Since the wreck coincided with the cormorant's arrival, not with its death, it is clear that the shooting was not the unlucky factor here. Unlike storm petrels and other gulls (*see* PORLOCK, South-West England & Channel Islands), the cormorant was not seen as a bird whose killing would bring misfortune, but was itself the bad omen. Cormorants had a

With their serpentine necks and dark plumage, cormorants are ominous-looking birds, and were traditionally regarded as portending storm and shipwreck.

poor reputation, and were regarded from Roman times onward as harbingers of storm. In the first century BCE, Virgil wrote in *The Georgics* that it was a sure sign of rough weather at sea when cormorants sported on dry land, and the same belief was held by nineteenth-century fishermen of the Côtes-du-Nord, who said that when cormorants were seen on the rocks, ships would be lost.

The name 'cormorant' comes from the medieval Latin *corvus marinus*, 'sea raven', and the bird is known as a sea-crow or water-crow on parts of the English coast. Corvids generally have long been regarded as ominous birds, and the cormorant's appearance and habits lend themselves to malign interpretation. One of its most noticeable features, especially when swimming, is its snake-like head and neck, and when it lands and spreads its wings to dry – because they are deficient in waterproofing oils – it looks not unlike a small pterodactyl. Its nests are untidy and stink of fish, and in the sixteenth and seventeenth centuries it was a symbol of insatiable greed, a role more often allotted today to the gannet.

The cormorant's dark hue, its reptilian head and its characteristic posture with bat-like wings hung out to dry suggested an image of the Devil, whose iconography by the Middle Ages had come to include the now traditional leathery wings. In Andrea Mantegna's *Agony in the Garden* (1458–60), a cormorant perches on a bare tree staring down avidly at the praying Christ, and John Milton uses the same image in *Paradise Lost* (1667) as a simile for Satan preparing to bring about the Fall of Adam and Eve. Having leapt into Paradise like a burglar breaking into a house, the Devil contemplates his plans:

> Thence up he flew, and on the tree of life,
> The middle tree and highest there that grew,
> Sat like a cormorant; yet not true life
> Thereby regained, but sat devising death
> To them that lived.

Cormorants plunge deep after their prey, and in the seventeenth century, Thames fishermen used to capture them and train them to help with their catches. Their diving habits are explained in a nineteenth-century version of one of Aesop's fables, which relates that the cormorant was a wool merchant who entered into partnership with a bramble and a bat. Their ship with its cargo of wool was wrecked, leaving the partners bankrupt:

> Since that disaster the bat skulks about till midnight to avoid his creditors, the cormorant is for ever diving into the deep to discover its foundered vessel, while the bramble seizes hold of every passing sheep to make up his loss by stealing the wool.

CRIMDON DENE, COUNTY DURHAM

A coastguardsman outwits the smugglers

One fine moonlit night in the 1860s, a coastguardsman named Murray was patrolling his beat from Hartlepool to Black Hall Rocks, and had reached Crimdon Dene, a few miles south of what is now Blackhall Colliery, when he noticed two figures on shore signalling to a small boat a little way out to sea. One of the men was Matthew Horsley, well known to Murray as a smuggler, and he realised that a contraband cargo was about to be landed. He was alone, however, and there were five or six in the boat as well as the two on shore. Thinking that it would be futile and even dangerous to take them on single-handed, he came up with an ingenious solution. Calling at a nearby cottage, he borrowed a white sheet, in which he draped himself, and then waded into the sea under cover of some rocks and approached the boat, which by now was being unloaded.

One of the crew caught sight of the apparition, screamed, jumped into the water and ran for the beach. He was too terrified to tell his friends what he had seen, but Murray got their attention with a penetrating wail: 'Matthew Horsley!' he cried. 'Matthew Horsley!' The men all turned to look, and saw a pale figure rising from the waves and moving towards them, lit by the spectral moon. Tough though they were, this was too much for them, and they all took off for safety on shore.

Left in peace, Murray examined the vessel's contents, which proved to be several hundredweight of tobacco. He marked the packages with the arrow sign of government possession, and waited for a colleague who was due to relieve him. He had forgotten, however, to take off his sheet, and when Coastguardsman Hicks arrived, he was appalled at the boat's ghostly occupant, especially when it called to him by name.

This story, from Athol Forbes's entertaining book *The Romance of Smuggling* (1909), reverses the common report that free-traders would dress as phantoms to discourage the curious. Several ghost stories have been explained by rationalists as smugglers' propaganda. The folklorist W. Walter Gill, writing in 1932 of 'Death-coach' apparitions on the Isle of Man, comments that he had come across suggestions made as early as 1815 that spooks might be reported or impersonated by the smugglers for their own purposes.

CULLERCOATS, TYNE AND WEAR

King of the herrings

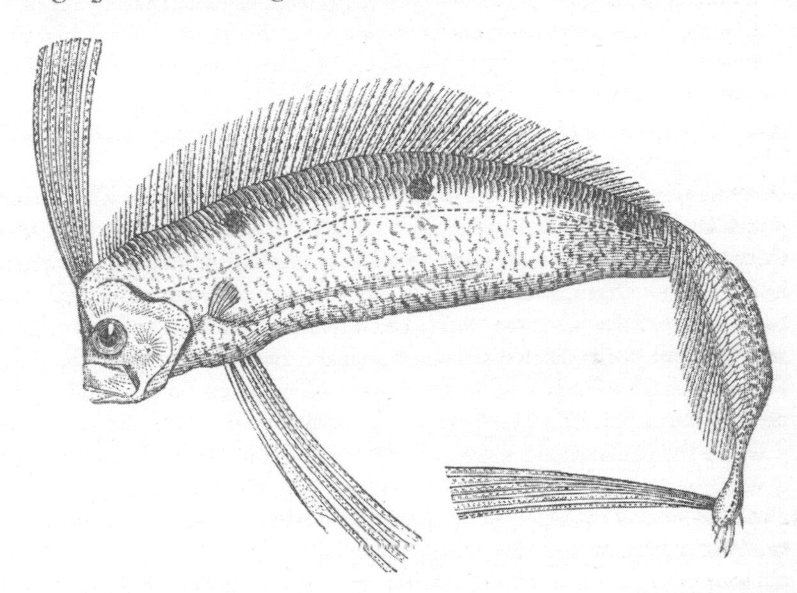

The silvery, serpentine oarfish has a coral-red crest along its spine, growing into a tall plume at the head.

A marvellous fish, silvery, crested, slender, and very long indeed, was caught off Cullercoats near Newcastle in 1849. It caused a sensation, some observers claiming that here at last was the great sea-serpent of legend (*see* SOUND OF SLEAT, Scottish Highlands & Islands), but the creature proved to be one known to science although rare in British waters, the ribbonfish or oarfish (*Regalecus glesne*). This specimen was over twelve foot long and less than a foot wide, easily qualifying for the 'serpent' description.

Quite apart from its occasional identification as a sea-snake, the oarfish has its own legends. It has been called 'King of the herrings', and was long thought to lead or accompany herring shoals. When one was caught, it was generally thrown back into the sea, since if that was not done, the leaderless shoals would then desert their waters, and the man who had netted the herring-king might be drowned. This belief was still held in Sussex in the twentieth century, as it had been in Skye, according to Martin Martin in his *Description of the Western Isles of Scotland* (1703), when he visited:

> The Fishers and others told me that there is big Herring almost double the size of any of its kind, which leads all that are in a Bay, and the Shoal follows it wherever it goes. This Leader is by the Fishers called the King of Herring, and when they chance to catch it alive, they drop it carefully into the Sea, for they judge it Petty Treason to destroy a Fish of that Name.

Martin's double-size herring does not sound particularly like the long thin oarfish, and it may be that another species was meant.

All kinds of fish and marine animals were supposed to have a leader. In Ireland in the 1940s, David Thomson was told that 'all the creatures of the water do have their own king', and as well as hearing of the king of the seals (*see* GREAT BLASKET ISLAND, Southern Eire), he talked to a man who had seen the king of the mackerel, a vast fish that had outswum two seals and beached itself, and which might perhaps have been a tuna.

DUNSTANBURGH CASTLE, NORTHUMBERLAND

Guy the Seeker

The ruins of Dunstanburgh Castle stand high above the sea. To the north and north-west, steep cliffs plunge down to the waves, and at the eastern end of the cliffs, under the main entrance to the castle, is a gully known

Ruined Dunstanburgh Castle is the setting for M. G. Lewis's tale of 'Sir Guy, the Seeker', who failed to free a beautiful lady from magical captivity.

as Rumble Churn. Even when it is calm, the sea mutters in the hollow rock, and in stormy weather it crashes and booms in confinement, sending up clouds of foam and spray.

In the nineteenth century, it was said that the Churn was 'ever sounding with the wail of malignant spirits', an idea that may have originated in the spectre-haunted brain of M. G. Lewis, best known for his ghost story *The Monk* (1796). While staying near the castle, Lewis wrote a poem supposedly based on local tradition, in which a knight, Sir Guy, roams one stormy night on the steep cliffs:

> Loud was the roar on that sounding shore;
> Yet still could the Knight discern,
> Louder than all, the swell and the fall
> Of the bellowing Rumble Churn!
>
> With strange turmoil did it bubble and boil,
> And echo from place to place;
> So strong was its dash, and so high did it splash,
> That it washed the castle's base:
>
> The spray, as it broke, appeared like smoke
> From a sea-volcano pouring;
> And still did it rumble, and grumble, and tumble,
> Rioting! raging! roaring!

At midnight, the gate of the castle bursts open to reveal a gigantically tall old man with a white beard, flames flickering around his bald head. He carries a staff of 'burning iron' and has a red-hot chain round his waist. The old wizard tells Sir Guy that a lady is waiting for him to rescue her, but warns him that he must not look at her.

Guy declares that he knows no fear, and follows the apparition through dark passages to a winding stair. As he climbs, he hears sobbing and whispers from above, and the tolling of a bell. At the top he finds a brass gate fastened with a live snake, which threatens to bite but is rendered powerless by the wizard's staff. The gate opens to reveal a huge marble-paved hall where a hundred knights and horses are asleep, the scene lit by candles held in dead men's hands which still drip blood. Two giant skeletons stand each side of a crystal tomb, in which a beautiful woman seems to whisper to Guy. Although he cannot hear her, he takes her words to be murmurs of love, and is enchanted.

The wizard points to a sword and a horn, held by the two skeletons, and tells Guy that he may either draw the weapon or sound the horn, but must choose only one. He blows the horn, at which the horses stamp and the sleeping warriors awake to attack him. Dropping the horn, he seizes the sword, and hears a piercing cry of grief. The wizard is furious:

> And he stamped in rage, and he laughed in scorn,
> While in thundering tone he roared,
> 'Now shame on the coward who sounded a horn,
> When he might have unsheathed a sword!'

Guy falls in a faint, and wakes to find himself back in the castle porch. For years afterwards he searches for the stairway, but never again can he find it, and at last he dies, a white and withered old man, still pining for a glimpse of his beautiful lady.

Shorn of its more baroque elements (skeletons, blood-dripping sconces, serpents, and indeed the lovely lady), the story is fundamentally the same as many 'sleeping hero' legends – usually linked to King Arthur in England, and in Scotland to Thomas the Rhymer and Fionn mac Cumhaill – in which somebody finds a band of sleeping warriors surrounding their leader, and is offered a choice between a sword and a horn, a choice he always makes wrongly. A local legend along these lines was the foundation on which Lewis built his narrative, a tour de force of Gothic detail, which by now has been retold enough to qualify as a tradition.

FARNE ISLANDS, NORTHUMBERLAND

St Cuthbert and the demons

The largest of the Farne Islands, Inner Farne, was once infested with demons. In the seventh century, St Cuthbert, having found his monastery at LINDISFARNE too sociable, retreated there in search of deeper solitude, but before he could settle down in peace he had to banish the evil spirits to the nearby island of Wideopens. There they remained for centuries afterwards, according to a manuscript life of the hermit Bartholomew, who stayed on Inner Farne in the twelfth century. The account describes the demons as dwarfish bearded beings, wearing black cowls, riding on she-goats, and brandishing their lances in a warlike manner. The Farne monks, who used Wideopens as a graveyard, were at first able to repel the apparitions by showing them a cross, but this soon lost its effect, and before the monks could bury their dead, they had to hem in the plot of ground with a barrier of straws marked with the sign of the cross.

It has been suggested that these stunted 'spirits' were human, aboriginal inhabitants of the islands who had never been in contact with the mainland and remained in a condition of savagery. This might be so, but reports remained current into the nineteenth century that the Farne Islands were haunted, either by demons or by the ghosts of the many sailors wrecked on the dangerous reefs.

Inner Farne itself remained under the special protection of St Cuthbert. The twelfth-century monk Reginald, writing of the saint's miracles, describes an occasion when several ships were obliged to shelter in the island's harbour from a storm. For days the bad weather persisted, and in the end the crews prayed to St Cuthbert for calm. That night, one of the men went outside and saw 'a light, above the brightness of the sun, and forthwith turning his eyes towards the ships, saw near them a man, tall and graceful, and clad in the robes of a Bishop'. Several of the sailors shared this vision, although a few of their womenfolk 'could see nothing, and turned the thing into ridicule'. Reginald implies that there was, in fact, nothing wonderful about the strange glowing light, since it was only visible around a large heap of fish – in other words, it was natural phosphorescence. Next morning, however, the storm had dropped, the sea was still and the ships went on their way in safety.

FILEY BRIG, NORTH YORKSHIRE

The haddock's spots

Filey Brig is a reef of rocks jutting out a mile from the Carr Naze headland at the top of Filey Bay. Exposed at low water but covered at the tide's flood, it was a danger to sailors, and the tradition among the Filey fishermen was that it was the work of the Devil, determined to hinder their trade and destroy their boats. While building the Brig, he dropped his hammer into the sea and made a grab for it, but caught a haddock instead, leaving a black mark on each side of the fish, either because his fingers were so hot they burnt its skin, or because they were sooty and black like the rest of him.

The two large black spots close to a haddock's gills are more usually said to be the marks of Jesus's fingers when he divided the fishes and loaves at the Feeding of the Five Thousand (Matthew 14:17–21), a legend that has led to the haddock being known as the Lord's Fish. Alternatively, they were supposed to be those of St Peter's hand, the haddock being identified as the fish caught in the lake of Gennesaret (Matthew 17:27) when Jesus, after tribute was demanded of him and his disciples at Capernaum, directed Peter to go to the sea, 'and cast an hook, and take up the fish that cometh

Filey Brig, said the local fishermen, was built by the Devil to wreck their boats.

up; and when thou hast opened his mouth, thou shalt find a piece of money: that take, and give unto them for me and thee.' The miraculous finding of the coin in the fish echoes the ancient Greek tale of the ring of Polycrates, a motif that appears in other traditions of strange things found in fish (*see* KING'S LYNN, East Anglia).

Pedantically speaking, the fish caught by Peter in the lake could not have been a haddock, since haddock live in salt water, but scientific accuracy is not an issue here. The legend is a folk explanation of the *spots*, not of the type of fish caught, and the same tale, part of the common stock of Christian legend, is told of the John Dory (*see* FOWEY, South-West England & Channel Islands).

GUNTHORPE, LINCOLNSHIRE

A cloud called Noah's Ark

Storm omens often came in the form of supernatural sounds or sights like those reported at SHERINGHAM (East Anglia) or LAXEY (North-West England & Isle of Man). There were, however, more everyday signals of bad weather on the way, such as unusual clouds. These continued to be observed and interpreted until modern meteorology took over, and are doubtless still taken into account today by weather-wise sailors and farmers.

One sign was the 'Noah's Ark' or 'Noe ship', a formation variously described as a single cloud looking like a ship's hull turned upside down, or as lines of cloud parted into wave-like ripples. The outlook, according to some writers, depended on their orientation. If the end of the clouds pointed towards the sun, there would be rain; if not, fine weather could be expected. Others held that it was their position relative to the prevailing *wind* that mattered. In Gunthorpe, in the early twentieth century, it was believed that:

> The Ark athwart the wind means rain.
> The Ark end-on, or bow-on, means that the Ark will have fair sailing.

Descriptions from various sources tend to contradict each other. Mrs Lubbock, an old woman of Irstead, Norfolk, in the mid nineteenth century, said nothing about which way the clouds were pointing, but maintained that if Noah's Ark were seen for several days on end there would be foul weather: 'It *shew* last February, forty days; and after that, the weather was very bad, wet, and stormy.'

The reference to 'forty days' here has a familiar ring, as the age-old formula for 'a very long time' used in the Old Testament for the duration of the original Noah's Flood.

HARTLEPOOL, COUNTY DURHAM

Hanging the monkey

When fear of Napoleon's forces was at its height around the coasts of Britain, a ship struck Long Scar near Hartlepool and went to pieces. One creature alone was saved, a monkey wearing a red cap, and the beast was immediately suspected of wicked intentions. It spoke no English, it gesticulated in a foreign manner, its behaviour was rude and outlandish, and its headgear was a radical symbol. Taking all this into account, it seemed obvious that this was a French spy. A trial took place, at which the monkey refused to state who he was, his nationality, or the name of his ship. 'When warned of the enormity of his conduct, he amused himself with the pursuit of petty difficulties lurking in his coat . . . He snatched at the Testament and bit it in halves, which established the fact that he was an Atheist and a revolutionary.' The verdict was plain, a gallows was erected on the sands where the attempted invasion had taken place, and the unfortunate creature was executed on the spot.

'It is considered bad taste to ask a native of Hartlepool why the monkey was hanged, or to seek further information,' writes Athol Forbes in *The Romance of Smuggling* (1909), making it clear that this was a story repeated *outside* Hartlepool in mockery of the residents' supposed gullibility. Although it is sometimes related as a historical incident, the same story is told of several places in Scotland, including Greenock, Cullen, and

Boddam, where a popular song with the refrain 'And the Boddamers hung the monkey O!' is cited in support of Boddam's claim to be the original town in question. This dubious honour is not theirs, since earlier printed sources for the legend link the story with Cullen and with Hartlepool.

Some versions give a different motive for the monkey-murder, in that the animal was the only survivor of the disaster. In order for the wreck to be legally claimed by the Hartlepool citizens, it was believed, *no* living creature could be left on board (*see* SMUGGLERS AND WRECKERS, p. 56), and so the monkey had to go.

HAUXLEY, NORTHUMBERLAND

Sea-fire

The Hauxley fishermen used to speak of 'Mar Fire' or 'Sea Mare', the light sometimes produced by tiny phosphorescent jellyfish and plankton. They said they saw 'a vast o' fire' at sea on nights before a storm. As well as being a sign of bad weather, the glow was useful in guiding the fishermen to shoals of herring, since the larger fish would often be found feeding on the smaller.

Elsewhere the phenomenon went under various names: 'Fire-burn', 'Water-burn', 'Sea-lamp', or 'Brimming'. In a nineteenth-century book on deep-sea fishing, Edmund Holdsworth writes:

> When the water is in this condition the slightest agitation, as is well known, produces sparks and flashes of light; and the presence of fish is often indicated by the streaks of light which are caused by their suddenly darting through the water, and even when stationary, by the luminous patches which then result, we are inclined to believe, from the constant movements of their fins.

Holdsworth had seen fishermen leaning over the water and tapping a piece of wood against the side of their boat to alarm the fish into movement. 'Now and then a fish betrayed itself by the line of light it produced as it darted away, and when these indications became more numerous it was decided to shoot [cast] the nets.' He adds, however, that the water was usually too clear on these occasions for good catches to be made, since the fish could see and avoid the nets, which were beautifully illuminated as they were drawn through the water, so that the fishermen generally did not make large hauls on nights of Mar Fire.

In *Folklore and the Sea* (1973), Horace Beck records a splendid legend accounting for the lights. The Devil once built a huge ship and sailed about in it, collecting up damned souls, wrecking other boats and making merry. St Elmo disapproved of his behaviour, and first set the Devil's ship on fire, then knocked a great hole in her side and sank her. The Devil is still looking for the remains of his vessel, and lights seen underwater are the torches with which he is searching. The mention of St Elmo suggests association with St Elmo's fire, also known as the corposant (*see* CROMER, East Anglia), but that light was seen around the rigging of ships, whereas here the phosphorescence is below the surface of the sea.

HESLEDEN, COUNTY DURHAM

Going out with the tide

It was once a very common superstition that life comes in with the tide, and goes out with it, so that births happen at the flow and deaths at the ebb. Sixteenth-century parish registers of Heslidon (now Hesleden) sometimes noted the state of tide at the time of death, so we learn that on 11 May 1595, at six in the morning, 'being ful water', Henrie Mitford of Hoolam died, and that on 17 May of the same year, at noon, 'being lowe water', Mrs Barbarie Metford died. The first entry does not support the theory that death only occurs on the ebb (although the tide might have been turning when Mr Mitford breathed his last), but such records do show the importance attached to the state of the tide at the moment of death.

The belief goes back to the time of Aristotle, who held that no animal died except on the ebb tide, although Pliny added the comment that this had been proved true only for human beings. Shakespeare refers to the notion in *Henry V* (1599), Act II scene 3, where Mistress Quickly reports the death of Falstaff: 'A parted even just between twelve and one, even at the turning o'th' tide,' and in *David Copperfield* (1849–50), Dickens's Norfolk fisherman Mr Peggotty explains that 'People can't die, along the coast . . . except when the tide's pretty nigh out. They can't be born, unless it's pretty nigh in – not properly born, till flood.'

Even in the twentieth century the tradition persisted. From David Thomson's *People of the Sea* (1954) comes a quote from a Mayo ferryman, talking of his wife's death:

> She went out with the ebb tide . . . the way I knew she would, for I was looking from the window when God took her and I never saw the water lower than it was that minute. I thought to myself, and I still praying, if God spares her now for these few minutes, and the tide to turn, she will be safe.

Jonah and the Whale

Summoned by God to preach destruction against the wicked city of Nineveh, Jonah was terrified by his destiny and fled to sea, but his ship was pursued by a mighty tempest. The sailors said:

Come, and let us cast lots, that we may know for whose cause this evil is upon us. So they cast lots, and the lot fell upon Jonah.

Then he confessed that he was trying to escape from the Lord's wrath, and they knew that it was on his account that the storm raged. He told them to throw him overboard, and as soon as they had done so, the sea was calm. Jonah was swallowed by a 'great fish', and stayed inside it three days and nights, praying for deliverance, until 'the Lord spake unto the fish, and it vomited out Jonah upon the dry land,' after which he went on to fulfil his mission.

A man devoured by a whale, imagined in this copy of a sixteenth-century drawing as a fanged monster.

Told in four short chapters of the Old Testament, the story of Jonah has given rise to a disproportionate amount of folklore. His name became used by sailors for any ill-starred member of crew, and even, in the corrupted form 'Jones', for a devil of the deep (*see* FIDDLER'S GREEN AND DAVY JONES'S LOCKER, p. 376). The tale that a prophet had once provoked a storm contributed to the superstition that a priest or minister was an unlucky passenger on board ship, although other factors were probably at work here (for instance, that because a clergyman could conduct a funeral, to have him so handy might be tempting fate).

As for the great fish, it was naturally assumed to be a whale (classed among the fishes until recent times). Although it would be quite impossible for a man to survive being swallowed by a whale, or any other animal, Jonah's adventure was said to have been literally repeated in 1891 (*see* HULL, North-East England). More often, however, his incarceration has been interpreted symbolically. In Christian thought, the time Jonah spent in the fish's belly prefigured the three days and nights Jesus spent in the tomb, and thus the whale can appear as a reference to the Resurrection (*see* ELACHNAVE, Scottish Lowlands). Alternatively, however, it may represent Satan, since the other great ocean-dweller of the Bible, Leviathan, also generally taken to be a whale, is described in Isaiah 27 as 'the dragon that is in the sea', a monstrous embodiment of chaos who will be punished by God at the world's end.

The whale's sheer size has led to its veneration and its persecution. Something on so grand a scale seems to present an irresistible challenge to competitive, supremacist man. Whale-hunting as a cosmic battle against Nature itself, a simultaneously heroic and barbaric struggle, is epitomised in Herman Melville's *Moby-Dick* (1851) by Captain Ahab, doomed to destroy the great white whale and meet his own death as he does so. His triumph is his tragedy, and the same, on a wider scale of devastation, is true of humanity's entire relationship with the whale.

Fertility and health were supposed to come under the same influence, a rising tide favouring conception of a baby, and an ebb tide promoting recovery from illness (*see* MORECAMBE, North-West England & Isle of Man). In the north-east of Scotland, it was said that eggs brooded when the tide was going out would hatch hens, roosters being the result of brooding when the tide was coming in, and so some farmers' wives would only set their hens to brood on an ebb tide, hens being far more valuable than cocks.

HULL, EAST YORKSHIRE

Whales and whaling

From the sixteenth century to the nineteenth, whale fishing played a central role in British maritime life, and Hull was one of the nation's whaling capitals, others being Aberdeen, Peterhead, and Whitby. Whalers, both the ships and men, had their own mystique and their own customs. A surgeon serving aboard a whaler in the 1820s described May Day celebrations out in the freezing Greenland waters, when the sailors made garlands of ribbons, mementoes of their wives and girlfriends. A more macho ceremony, also performed on 1 May, was the 'shaving' of new recruits, a burlesque rite similar to that performed when 'crossing the line' (*see* ABERDEEN, Scottish Lowlands). One of the crew dressed as Neptune, and others disguised themselves in whatever grotesque costumes they could put together. Anyone on their first voyage was then 'soaped' with a disgusting mixture of oil, tar, paint, and suds, his face was scraped by the boatswain with a rough piece of iron hoop, and after that the day concluded with fiddle-playing, dancing, and a good deal of grog.

Whales were the subjects of many legends (*see* for example ELACHNAVE, Scottish Lowlands), some based on fact. A whale could wreck a vessel, and the sinking of the hunters' ship at the climax of Herman Melville's *Moby-Dick* (1851) was inspired by a real case in 1819, when the *Essex* was rammed twice, head-on, by an eighty-five-foot whale – quite deliberately, it appeared to the crew. Something still more freakish was said to have occurred in the same year, when the whaler *Star of the East* was cruising off the Falkland Islands. A harpooned whale, thrashing in its death throes, struck and overturned one of the ship's whaleboats, throwing the men into the water. One was dead before he could be picked up, and another, James Bartley, disappeared, so it was thought that he too had drowned. The whale's huge body was towed back to the ship and 'flensed' in the customary manner, its blubber stripped away to be melted down

A picture from the mid nineteenth century shows a man in the act of harpooning a whale, whose enormous flukes could easily upset the whalers' little boat.

for oil, and the whalebone taken out. The labour took all day and part of the night, and next morning the men resumed work, coming to the stomach.

> The workmen were startled while labouring to clear it and to fasten the chain about it to discover something doubled up in it that gave spasmodic signs of life. The vast pouch was hoisted to the deck and cut open, and inside was found the missing sailor, doubled up and unconscious.

Bartley was washed in sea water and partly revived, but was quite mad from his experience, and remained a raving lunatic for some while. He looked most peculiar, his face and hands bleached deadly white from the whale's gastric juices, and his skin wrinkled, 'giving the man the appearance of having been parboiled'. After three weeks, he began to get his senses back, and was able to tell people what he had undergone.

> Bartley affirms that he would probably have lived inside his house of flesh until he starved, for he lost his senses through fright and not from lack of air. He says that he remembers the sensation of being lifted into the air by the nose of the whale and of dropping into the water. Then there was a frightful rushing sound, which he believed to be the beating of the

> water by the whale's tail, then he was encompassed by a fearful darkness, and he felt himself slipping along a smooth passage of some sort that seemed to move and carry him forward. This sensation lasted but an instant, then he felt that he had more room. He felt about him, and his hands came in contact with a yielding slimy substance that seemed to shrink from his touch.

When it finally dawned on him that he had been swallowed, he was overcome with terror. Although he could breathe, the heat was dreadful, and he was so sick and faint that he lost consciousness, and remembered nothing more until he found himself in the captain's cabin.

Bartley made a fair recovery, apart from the skin on his face and hands, which remained yellow and wrinkled like old parchment, and apart from his dreams, which were haunted by angry whales and 'the horrors of his *fearful* prison'. The whaling captains said that men were quite often swallowed by whales, but they had never come across another who survived.

A detailed investigation into this episode has been made by Professor Edward B. Davis, his results published in a 1991 article, 'A Whale of a Tale: Fundamentalist Fish Stories'. Davis ascertained that the *Star of the East* was a real ship, but found that its crew list contained no James Bartley. He guesses that the story started as a publicity stunt by someone who promoted himself around the sideshows and music halls of the time as a real-life Jonah (*see* JONAH AND THE WHALE, p. 164), and that it was picked up and embellished by religious enthusiasts who wanted to prove the truth of the Bible. Bartley's horrible adventure, however, continues to appear from time to time as a factual report, its mixture of drama and convincing detail too good to let go.

LINDISFARNE (HOLY ISLAND), NORTHUMBERLAND

A miracle of St Cuthbert

Bleak, windswept Lindisfarne was chosen by St Aidan for his monastery in the seventh century, and a little later St Cuthbert became its bishop. Winding through quicksands, the causeway from the mainland is accessible to foot traffic twice a day at low tide, but the rest of the time Lindisfarne is truly an island.

When the Vikings attacked in the ninth century, the monks fled, taking with them the sacred body of Cuthbert, for which they eventually found a home at Durham. In 1069 the dead saint had to be removed again in a hurry, this time in fear of William the Conqueror. When the body was disinterred, according to the contemporary observer Simeon of Durham, it was found to be quite undecayed and fragrant. It was carried back to the eastern coast, the plan being to return it to Lindisfarne, but when the crossing point was reached, the tide was at the flood. It was late December, the weather was freezing, and the frailer members of the party, which included women, children, and aged monks, were desperate for shelter.

> 'What shall we do?' said they; 'we are prevented from crossing over to the island at this time by the full tide; nor is there any place of residence for us in which we can escape this nipping cold.' Whilst they were in the midst of these lamentations, the sea suddenly receding at that spot (but at no other), afforded them the means of passing over, whilst at every other point the tide was at the fullest.

All of them were able to cross to the island dry-shod, carrying Cuthbert's body with them, and as they advanced, the waves followed close behind them, protecting them from pursuit. Cuthbert's body was returned to Durham a little later, but Lindisfarne continued to be closely associated with his name and miracles, and is still known as 'Holy Island'.

For a while in the seventeenth century, however, the piety of its inhabitants seemed to have taken a strange turn. Captain Robin Rugg, governor of the island, reported that when a ship was seen in danger at sea, the 'common people' would pray for it 'very devotedly'.

> You seeing them upon their knees, and their hands joined, do think that they are praying for your safety; but their minds are far from that. They pray, not to God to save you, or send you to port, but to send you to them by shipwreck, that they may get the spoil of her. And to show that this is their meaning if the ship come well to port, they get up in anger crying 'the Devil stick her, she is away from us.'

Their rapacity was not unique. On ST AGNES (South-West England & Channel Islands), the islanders similarly prayed to their saint for wrecks.

LONGSTONE LIGHTHOUSE, NORTHUMBERLAND

Grace Darling

An illustration from Grace Darling or the Heroine of the Farne Islands *(1839), showing the lighthouse keeper and his daughter rowing to the rescue of passengers from the wrecked paddle steamer* Forfarshire.

On 5 September 1838, the *Forfarshire*, a passenger steamer, put out from Hull on her way to Dundee. She was in an unseaworthy condition, and as the weather grew worse over the next two days she found herself in serious difficulty. Forced to put back from Berwick Bay, she tried to make her way between the Farne Islands, but at 3 a.m. on 7 September, at the height of the storm, she struck the Harcar Rock, about half a mile from the Longstone Lighthouse.

Some of the crew took to the lifeboats. Most of the passengers were asleep when the steamer struck, but a Mr Ruthven Ritchie of Perthshire 'arose instantly, and seizing his trousers, rushed upon deck, from whence, observing the sailors leaping into the boat, he, with an extraordinary effort, by means of a rope, swung himself into it, and was thus miraculously preserved'. His uncle and aunt too tried to get into the boat, but fell into the sea and were drowned.

Near the Harcar Rock runs a tremendous current known as the Piper's Gut. Struck by this current, the *Forfarshire* broke completely in half, her stern quarterdeck carried away with everyone on it, including the captain with his wife in his arms. A few passengers were now left on the rock, crying for help but with small hope.

> And now one of those heroic actions occurred, which for their romantic daring are remembered for generations with admiration, and produced a burst of enthusiasm throughout Europe for the humble girl who performed it.

From her bedroom window, Grace, daughter of William Darling the lighthouse keeper, saw the pathetic figures clinging to the wreckage. She and her father set out in their small boat, and while William landed on the reef, Grace rowed to and fro among the crashing waves to keep the craft from being smashed to pieces on the rocks. Nine people – all that remained alive of the original sixty or so – were taken off, and the Darlings took care of them in the lighthouse for three days until they could be ferried to the mainland.

The details are taken from an account written in November 1838, and printed later along with Grace Darling's obituary. She died of consumption at the age of twenty-five, less than four years after the wreck and rescue that had made her a national heroine. Her exertions on the night of the storm may have injured her health, and so too may the fame that afterwards pursued her: she was fêted in poem and song, her portrait was painted, she received many proposals of marriage, all of which she declined, and she also refused to appear at the Adelphi Theatre in London. A modest woman, she might well have shrunk from her celebrity, which led to the most extravagant eulogies. 'Had Grace Darling lived in the remote ages,' her obituary enthuses, 'such heroism as hers would undoubtedly have been deified.' It concludes more soberly and truly: 'as it is her name even now will scarcely ever be forgotten.'

NORTH SHIELDS, TYNE AND WEAR

Press gangs

The verb to 'press' or 'impress', in the sense of enlisting men into the navy (or, more rarely, the army), comes from the Old French *prest*, a loan or advance on wages. Sailors were encouraged to join up by payment of the 'king's shilling', and it was said that many were tricked into taking the money unawares – finding a coin, for instance, in the bottom of a drink they had been treated to. Others were literally forced into the service, taken during raids on ships, fishing boats, inns or even private houses. The practice was not discontinued until 1833, and was at its height during the Napoleonic wars, when there was urgent need to man the ships, and volunteers were in short supply.

The keelmen of the Tyne were in theory exempt from impressment, but in practice their skills and experience, gained on the coal barges, meant that they were eagerly sought as crew for men-of-war. *The Peggy*, a naval tender used by the press gang, used to anchor close to the mouth of the Powburn river, west of what was once the North Shields fish quay and market, at a deep pool that became known as Peggy's Hole. Many traditional Tyneside songs deal with fear of the press gang:

> Here's the tender comin', pressing all the men:
> Oh dear hinny, what shall we dee then:
> Here's the tender comin', off at Shields Bar
> Here's the tender comin', full of men o' war.

The verses go on to urge the men to hide themselves away, since if they are taken, who will support their wives and children? Men who were pressed had to leave their families to rely on parish relief, the meagre social security of the time, and official records show that poverty in Tyneside seafaring communities reached starvation levels after the press gang's visits.

The novelist Mrs Gaskell relates in *Sylvia's Lovers* (1863) how in the 1790s, the press gang was driven south by the resolute resistance of the North Shields men.

> For on a certain Tuesday evening yet remembered by old inhabitants of North Shields, the sailors in the merchant service met together and overpowered the press-gang, dismissing them from the town with the highest contempt, and with their jackets reversed. A numerous mob went with them to Chirton Bar; gave them three cheers at parting, but vowed to tear them limb from limb should they seek to re-enter North Shields.

The press gang bided its time, and had its revenge a few weeks afterwards. Overnight, North Shields was cordoned off by soldiers while the pressers descended in force, and left with a haul of more than two hundred and fifty men, including sailors, mechanics and every kind of labourer.

Stories of the tragic consequences of pressing circulated widely. One cautionary tale of the press gang was told at SHEEPSTOR (South-West England & Channel Islands), and another appears in *About Yorkshire* (1883), whose authors, Thomas and Katharine Macquoid, were assured that the events had happened within living memory. Between Whitby and Robin Hood's

Bay lived two sisters, Hester and Dorothy Mossburn, who were engaged to two sailor brothers, Bill and Peter. On their return from a voyage to Greenland, the brothers were on their way to see their sweethearts when they met another girl, Polly, who had been involved with Bill before he fell in love with Hester, and who now tried to get him back again. When he refused, she cursed him, and did not stop at that, but instantly went to tell the press gang that two fine sailors could be found at the Mossburns' cottage. Bill and Peter, having spent the evening planning their weddings, had no sooner left the house than they were dragged off to fight for their country. As the months passed with no word from them, Hester and Dorothy feared that they must have been killed, and put on their mourning clothes.

Polly unwisely boasted that 'if a lad broke faith with her she knew how to punish him'. When it was realised what she had done, she was ostracised by the whole town. 'Polly's landlady turned her out of doors, and not a soul would give shelter or employment to the girl who had betrayed Whitby sailors to their natural enemies.' Unable to find work or friendship, she lost her wits, and ended up wandering over the moors telling anyone she met that she was 'waiting for her lad'. For Hester and Dorothy, however, there was a happy ending, since four years after their disappearance Bill and Peter returned alive, although both had been wounded in the wars, and the two couples were married at last.

ROBIN HOOD'S BAY, NORTH YORKSHIRE

Robin Hood at sea

Robin Hood's Bay has been known by that name since at least the sixteenth century. The traditional explanation is that the famous outlaw, tired of his usual hunting grounds in Sherwood Forest, took a fancy for a seaside holiday on the east coast. Unable to decide exactly where he should go, he shot an arrow at random and hit a spot on the cliffs overlooking the bay, which has been called after him ever since.

This tradition is elaborated in Lionel Charlton's *History of Whitby* (1779), which relates how, when Robin Hood had made Nottingham too hot to hold him, he retreated to the coast near Whitby, where 'he always had in readiness near at hand some small fishing vessels'. In these he could put to sea whenever he was pursued, and 'frequently went a fishing in the summer season, even when no enemy approached to annoy him'.

The tale of Robin's ventures on the sea was told as a ballad, 'The Noble Fisherman', popular in the eighteenth century. The song tells how he decides to enrich himself by taking to the waves:

> 'The fishermen brave more mony have
> Than any merchants two or three;
> Therefore I will to Scarborough go,
> That I a fisherman brave may be.'

He is employed as a fisherman by a widow who owns a ship, but at first is an unsatisfactory member of the crew, since when the others cast their baited hooks into the sea, he throws in a bare line, and is mocked for his ignorance.

> 'It will be long,' said the master then,
> 'Ere this great lubber do thrive on the sea;
> I'le assure you he shall have no part of our fish,
> For in truth he is of no part worthy.'

He proves his value, however, when a French warship comes in sight and bears down on the fishing boat. Robin asks for his bow, and declares that he will spare no Frenchman. The master thinks this a boast, but Robin shoots one of the enemy through the heart, and proceeds to dispatch the rest.

> Then streight they boarded the French ship,
> They lyeing all dead in their sight;
> They found within their ship of warre
> Twelve thousand pound of money bright.

Robin divides the riches, giving half to his fishermen colleagues and devoting the other half to building a refuge for the poor and oppressed.

SALTBURN-BY-THE-SEA, NORTH YORKSHIRE

The greedy sea

A stream running into the sea at Saltburn used to be treated by the local sailors as a sort of oracle, since its hollow banks sometimes amplified the sound of the sea to make a groaning noise audible six miles away or more, even when the waves were smooth as a pond and no wind blew. At such times, according to an early seventeenth-century manuscript 'Description of Cleveland', the fishermen dared not put forth, however eager for a catch they might be, 'houldinge an opynion that the ocean, as a greedy beaste rageinge for hunger, desyres to be sattisfyed with men's carkases'.

The image of the sea as hungry for human life was a common one, though not always so strikingly expressed. It was proverbial among Scottish and Irish fishermen of the nineteenth and twentieth centuries that 'the sea takes its own share' (*see* for example COLLIESTON, Scottish Lowlands), meaning that once the ocean had demanded a sacrifice, its prey could not, or should not, be saved. Its appetite, once aroused, had to be fed, and if deprived of one victim, it would take another instead.

SCARBOROUGH, NORTH YORKSHIRE

Lucky water

A rite that surely cannot have been unique, though seldom recorded by modest antiquarians, used to be performed on Scarborough pier. James Schofield, in his *Guide to Scarbrough* (c. 1787), describes 'a small circular cavity' about forty paces along the pier, where a sailor's wife or girlfriend would go when worried about the return of her loved one. This, he writes, would receive 'a saline and tepid libation' – in other words, she would piss into the hole – 'while the sacrificer, muttering her tenderest wishes, looks towards that quarter, from whence the object of her anxiety, is expected to arrive'. This was performed with a view to appeasing the waves, and obtaining a favourable breeze. A certain fisherman named Gradling was once given up for lost in a storm, but when his wife carried out the ritual, reports Schofield, 'strange to relate, the libation was scarcely cold, before the missing coble [boat] came in sight!'

SKINNINGROVE, NORTH YORKSHIRE

A Sea-Man

An old manuscript account of the Cleveland district gives an account of a 'Sea-Man' caught by some fishermen of Skinningrove on a shingle beach under high cliffs, 'sixty yeares since, or perhaps eighty or more', meaning some time in the mid sixteenth century. For many weeks the Skinningrove villagers kept their merman in a cottage, feeding him on raw fish, since that was all he would eat.

Although he expressed himself only in shrieks, he showed 'a curteous acceptance' of the attention he received, and seemed particularly to welcome the visits of pretty young women, 'whome he would behould with a very earneste countenance, as if his phlegmaticke breste had bin touched with a sparke of love'. His resigned behaviour convinced his guards that

Survival Cannibalism

A notorious episode of cannibalism at sea occurred after the wreck of the French ship Medusa *in 1816. Théodore Géricault's painting* The Raft of the Medusa *(1818–19) showed the sailors fighting for survival on their makeshift lifeboat, a horrific scene imitated in this nineteenth-century engraving.*

In September 1884, three men were rescued from a dinghy drifting in the South Atlantic. More than three weeks had passed since their yacht, the *Mignonette*, had foundered, and they had survived only by killing and eating the fourth member of crew. Two were convicted (the third having taken no part in the homicide, although he had shared the food), but sentenced to a mere six months' imprisonment, in recognition of the fact that they had had no alternative but starvation. There was widespread surprise that the case had even come to court, but the authorities were determined to demonstrate that murder and cannibalism, under whatever circumstances, were not acceptable behaviour in the late nineteenth century.

There were, however, many precedents. After the *Essex* was wrecked by a whale in 1819 (*see* HULL, North-East England), one boatload of survivors ate their comrade after he died naturally, but another had to adopt the even more painful decision to draw straws for who should die and who should shoot him. The casting of lots to decide on a sacrifice was ancient practice, mentioned in the biblical story of JONAH AND THE WHALE (p. 164), and for starving sailors cast adrift, survival cannibalism amounted to a 'custom of the sea'. The horrendous procedure is graphically described by Edgar Allan Poe in *The Narrative of Arthur Gordon Pym* (1838) – in which, by a macabre coincidence, the name of the man eaten is that of the *Mignonette*'s victim, Richard Parker.

Many ballads deal with the same subject, some tragic, some exploiting the black comedy inherent in the sailors' horrid predicament. A French song, '*La Courte Paille*' ('The Short Straw'), is an early example, and a Guernsey version concludes with dinner, the unfortunate cabin boy being eaten with white sauce and salad. A related English ballad, 'The Ship in Distress', tells the same story in a poignant vein:

> For fourteen days, heartsore and hungry,
> Seeing but wild water and bitter sky,
> Poor fellows, they stood in a totter,
> A-casting lots as to which should die.
> The lot it fell on Robert Jackson,
> Whose family was so very great.
> 'I'm free to die, but oh, my comrades,
> Let me keep look-out till the break of day.'

Here there is a happy ending, since a ship is sighted before the fatal deed has been accomplished.

William Thackeray's 'Little Billee' (1845) is a literary rendition that passed into the folk repertoire. His 'gorging Jack and guzzling Jimmy' are two voracious friends who agree to spare each other, but gang up on the 'young and tender' little Bill, who luckily spies a British vessel in time to avoid his fate. In W. S. Gilbert's 'Yarn of the *Nancy Bell*' (1866), things do not turn out so well for most of the men, who are devoured one by one, leaving only an elderly man who laments that, because he has taken part in eating all his former shipmates, he now embodies the entire band, cook and captain, mate and bo'sun, and the whole crew of the captain's boat.

The courteous and amorous 'Sea-Man' caught by the sixteenth-century fishermen of Skinningrove might have been a whiskered seal.

he was happy in captivity, and one day they left him alone, whereupon he made his escape back to the sea. Polite to the last, 'as one that would not unmanerly depart without takinge of his leave', before swimming away he raised himself several times shoulder-high above the waves, 'making signes of acknowledging his good entertainment to such as beheld him on the shoare, as they interpreted it', but finally plunged below the water and was seen no more.

No physical description is given of this 'Sea-Man', but Mackenzie Walcott, writing in 1861, firmly identifies him as a large seal.

STAITHES, NORTH YORKSHIRE

Burning fishbones

The village of Staithes is best known to the wider world as the early home of the eighteenth-century navigator James Cook, the first explorer to map part of Australia's coast, and the first to cross the Antarctic Circle. Locally, however, it boasts a variety of legends, one of which holds that it was first settled by survivors of a French ship wrecked on the coast. This is no more than a picturesque tradition, since the name 'Staithes' derives from Old Norse for 'the landing place', and the town originated as a Viking homestead.

More than one ghost used to haunt its shores. In the early nineteenth century, many people saw the wraith of James Harrison, a fisherman who had fallen to his death in the sea. His family not only recognised their dead relative by his appearance and clothes, but were apparently able to have conversations with him. James was eventually exorcised by a priest (Catholic clergy being quite often said to be more skilled at such rites than their Church of England counterparts).

Another apparition seen 'several nights in the year', according to a late nineteenth-century *Guide to Saltburn*, was that of a young woman who had died in a gruesome accident. When standing at the foot of Colburn Nab, slightly to the west of the town, a huge rock had fallen from the cliff above 'and, striking her on the neck, cut her head completely off, throwing it to some distance'. It was said that her headless body visited the place, crossing a bridge across a small brook entering the sea.

The Staithes fishermen used to carry out a ritual whenever their catches were bad for several nights running, keeping the first fish that swam into their net and burning it on their return home as a sacrifice. This rite, reported by the *York and Yorkshire Herald* in 1885, is rather surprising, since there was a common and widespread taboo against burning fishbones. In *Fisher Folk-Lore* (1965), Peter Anson, having spent most of his life among Scottish fishing communities, writes that 'Even to-day at the age of seventy-five nothing would induce me to throw fish bones on the fire.' It is unlikely that this prohibition would have been unknown in Staithes, but something forbidden as an everyday action may sometimes be allowed in a more ceremonial context, one example being HUNTING THE WREN (p. 194).

WHITBY, NORTH YORKSHIRE

St Hilda and the serpents

The cliffs around Whitby are thickly embedded with fossil ammonites, extinct marine molluscs of the Jurassic age with a coiled spiral shell. Before anything was known about fossils, these were thought to be petrified snakes, and traditionally called 'snakestones'. They were popular talismans, carried for luck and thought to cure cramp (*see also* WHITSTABLE, South-East England).

The local legend accounting for their origin is that when St Hilda founded Whitby Abbey in the mid seventh century, the place was crawling with serpents. Hilda was terrified, and prayed that all the snakes should crawl or tumble down the cliff and turn to stone, which they did. Although ammonites do look rather like curled-up snakes, of course they have no heads, and it was said that they had lost them in their fall towards the sea. Inventive salesmen sometimes 'restored' the missing feature by carving heads on the fossils, and examples with added heads can be seen in the Natural History Museum in London, the Yorkshire Museum in York, and the National Museum Cardiff. Although the miracle was not unique to Hilda – the same story was told of St Keyne of KEYNSHAM (South-West England & Channel Islands) – the tale was so famous that snakestones came to be the emblem of Whitby, and three of them appear on the town's coat of arms.

Ammonites are really the fossilised remains of shellfish, but those found near Whitby Abbey were traditionally supposed to be snakes turned to stone by St Hilda.

So holy was Hilda, apparently, that seabirds flying over the abbey swooped low to do her honour, a legend that developed in a curious fashion. The sixteenth-century historian William Camden wrote that he had heard from several credible sources that wild geese, migrating from colder northern lands, 'to the great amazement of every one, fall down suddenly upon the ground, when they are in their flight over certain neighbouring fields hereabouts'. An eighteenth-century observer, commenting on this tradition, noted that seabirds did indeed land in great numbers in some fields near the abbey, and suggested that they were taking refuge from storms at sea or resting in a long migratory flight, but such rational explanations cut no ice in nineteenth-century Whitby, where according to the *Athenaeum* (1899), the local superstition was that 'all sea-birds drop down dead if they cross the abbey' because of a curse laid upon them by Hilda. No reason for her cursing the birds is offered, and it seems that this idea resulted from confusion between two different tales, one relating how birds would bow in their flights to show respect for the saint, the other telling of Hilda's petrifaction of the snakes.

See also HULL.

YORK, NORTH YORKSHIRE

'The Cruel Knight and the Fortunate Farmer's Daughter'

Although quite far inland, York has a magnificent sea story, a highly elaborated version of 'The Ring and the Fish'. At its simplest, that motif deals with a ring thrown into water and later found inside a fish that has swallowed it. The tale, which goes back to ancient Greek legend, was told in Glasgow as a miracle of St Mungo, and forms part of many local legends, sometimes involving odder objects than rings (*see* KING'S LYNN, East Anglia).

The York story takes some time to get to the fish and the ring, incorporating many other traditional elements along the way in an entertaining black comedy. Popular in eighteenth-century chapbooks, it has been told in various forms, of which the nineteenth-century ballad of 'The Cruel Knight and the Fortunate Farmer's Daughter' is one of the best.

> In famous York city a farmer did dwell,
> Who was belov'd by his neighbours well;
> He had a wife that was virtuous and fair,
> And by her he had a young child every year.
> In seven years six children he had
> Which made their parents' heart full glad.

After that, however, things started to go badly for this large family. The farmer fell on hard times, and could hardly keep the children fed, so when his wife became pregnant again he was not so glad as before. Her time came upon her, and as she cried out, a passing knight heard the sound of her pain.

This knight was an astrologer, and he paused to consult his book and read the newborn child's future. To his astonishment and displeasure, he found that the baby was destined to be his own wife. It was no part of his plans to marry into the peasantry, and so he visited the house next day to investigate. He offered the farmer £3,000 for the little girl, promising to make her his heir, and the parents agreed, begging him to be kind to the baby. As soon as he was out of their sight, however:

> Being cruelly bent he resolv'd indeed
> To drown the young infant that day with speed,
> Saying, 'If you live you must be my wife,
> So I am resolved to bereave you of life;

For till you are dead I no comfort can have,
Wherefore you shall lie in a watery grave.'

Throwing her into a river, he rode away. Luckily the tide carried the baby in sight of a fisherman, who retrieved her, and for the next eleven years she lived happily with him and his wife.

One day she came to call her foster-father home from where he was drinking in the company of several gentlemen. When they commented on how beautiful she was, the fisherman told them how he had found her in the river. Now, among the drinkers was the very knight who had tried to drown the baby, and when he found out that he had failed he was furious. He visited the fisherman and his wife, once more offered to adopt the child, and again was allowed to take her away, since the humble couple thought that she would have great advantages with such a rich man.

This time, the knight sent the girl to his brother in Lancashire, with a sealed letter instructing him to kill her at once. On her way, however, she stayed at an inn, which was burgled during the night. The thief found the knight's letter, and deciding that the girl was too young and lovely to die, he substituted a note asking that she be given a good education, and departed from the scene.

A year later, when the cruel knight came to visit his Lancastrian brother, there he found the girl, fit as a fiddle and prettier than ever. In a passion, he cried:

'Why did you not do as in the letter I writ?'
His brother replied, 'It is done every bit.'

Shown the thief's faked letter, the knight was greatly puzzled, but resolved again to get rid of the troublesome girl. Taking her with him to the sea-shore, he told her that her last hour had come. She begged for mercy, and promised never to come near him again if her life was spared. Relenting only slightly, the knight pulled a ring off his finger, and told her to look at it carefully, for never again should she come in his sight unless she had it with her. Then, throwing the ring into the depths, he rode away.

The girl, forbidden to go back to the brother in Lancashire, took a job as cook at a nobleman's house. Preparing dinner one day, she cut open a cod and found inside it the very ring the knight had dropped into the sea. Smiling to herself, she said nothing to anyone about her discovery, but continued to work so well, and made herself so popular in the household, that after a while the lady of the house took her as a companion.

Sometime later, who should come visiting the nobleman and his wife but the cruel knight. When he saw the girl, richly dressed and happy, he flew into a terrible rage, and as soon as he got her alone he demanded what she thought she was doing. She knew very well, he said, that now she must die.

> Said she 'In the sea you flung your ring,
> And bid me not see you unless I did bring
> The same unto you. Now I have it,' cries she,
> 'Behold, 'tis the same that you flung into the sea.'
> When the Knight saw it, he flew to her arms,
> And said 'Lovely maid, thou hast millions of charms.'
> Said he, 'Charming creature, pray pardon me,
> Who often contrived the ruin of thee;
> 'Tis in vain to alter what heaven doth decree,
> For I find you are born my wife to be.'

The pair paid wedding visits to the girl's parents and foster-parents, and great was everyone's joy when they saw the farmer's daughter become a great lady.

A man who has tried to kill you a few times would not be everyone's ideal husband. In some versions, the girl is destined to marry the knight's son rather than the knight himself, a change that perhaps makes better sense, but rather spoils the surprise denouement.

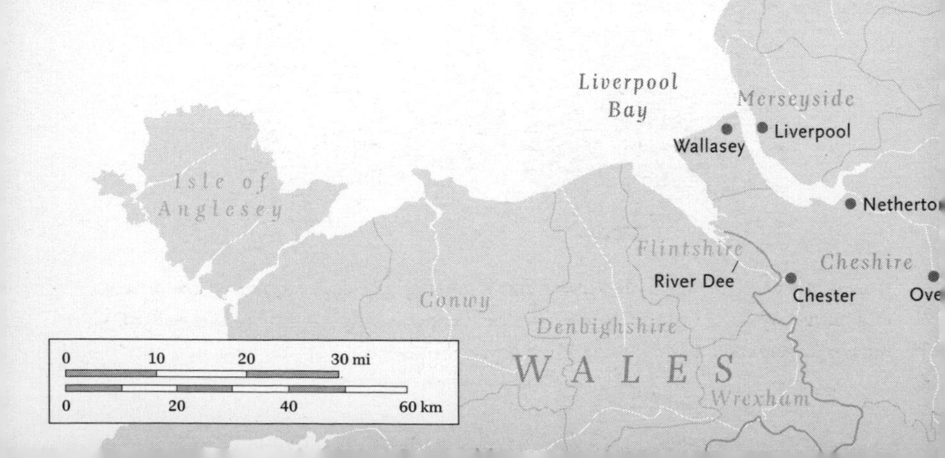

SCOTLAND
Dumfries and Galloway
Bowness-on-Solway
Solway Firth
Luce Bay
Brigham
Workington
Cumbria
King William's Banks
ENGLAND
Drigg
Ravenglass
Kirk Michael
Maughold Head
Peel
Isle of Man
Laxey
Niarbyl
Millom
Port Erin
Walney Island
Priest Skear
Spanish Head
Morecambe
Morecambe Bay
Rossall Point
Irish Sea
Lancashire
Blackpool
Liverpool Bay
Merseyside
Wallasey
Liverpool
Isle of Anglesey
Netherto
Flintshire
Cheshire
River Dee
Chester
Ove
Conwy
Denbighshire
WALES
Wrexham
0
10
20
30 mi
0
20
40
60 km

NORTH-WEST ENGLAND & ISLE OF MAN

Cheshire, Cumbria, Lancashire, Isle of Man, Merseyside

BLACKPOOL, LANCASHIRE

Drowned bells

A couple of miles off shore from Blackpool, according to tradition, is a submerged cemetery and church called Kilgrimol, said to have been swallowed up by an earthquake. According to a history of Blackpool written in 1837, rumour remained active about the site at that time. Fearful visions

It was said in the nineteenth century that a ringing church bell could be heard from underwater, off the coast of Blackpool. In this illustration to one of Hans Andersen's fairy tales, a merman wonders at a huge bell lying on the seabed.

had been seen, and 'many a benighted wanderer has been terrified with the sound of bells pealing dismal chimes'. A little later, in 1885, the collector of sea legends Fletcher Bassett makes this a local maritime tradition. He writes that 'mariners say the sound of a submerged bell is often heard near Blackpool', implying that they hear it not on land but at sea.

Stories of drowned church bells, such as those told at BOSCASTLE (South-West England & Channel Islands) and BOWNESS-ON-SOLWAY, generally start with an act of sacrilege – swearing, or sale or theft of the bells – but no tale survives to account supernaturally for the sinking of Kilgrimol.

BOWNESS-ON-SOLWAY, CUMBRIA

'When ours come back from the sea'

An old rivalry is still kept up between St Michael's church in Bowness-on-Solway and the parish churches of Dornock and Middlebie in Dumfries and Galloway, south-west Scotland, a quarrel that is explained differently depending which side of the border you are on. In Scotland, the story goes that in the mid seventeenth century, Bowness robbers stole two bells, one from Dornock and one from Middlebie, and in retaliation, the Dumfries men went and took a bell from Bowness. In Bowness, you are told that the Scots were the raiders, and that the Cumbrians only took the Dumfries bells to replace their own. Whoever started it, both parties agree that the Bowness bell was dropped in the Solway Firth, between Bowness and Dumfries, at a place known afterwards as the Bell Pool. Dangerous currents in the firth have prevented anyone from finding out if the bell is really there, but it is said to ring from underwater on evenings of still weather.

The thefts are clearly remembered in some quarters. The retired minister of Dornock reports that for the bicentenary of his church in 1993, they borrowed 'their bell' back from Bowness. It was decorated on one side with a traditional English flower display, arranged by a Bowness parishioner, and on the other with a Scottish arrangement by a lady of Dornock.

Whenever Bowness changes its vicar, it is traditional for the ministers of Dornock and Middlebie to ask for their bells back, and to receive the answer, 'When ours come back from the sea.' In 2001, when a new incumbent took over St Michael's in Bowness, he was invited to travel by boat across the Solway Firth to meet one of the Scottish ministers at the Bell Pool. In the end he could not go, because the outbreak of foot-and-mouth disease at the time meant that travel in rural areas was restricted, but had he done so, he would presumably have been confronted with the usual request, and been expected to give the usual reply.

BRIGHAM, CUMBRIA

Mutiny on the Bounty

Captain Bligh and his companions are cast off in a small boat by Christian Fletcher and the mutineers of the Bounty.

Brigham's most notorious son was Fletcher Christian, born in about 1764 at the farmhouse of Moorland Close. After attending Cockermouth Grammar School – where a younger pupil was William Wordsworth – Christian joined the navy. He served with William Bligh on a couple of voyages, and under the same commander joined the *Bounty*, which sailed to Tahiti in 1787 on a mission to collect breadfruit seedlings.

After several months in Tahiti, the ship left for the West Indies, and in April 1789 a mutiny broke out under Christian's leadership. Bligh and eighteen men were set adrift in a launch, with minimal provisions, and almost miraculously survived a journey of nearly four thousand miles before arriving in Timor, from where they returned to England. The mutineers, meanwhile, first returned to Tahiti, where some of them stayed, the rest then sailing with some Tahitians to settle on uninhabited Pitcairn Island. Almost all the men of the colony, it seems, were later massacred by the Tahiti men, so that by 1808, when a ship called at Pitcairn, only one member of the *Bounty*'s crew survived, together with several women and many of the crew's offspring, some of them quite grown-up by that time, including a son of Christian's named Thursday October Christian. It is almost certain that Fletcher Christian had met his end at the same time as the other settlers, but conflicting rumours circulated: some said that he had gone mad and thrown himself into the sea, and it has even been said that he returned to England and met the poet Wordsworth in Bristol. The last, at least, is highly unlikely, a report that seems to have been inspired simply by the fact that the two had once been schoolfellows at Cockermouth, although the supposed friendship between the men has been made the subject of a thriller, *The Grave Tattoo* (2006), by Val McDermid.

Bligh's further career alternated advancement and setback. In 1801 he commanded a ship at the battle of Copenhagen, and in 1805 he was appointed governor of New South Wales, where he made himself so unpopular that another mutiny took place. He was imprisoned for two years, but after being freed was promoted to rear admiral.

As far as the *Bounty* episode was concerned, sympathy at the time was all on Bligh's side, naturally enough, since his was the only available version of events. According to his own report, printed in 1792, he was blameless, and suspected no ill up to the moment, at dawn on Tuesday 28 April 1789, when he was roused from sleep by Christian and some other men and hauled out of bed at cutlass-point. He said that Christian appeared depressed to the point of insanity, 'as if meditating destruction on himself and everyone else'.

> When they were forcing me out of the ship, I asked him, if this treatment was a proper return for the many instances he had received of my friendship? he appeared disturbed at my question, and answered with much emotion, 'That, – Captain Bligh – that is the thing; – I am in hell – I am in hell.'

It was a detail that stuck in many people's minds, including Byron's:

> When Bligh in stern reproach demanded where
> Was now his grateful sense of former care?
> Where all his hopes to see his name aspire
> And blazon Britain's thousand glories higher?
> His feverish lips thus broke their gloomy spell,
> ''Tis that! 'tis that! I am in Hell! in Hell!'

Evidence given at the court martial in 1792 of the surviving mutineers taken off Tahiti, however, suggests that Bligh misremembered or misrepresented Christian's words. What Christian really said, apparently, was, 'I have been in hell for weeks past,' referring to the frequent quarrels he and the other officers had had with Bligh. The captain had flogged the men, stopped their rations, and clapped more than one of them in irons, and an argument over coconuts, which Bligh insisted Christian and the others had stolen, had proved the final straw. Public opinion eventually swung to the view that Bligh had been an intolerable tyrant and Christian's mutiny a heroic act – the angle taken by Hollywood in the classic *Mutiny on the Bounty* (1935), starring Clark Gable and Charles Laughton.

Whether Bligh or Christian was the greater villain is a matter of opinion. That Bligh was irritable, unreasonable, and arbitrary is well established, and by modern standards Christian was justified in rebelling, but beyond question it was inhuman to cast nineteen men adrift in an open boat, thousands of miles from the nearest landfall.

CHESTER, CHESHIRE

The Virgin Mary drowned for murder

An open space in south-west Chester, encircled to the south by the river Dee and now the site of the city's racecourse, was once flooded, leaving only a small islet surmounted by a stone cross, which perhaps originally marked a grave, above the level of the surrounding water. It must have remained like this for some time, long enough for the area to become known as 'Roodee' or 'Roodeye', meaning 'the Island of the Cross', a name that is still used, although the cross has now vanished and the fields have been drained.

A legend accounting for the cross's presence in the middle of the water is given in Thomas Hughes's *Stranger's Handbook to Chester* (1856). The story starts in Hawarden, a few miles west of Chester towards the mouth of the Dee, where long ago there was a severe drought. The people prayed earnestly for rain before an image of the Virgin Mary in their church, and among the supplicants was Lady Trawst, wife of Hawarden's governor, who 'prayed so heartily and so long, that the image, grown desperate we suppose, fell down upon the lady and killed her'. The enraged townspeople held a trial, and the statue was convicted of murder, but it was feared that if they went so far as to execute the Virgin they would suffer divine punishment. Accordingly it was decided to deposit the image on the bank of the river, from where it was carried away to Chester by the incoming tide.

Finding the statue and learning why it had been washed there, the Chester citizens buried it and erected a stone cross bearing the words:

> The Jews their God did crucify,
> The Hardeners theirs did drown:
> Because their wants she'd not supply, –
> And she lies under this cold stone.

It is unlikely that the cross ever bore such an inscription (although as it has long gone, one cannot be certain), but it would not have been unique for a statue to be accused of a crime, or of failing to bring rain in answer to prayer. According to Christina Hole's *Traditions and Customs of Cheshire* (1937), in late nineteenth-century Sicily, after a six-month drought, the enraged locals took their revenge on their saints' images 'by ducking them in horse-ponds, tearing away their garments and dumping them in parched gardens to see for themselves what the weather was like'. Hughes's story, however, is more like that of the monkey hanged at HARTLEPOOL (North-East England), an anecdote told by one community (Chester, in this case) to mock their supposedly ignorant and credulous neighbours (the 'Hardeners' or Hawardeners).

RIVER DEE, CHESHIRE

A ghost aboard

In 1850, the *Asp*, an old paddle-wheel steamer recommissioned as a naval surveying vessel, was put under the command of Lieutenant George Manley Alldridge. The superintendent of the dockyard told him that the ship was supposed to be haunted, but Alldridge replied that he didn't mind ghosts, and although after a few days repairing the steamer the shipwrights begged him to abandon it, he held firm, and the *Asp* was made ready for sea. What followed was related by Alldridge in 1867, some years after the events he describes. 'Call it a ghost or what you will,' he wrote, 'still I assure you that which I am about to relate is what really did take place.'

It was when the *Asp* arrived at the mouth of the Dee that he began to notice strange things about his ship. While it was at anchor there, he used to sit in his cabin after supper, and he and his officers often heard strange noises, 'such as would be caused by a drunken man or a person staggering about', which seemed to come from the aft cabin, formerly the ladies' quarters when the boat had been in civilian service. On one such occasion Alldridge called out, and the noise stopped, only to start again immediately.

He looked into the cabin, which was directly opposite his own, and found it empty. There was, he noted, no communication between it and the rest of the ship except by a companion ladder which was clearly in his view.

The strange sounds became more frequent after this. 'Sometimes it was as though the seats and lockers were being banged about, sometimes it sounded as though decanters and tumblers were being clashed together.' While he was in bed, Alldridge would hear the noises close to him, and felt the presence of somebody or something in the room.

> One night, when the vessel was at anchor in Martyn Roads I was awoken by the quartermaster calling me and begging me to come on deck as the look-out man had rushed to the lower deck, saying that a figure of a lady was standing on the paddle box pointing with her finger to Heaven.

When Alldridge ordered the lookout man to go back on deck he went into a 'violent convulsion', and Alldridge himself had to go and stand on watch until morning.

From this time on the apparition was visible as well as audible, and always appeared pointing at the sky. One Sunday, when the boat was in Haverfordwest river opposite Lawrenny, Alldridge and his steward were the only two on board. The steward suddenly appeared below decks, having fallen down the ladder in his terror. He said that he had been spoken to by something unseen, and refused to stay another night on board the *Asp*. Alldridge consulted the vicar of Lawrenny, who concluded that 'some troubled spirit must be lingering about the vessel'.

One night Alldridge was woken by something touching his bedclothes. He grabbed for it, but caught nothing. On another occasion, a cold hand was laid on his forehead. He must have been a strong-minded man, because he stayed in command of the *Asp* for several years, although many of his crew resigned or deserted.

In 1857, the vessel put in for repairs at Pembroke. The first night she was anchored there, one of the sentries on duty saw a woman mount the starboard paddle-box, pointing upwards, before stepping ashore and advancing towards him. 'Who goes there?' he demanded, levelling his musket, but the phantom walked on, the gun apparently passing right through its body. The man dropped the weapon and bolted, and another sentry then saw the ghost walk into the graveyard of Pater Old Church, where she climbed on a tombstone, again raised her hand towards heaven, and then vanished. This was the ghost's final appearance, and Alldridge winds up his story with the most dramatic detail of all:

> The only clue I could ever find to account for my vessel being haunted is as follows: – Some years previously to my having her, the *Asp* had been engaged as a mail packet between Port Patrick and Donaghadee. After one of her trips, the passengers having all disembarked, the stewardess on going into the ladies' cabin found a beautiful girl with her throat cut lying in one of the sleeping berths quite dead! How she came by her death no one could tell and, though, of course, strict investigations were commenced, neither who she was or where she came from or anything about her was ever discovered.

DRIGG, CUMBRIA

Herdwick sheep

In 1588, when the Spanish Armada was retreating around the west coast of England from Drake and the English forces at PLYMOUTH (South-West England & Channel Islands), it is said that one of the vessels was wrecked at Drigg, and forty sheep swam ashore from the wreck. (Ships often sailed with live animals on board, a living larder in the days before refrigeration.) From these sheep came the famous, though now rare, Herdwick breed, remarkably hardy animals which can survive burial under snow for days at a time. Farmers say that they will always return to the land on which they were reared.

Tales of people, animals, or customs surviving from the Armada are quite common (*see* for example FAIR ISLE, Scottish Highlands & Islands). The Herdwicks, however, have an alternative origin-legend, that they came not from a sixteenth-century Spanish galleon, but centuries earlier from a Viking longship. Both tales were familiar to Beatrix Potter, who bought a farm in the Lake District in 1923 complete with a stock of Herdwicks. In *The Fairy Caravan* (1929) she writes of the 'proud ancient heritage of our Herdwick sheep', which she maintains pre-date the Armada and the Vikings, although animals from far lands have added their genes into a prehistoric breed:

> Wild and free as when the stone-men told our puzzled early numbers; untamed as when the Norsemen named our grassings in their stride. Our little feet had ridged the slopes before the passing Romans. On through the fleeting centuries, when fresh blood came from Iceland, Spain, or Scotland – stubborn, unchanged, UNBEATEN – we have held the stony waste.

She was probably right that the stock was indigenous: their ancestors may have been bred in the Bronze Age.

KING WILLIAM'S BANKS, ISLE OF MAN

The moving island

> I believe the old Arms of the Isle of Man was a ship – yea, and that most meet and fitting . . . because floating in the ocean, it resembleth a moveable island.

This passage from William Blundell's seventeenth-century *History of the Isle of Man* hints at an old belief that the island once moved freely about the Irish Sea, and was only later anchored in its present position (as in classical myth the island of Delos floated around the Aegean until Zeus moored it). That this was so was stated as a fact in 1620 by a Hungarian traveller, who wrote that Monia (Man) is the most celebrated of the small islands around the English coast because it has no foundation. Blown hither and thither by the winds, he reported, its position can vary by hundreds of miles.

It was often said that the island had once been a different shape or size. In the eighteenth century, George Waldron heard that long ago it had been much larger than in his day, 'but that a magician, who had great power over it, and committed many wonderful and horrible things', had raised a storm 'not only in the air but also in the bosom of the earth', implying some kind of earthquake or volcanic eruption. Several pieces of the island were torn off and became dangerous rocks in the sea.

> The smaller fragments, they say, are sands, which, waving up and down, are at some times to be seen, and at others shift themselves far off the coast. They maintain that it was on one of these that the late King William was like to have perished.

King William's Banks, about twelve miles north-north-east of Ramsey, are named after William of Orange, whose ship nearly foundered on the sandy reef in 1690, when he was on his way to the battle of the Boyne.

Others said that Man had anciently been formed of two or three distinct islands. John Welch, visiting in the 1830s, deduced that the dividing channel or channels must have run roughly east–west between Douglas and Peel and perhaps between Ramsey and Ballaugh, supporting his theory with 'the discovery of ancient anchors and remnants of vessels half-way between Douglas and Peel'. This tradition persisted into the twentieth century, when W. Walter Gill was told that the island had formerly been

Hunting the Wren

A remarkable custom, common in England until the mid nineteenth century and surviving longer in Wales, Ireland, France, and the Isle of Man, was the annual 'Wren Hunt', when people pursued and killed the tiny birds, keeping their bodies or feathers as charms against drowning.

The ceremony, which happened around the winter solstice in late December or at New Year, was sometimes explained by the legend of a sea-nymph so beautiful that no man who had seen her could stop yearning for her love. She led many to a watery death before at last a counter-spell was found to destroy her, which she escaped only by taking to the air as a wren. Once every year she was said to resume that form, and therefore at that time all wrens were legitimate prey, although it was thought unlucky to kill a wren on any other day.

The wren was sacred to the pagan Celts, who believed that predictions of the future could be made from its flight and song, and it may therefore have been seen as an evil influence by Christian clergy trying to stamp out Druidic practices of augury. Belief in the bird's protective power, however, lingered, leading to its corpse or feathers being cherished as talismans that preserved sailors from shipwreck. A blend of these two factors – demonisation of the wren, and faith in its potency – is the most likely explanation for its pursuit and sacrifice. The tale of the nymph's fatal charms seems to have begun as a separate legend, current from the early eighteenth century, about a siren who escaped death by turning not into a wren but into a bat, and making no mention of any subsequent hunt. Later versions in which she becomes a wren, making this a reason for the bird's destruction, are probably rationalising attempts to justify an ancient practice whose roots had been forgotten.

Writing in 1824, the Galloway antiquarian John MacTaggart remarks that Manx herring fishers dared not go to sea without the body of a wren:

Their tradition is that of a '*sea-spirit,*' that haunted the '*herring-tack,*' attended always by storms, and at last it assumed the figure of a wren and flew away. So they think when they have a dead wren with them, all is snug. The poor bird has a sad life of it in that singular island; when one is seen at any time, scores of Manxmen start and hunt it down.

He is mistaken in supposing that the practice occurs 'at any time', and that it is unique to Man, but it was virtually unknown in Scotland, and must have seemed particularly outlandish to a Scottish commentator.

The actual slaughter of wrens seems to have stopped at around the time of the First World War, but 'Wren-bushes' made of greenery were still carried in procession in the 1930s, and a traditional Manx 'Wren Song' continued to be performed by groups of boys on St Stephen's Day (26 December) in the 1970s, mainly as a pretext to collect money.

It was said that a wicked mermaid had once become a wren, and that she might be found in that shape again on the anniversary of her transformation. On one day each year, wrens were therefore hunted and killed in great numbers.

two separate countries, the northern part called 'Sodor' and the southern 'Man'. Gill notes an Irish legend that Scotland, Ireland and Man had once been united, with Man presumably forming an isthmus between the Mull of Galloway and the Ards Peninsula, but had been separated into three by magical or Druidical power, a tale perhaps suggested by Man's armorial symbol of the three legs (*see* SPANISH HEAD).

Related to these legends is the widespread tale that the entire Isle of Man was once sunk beneath the waves, and when it rose to the surface was hidden under a magical mist. In John MacTaggart's *Scottish Gallovidian Encyclopedia* (1824) it is recorded that the spell was broken when a burning ember fell on the land from a sailor's pipe. If all the fires on the island were extinguished at the same time, he writes, the enchantment would return and the island would vanish. Once, indeed, every flame went out except a tiny blaze at Laxey, 'and faith, the Isle o' Man was begun to shog and quake'.

KIRK MICHAEL, ISLE OF MAN

Manx charmers

The Manx were once largely dependent on fishing for subsistence. 'Life to man, and death to fish' was a regular toast at public dinners on the island, a wish for abundant catches without loss of life to the fishermen, and 'No herring, no wedding' was a common proverb, meaning that if the herring fishing were to fail, the young men would have no money to marry. This was literally true, as proved by comparisons between the marriage registers and the fishing records.

Given the central importance of fishing to the economy, it is natural that much folk magic on the island was devoted to securing good catches. The term 'charmer' was used both for the practitioner and the practice ('doing the charmer') of a particular class of spell involving herbs. In the nineteenth century, a charmer who lived near Kirk Michael was often consulted by the mariners. One fisherman, who paid the wise man several times for help, was told that 'I cannot put the fish in your nets for you, but if there is any mischief in the way of your luck, I can remove that for you.' The charmer supplied packets of finely chopped herbs, which then had to be boiled, the infusion strengthened with spirits, and the mixture partly drunk by captain and crew, partly sprinkled over the boat and nets.

On one occasion, this fisherman confessed, he went drinking with the money he was supposed to give the charmer, and to avoid getting found out he went and collected some random herbs. The captain went through the prescribed ritual, and the boat had a most successful expedition, so much so, in fact, that the rest of the crew wished the charmer had distributed the catch over two nights, rather than cramming their nets so full that they almost burst.

Exceptionally good catches could provoke jealousy. From the early twentieth century comes a report that if one or two boats did much better than others, the less fortunate men would go in the night to the houses of the successful crews and pull some straw out of their thatch, 'and that took away all their luck with it'. Straw was traditionally significant in spells and charms, used for instance to make crosses against demons at the FARNE ISLANDS (North-East England), and the material of somebody's roof would have been thought to be particularly powerful over the fortunes of those who lived underneath.

Winds could be bought from witches who supplied knotted cords, a practice that lasted from at least the sixteenth century until the twentieth. The first two knots could be undone when wind was needed, but the third was never to be loosened except in great emergency, and should under no circumstances be cut with a knife. If it was cut, or untied needlessly or by any unauthorised person, a terrible hurricane would result. Witches might also curse boats, and a procedure to disenchant a vessel was to go over it with a flaming torch to 'burn out' the evil influence, a ceremony also performed at BURGHEAD (Scottish Lowlands).

LAXEY, ISLE OF MAN

The Night Man

In the eighteenth century, there were reports of a spirit that haunted a rock somewhere around the coast of Man. He was vocal at all times, and people did not mind his customary cry of 'Hoa! hoa! hoa!' but the prospect of a tempest excited him tremendously. Having died in a shipwreck himself, the ghost bellowed so loudly before any other such disasters that the houses shook with the noise, and not only the islanders but even their animals were terrified.

There are many tales of 'this useful precursor of the barometer', as the folklorist Walter Gill calls the storm omen, known in Manx as the *Dooiney Hoie* or 'Night Man'. One place he was supposed to inhabit was Ballaconnell valley in Malew, an area where, perhaps coincidentally, apparitions were seen of 'a strange and nameless figure' wearing a big hat and a long grey cloak with shining buttons, but mostly the Night Man was to be heard near the sea.

In *Fairy Tales from the Isle of Man* (1951), Dora Broome gives a story set on the cliffs near Laxey, where the Dooinney-Oie, as she spells it, made his home in a cave. Whenever he saw the clouds gathering, he would blow on his bugle 'How-la! How-la!' Then the Laxey fishermen would say, 'Theer's the Dooinney-Oie blowin' on his horn,' and they would know better than to put to sea that day, so they would bring in their boats and haul up the nets.

The Dooinney-Oie was proud of his horn-blowing, and after a while he got conceited and blew 'How-la! How-la!' all day long, when there wasn't a cloud in the sky. An old woman called Nance Corlett got tired of the uproar, and plugged the horn with goose grease so that it made no noise for a while. The Dooinney-Oie left it by a fire that melted the grease, but he was not very bright, and tried to blow it while it was still hot, burning his lips. He threw it from him, and it went over a cliff into the sea, and then the Dooinney-Oie was so upset he stamped and cried, making a worse noise than the horn itself.

The horn was cast out of the sea by a merman, and landed outside Nance Corlett's cottage. She thought she had better give it back, so she polished it up and left it by the cave in the cliffs with a note tied to it. The Dooinney-Oie couldn't read, but a passing seagull from Douglas could, and told him what the note said:

> If ye blow every day,
> It'll be took away,
> By Misthress Nance Corlett.

After that, the Dooinney-Oie was so frightened of losing his horn again that he never made a sound unless a storm was brewing.

LIVERPOOL, MERSEYSIDE

A bird of ill omen

Hearing a flock of seagulls crying, a character in Rumer Godden's novel *In This House of Brede* (1969) remembers 'how sailors say they are the souls of Liverpool seamen – Liverpool men, because, legend says, they are the worst and so their souls are lost'.

That gulls embodied the spirits of dead mariners was a widespread tradition (*see* PORLOCK, South-West England & Channel Islands), linked in some cases to the belief that it was unlucky to interfere with the birds. The related idea that a curse would rest on anyone who killed an albatross, although of relatively recent and literary origin (*see* WATCHET, South-West England & Channel Islands), gained broad currency, and by the mid twentieth century it was an article of faith, if not among sailors, then certainly among the media, as demonstrated by events in Liverpool in 1959.

The drama was set on board the cargo liner *Calpean Star*, described in the press as an 'Ark', since she was carrying a cargo of seals, penguins, and other animals from the Antarctic destined for zoos. When the ship stopped off at the Huskisson Dock in Liverpool, the crew, who had understood that they were to be paid off at this point, learned that they were expected to take the ship on to Oslo, and staged a sit-down strike in protest. This might not have made the national press had it not been for the fact that, as reported in the *Daily Telegraph* on 7 July, on board was an albatross, 'bird of ill omen to seafarers', lying dead in its cage.

According to Warren Armstrong's *Sea Phantoms* (1963), it had been shot by an amateur marksman with a home-made bow and arrow, an unconvincing detail since the bird died while already in captivity. Others said that it succumbed after eating a sausage roll. In any case, it was deceased, and the *Telegraph* claimed that 'Superstitious seamen, remembering Coleridge's "The Rime of the Ancient Mariner," were asking among themselves: "Are our troubles only beginning?"' This was fantasy, as made clear in the *Liverpool Echo*: 'It had been suggested that superstition about the "bird of ill-omen" was a part cause of the trouble but this the men denied.' Obviously the *Telegraph* was doing its best to create a presentiment where none existed.

It should, however, be recorded that two days after the *Calpean Star* anchored in Liverpool there was a violent explosion and fire at the Huskisson Dock, and that although the ship survived the blaze, she met her fate less than a year later, in June 1960, when she was wrecked in Montevideo.

MAUGHOLD HEAD, ISLE OF MAN

The Carrasdhoo Men

A gang of wicked wreckers known as the 'Carrasdhoo Men' used to infest the cliffs of Man. According to a woman of Northside who talked to the folklore collector Walter Gill in the early twentieth century, and had been told the story by her grandfather, they lured ships on to Maughold Head and pillaged the wrecks, hiding their loot further up the glen.

Gill believed that there was a historical basis for the tale, and certainly real-life wreckers did operate on the island by means of false lights designed to mislead benighted ships. In the early nineteenth century, it was recorded that Archdeacon Philpott used to patrol on stormy nights expressly to find any fires lit for this purpose, and to put them out. The Carrasdhoo Men, therefore, may have started out as human villains, but in the folk imagination they became more terrible figures. Gill's informant had been told as a child, when she was naughty, 'I'll give you to the Carrasdhoo Men,' and in a poem of 1839 by Esther Nelson they figure as a tribe of uncanny murderers:

> The Carrasdoo men were a fearful race; –
> A band of borderers none might trace;
> Whose land or lineage no one knew
> In the wild lone isle wherein they grew.

Crashing seas drive a ship on to fearsome rocks. Once stranded, she would be easy prey for 'wreckers' such as the wicked Carrasdhoo Men of the Isle of Man, who might even kill surviving sailors in order to rob the vessel.

They usually drugged their victims and drowned them in peat bogs, but they could also fling pebbles from a distance with 'fatal power', an ability that links them with the fairies and their stone arrows. The Men's hearing was also preternaturally acute, the slightest whisper enough to alert them. The poem locates the Men's hideout in Jurby:

> In Yorby glen the pools are black; –
> But dead men's spirits will come back,
> And, shrieking, point to far beneath
> Where the dark men hurl'd them unto death.

Walter Gill points out that in fact there are no glens in Jurby, a fact of which Nelson, a local writer, must have been well aware, and suggests that she wanted to conceal the true locality, as well as suppressing references to the Carrasdhoo Men's wrecking activities. On this subject, he says, 'Manx oral tradition preserves a judicious silence.' The implication is that it came too close to the bone, since in the early nineteenth century, when Nelson composed her verses, the practice of enticing ships on to the rocks had not yet died out.

MILLOM, CUMBRIA

A rainy buzzer

In the heyday of the Cumbrian shipyards and steelworks, residents of Millom and some parts of Barrow-in-Furness said that the sound of the factory whistles, known as 'buzzers', foretold rain or cold weather. This very localised portent found its way into regional dialect, in the phrase 'a rainy buzzer', and into verse. The twentieth-century poet Norman Nicholson recorded the belief in 'Weather Ear':

> Lying in bed in the dark, I hear the bray
> Of the furnace hooter rasping the slates, and say:
> 'The wind will be in the east, and frost on the nose, today.'

As the poem makes clear, whether the buzzer can be heard or not depends on the direction of the wind, so that this is an essentially rational prediction. The same can be said for other weather omens, such as gulls flying inland (see THE POWER OF SEAGULLS, p. 304), and even for the howling of certain elemental spirits such as the *Dooiney Hoie* (*see* LAXEY), whose voices were the sounds of wind and waves.

A more pessimistic outlook is recorded in William Rollinson's *Cumbrian Dictionary* (1997). Generations of Cumbrians, he writes, have looked westward across the Irish Sea to predict the weather: 'if the Isle of Man can be seen, it will rain within twelve hours; and if the island can't be seen – it's raining already!'

MORECAMBE, LANCASHIRE

Cures from the tide

Ebb tides were often said to draw with them the life of a dying patient (*see* HESLEDEN, North-East England), but in some cases they might provide a cure. A letter printed in the *Manchester Guardian* in 1956 records the writer's childhood experience of whooping cough in what the paper calls 'pagan Lancashire':

> I don't remember the details of my cure, but I do know that an aunt, an uncle, one set of grandparents, and my parents walked in solemn procession at 7.30 a.m., clad in I know not what, out with Morecambe's tide, bearing one whooping infant like an offering to the sea gods. My mother is now strangely reticent about this episode, but as my grandmother tartly puts it, 'Ah, but it cured you!'

It was a quite common suggestion, well into the twentieth century, that whooping cough could be cured by the tide. In Norfolk, sufferers were sent to meet the incoming tide, and as it ebbed, so, it was thought, the disease went. Similar ideas were held in Wales, although some recommended a daily walk before the *flowing* sea, rather than the ebb. A case was recorded in 1916 of a mother taking her sick child back and forth on the Severn ferry near Chepstow until its cough had gone, although this may have been with a vaguer idea of the benefits of sea air.

The link between whooping cough and sea perhaps comes by association of the sound of the waves with the gasping of the sufferer, but other illnesses could be cured in this way. Describing Irish customs in the late nineteenth century, 'Speranza' Wilde (mother of Oscar) wrote that a feverish patient should be placed on the shore when the tide was coming in, and when it went out, the retreating waves would carry away the fever and leave the sufferer well.

NETHERTON, CHESHIRE

A blessed harbour

The 'Cheshire prophet' Nixon was a mysterious figure, whose apocalyptic forecasts first appeared in pamphlets of the early eighteenth century. By the nineteenth century Nixon was nationally famous, his pronouncements much debated, and mentioned by Dickens (who calls him the 'red-faced Nixon', since portraits show the sage with a very ruddy complexion).

One of the seer's 'occult deliverances', discussed in the *Cheshire Sheaf* in 1878, related to 'God's croft, between the rivers Mersey and Dee', said by Nixon to be the only refuge on the Day of Judgement. The *Sheaf's* correspondent, J. Rogers, wanted to know whether any such field or place in fact existed in that area, and received two replies a fortnight later. M. Blezard wrote that in Netherton, midway between Frodsham and Helsby, there was a farm called God's Croft, which could not, despite its name, be considered to occupy any peculiarly fortunate position. M. Harrison's letter interpreted the phrase as applying to a blessed harbour for sailors:

> Considerable change has taken place in the River Mersey from Eastham by Thornton and Frodsham. At some remote period the river and tide must have flowed over Trafford, under Dunham and Helsby. Under the latter during a storm, the navigation must have been not only uncertain but dangerous. . . . Approaching Frodsham is a spot safer for vessels and sheltered; this is called God's Croft.

The next communication came from Thomas Hughes, editor of the *Sheaf*, pointing out that the 'mystic utterance' was probably a pun. Asked by a friend where a man should find safety on Judgement Day, Nixon had replied, 'In God's Croft, betwixt Mersey and Dee.' Although, Hughes wrote, this had long been accepted as literal, and had encouraged generations of Wirral-dwellers to anticipate a happy afterlife, Nixon's words could be interpreted as meaning that between mercy (Mersey) and damnation (D—, as the word was often politely written), sinners could hope for salvation. Therefore, although the names Godscroft Lane and Godscroft Hall are still in use, a mile or so west of Netherton, their inhabitants cannot count on any special favours.

NIARBYL, ISLE OF MAN

Mermaids of Man

One of the mermaids' favourite haunts on Man was Niarbyl or *yn Arbyl*, 'the Tail', a descriptive term for the long reef of naked rock jutting out on the south-western coast between Peel and Rushen. Here, Walter Gill was told in the 1920s, people used to see 'an old man with long white hair, seated in a boat which seemed to be part of himself, so that they couldn't tell where one ended and the other began'. He was generally coming north from Fleshwick at the southern end of the bay, and he sang so beautifully that everyone used to gather on the shore to hear him.

The boat that seemed to be part of the sailor sounds very like the craft piloted by the Finn-men of EDAY (Scottish Highlands & Islands), and Gill reports that belief in 'a race of people who lived on, in, or under the sea as ordinary beings do on land or in boats' was not in his time wholly abandoned in the Isle of Man, 'though now seldom openly acknowledged'.

The mermaid was dangerous if her affections were spurned. One pretty siren of the Isle of Man fell in love with a shepherd, but when he drew back from her embrace, she threw a stone that caused his agonising death.

He had heard them called 'fish-people'. They were said to talk among themselves, and also to sing and whistle, an ability that suggests a link with seals since these animals can make very musical sounds (*see* BARRA, Scottish Highlands & Islands), although Gill says that they were not, as far as he knew, associated.

More traditional mermaids were known among the Manx, who told stories of the *Ben-varry*, 'Woman of the sea', and her liking for handsome young men. In the eighteenth century, George Waldron heard one tale of a beautiful mermaid who fell in love with a comely shepherd and offered him gifts of pearls, coral, and shining shells, her presents 'accompanied with smiles, pattings on the cheek, and all the marks of a most sincere and tender passion'. One day she threw her arms fondly around him, and he withdrew, frightened that she meant to pull him into the sea. Offended, she threw a stone at him and disappeared in the waves. Although only slightly wounded, the shepherd 'felt from that moment so excessive a pain in his bowels, that the cry was never out of his mouth for seven days, at the end of which he died'.

OVER, CHESHIRE

Hay from Merton Sands

For such a powerful personage, the Devil was easily fooled. A Cheshire story tells how he tried to bribe a monk from Vale Royal Abbey to give him his soul. In return, the monk wanted an endless supply of ale to drink and venison to eat, and as his third gift he demanded six bundles of hay from Merton Sands near Over. Christina Hole comments in *Traditions and Customs of Cheshire* (1937) that 'As the monk's chief characteristics appear to have been gluttony and trickery, the Devil might reasonably have hoped to get this soul for nothing by merely waiting a little,' but he agreed to the requests before realising that no grass grew on Merton Sands and therefore no hay could be harvested from them. The monk escaped damnation, the Devil retired frustrated, and to make sure he stayed away, the men of Over used to plough the sands every year to stop any blade sprouting there.

Similarly impossible seaside tasks defeated many demons, including Jan Tregeagle at GWENVOR SANDS (South West England & Channel Islands).

PEEL, ISLE OF MAN

The Mauthe Doog

A fishing boat was due to set out from Peel one evening, but although the crew was ready and waiting, their captain did not arrive. Eventually the other men gave up and went home, and it was just as well they did, because in the small hours a gale blew up, so sudden and violent that if they had been out in it, they might not have survived. The next day, the captain told them what had kept him from the harbour. He had been on his way down from his house at the top of the town when he had met a big black dog that would not let him pass. He had threatened it with a big stick, to no avail, and had tried other routes, but whatever path he chose, the dog was there first. Finally he went home, having decided that this was a 'sign' – as events proved it to have been.

Telling this story in *A Manx Scrapbook* (1929), Walter Gill comments that black dogs were 'long and conspicuously associated with Peel'. He means the Mauthe Doog or Moddey Doo, a big black spaniel with a curly coat that was said to have haunted Peel Castle in the seventeenth century. The soldiers got quite resigned to it coming into their guardroom, where it used to lie in front of the fire all night, but they were still reluctant to venture alone through a dark passage which they regarded as being the phantom dog's special territory. Unfortunately this was the only way to get to the captain's room, where they had to go every evening to deliver the keys after locking the castle gates, and so they used to go in pairs. One night, however, a drunken soldier insisted on going by himself. A fearful noise was heard, and when he returned to the guardroom he could not tell them what had happened, for he was struck dumb with horror, and died three days later. After this the passage was walled up.

The tale was made famous by Walter Scott, who included a version of it in *Peveril of the Peak* (1823). In the nineteenth century, Peel Castle was visited by hundreds of tourists every year, all eager for a sight of 'the haunted guard-room', and such was Scott's influence that ghostly black dogs all over the British Isles became known, for a while, as Mauthe Doog or Moddey Doo. The original spaniel was a malign apparition, and other canine spectres on Man were, Gill says, 'seen chiefly by other dogs, and not admired by them', but black dog apparitions elsewhere, such as Shuck (*see* OVERSTRAND, East Anglia), often performed as both guardians and storm omens, functions combined in Gill's story of the Peel fishermen.

It was thought that other supernatural beings inhabited the castle ruins, and alerted the local sailors to storms at sea by shouting. The spirits in question, described as 'big fairies' in an account given to Gill in the 1920s, may be another manifestation of the Dooinney-Oie or Night Man (*see* LAXEY), who was also said to yell before a tempest. Once, an old Ballaugh woman said, some sailors were putting to sea, 'and the Man at the back of Peel Castle shouted to them there was a storm coming and they were not to go out, but they went, and the fleet was lost off the Calf.'

Other kinds of warning were recorded, one from a mermaid who called to some Peel boats to make for land when they were fishing off Spanish Head. Those who obeyed her were safe, but the boats that stayed out lost their tackle and some of their crew. On another occasion the men in two boats at some distance from each other all distinctly heard a voice telling them to weigh anchor, and did so just in time to escape a tempest.

These tales, Gill remarks, 'seem to describe a kind of collective second sight', and it seems that the Manx, like the people of the Hebrides, were (and perhaps still are) particularly gifted in this line. When away from home, they often visited their families in spirit, a sort of 'sleep-walking of the soul', as Gill describes it. One Peel man, taking a long fishing trip off the south of Ireland, claimed that he regularly did so. 'I was back home again last night, boys,' he would say in the morning, but his friends said he was lying, or dreaming. One day he looked worn out, and said he had been back yet again. The next letter that came from his wife complained

Peel Castle was said to be haunted by a spectral dog, and by other spirits that shouted to sailors before a storm.

that on that very night she had been disturbed by his apparition, making a lot of noise and waking up the whole family: 'it was not the first time, and she wanted him to stop it.'

PORT ERIN, ISLE OF MAN

The undersea world

One of the most circumstantial accounts of a visit to the undersea world was given to George Waldron at Port Erin, in the early eighteenth century. He was assured that it was vouched for by a whole ship's crew, and had happened within living memory. Forty or fifty years earlier, his informants told him, there had been a deep-sea treasure-hunting project. A ship anchored near Man and let down a diving bell made of glass cased in leather. Vast amounts of rope were paid out, but the diver continued to pull, indicating that he wanted to go further. Eventually a mathematician on board the ship calculated that so much cable had been let down that the man below 'must have descended from the surface of the waters more than twice the number of leagues that the moon is computed to be distant from the earth'. The rope ran out, so they had to bring the man back to the surface, where he complained bitterly, saying that if they had been able to let him have just a little more slack, he could have brought them wonderful things.

Having recovered himself with a dose of brandy, he told them what he had seen. After passing the region of the fish, he had come into 'a pure element, clear as the air in the serenest and most unclouded day', and found himself in a wonderful city, paved in coral and glassy pebbles. Mansions of mother-of-pearl and crystal embossed with shells were on every side, and he managed to manoeuvre his diving machine through a doorway into a room floored in jewels, which he tried to prise loose, but they were too firmly set. At this point he was pulled back by the crew above, seeing on his return 'several comely mermen and beautiful mermaids', who were frightened at the sight of his glass diving bell.

Having undergone this remarkable experience, the man pined away with longing for the submarine kingdom – which he could never visit again, as the diving project had been abandoned – and he died not long afterwards.

Before visiting Port Erin, Waldron had never met anyone who believed in mermaids and mermen, but there he met plenty. People told him of some fishermen who had actually caught a mermaid, 'in all parts above the waist resembling a compleat young woman, but below that, all fish, with fins, and a huge spreading tail'. They had kept her for three days, but then, seeing that she refused to eat, drink or speak (although they had heard the mer-people talk to each other by the sea), they let her go back to her element. When she plunged in, they heard one of her own people ask her what she had observed among the land-dwellers. Nothing very wonderful, she replied, 'but that they are so very ignorant as to throw away the water they have boiled eggs in'.

The remark about egg water gained wide currency, and was reported as having been said by other sea-people. Perhaps it was a mermaid joke.

PRIEST SKEAR, LANCASHIRE

A sea phantom

The coast near Priest Skear or Priest Keer, a tiny island in Morecambe Bay, was the site of a potentially murderous apparition in the early twentieth century. As reported by an anonymous contributor to *Lancashire Lore* (1971):

> My father and his friend were making their way out to Priest Keer, or Perry Winkle Island, duck shooting under the moon, when a figure appeared and waved them on. As this figure was nearing the Keer Edge it disappeared into the sea. If the men had followed they would have been drowned.

The implication is that this was a marine spectre that lured people to their deaths. More frequently, such phantoms appeared as shining lights, like the 'fiery devil' that haunted BUCKHAVEN (Scottish Lowlands), or the Lantern Man of Norfolk, a particularly savage manifestation of the will-o'-the-wisp.

RAVENGLASS, CUMBRIA

Pearls

Pearls are proverbially linked with oysters, but can be produced by other shellfish. Pearl mussels are still found, among other places, at Ravenglass, at the mouth of the river Irt, and in the late seventeenth century a company was set up to collect the mussels and harvest the pearls. At least a hundred years earlier, the locals had been accustomed to search for the pearl-bearing shells, marketing their finds to jewellers who made a good profit by buying cheap and selling dear. The sixteenth-century historian William Camden describes how, at the point where the Irt runs into the sea, 'the shell-fish, gaping and eagerly sucking in its dewy streams, conceive and bring forth Pearls, or (to use the Poet's name) Shell-berries.'

The notion that mussels or oysters 'conceived' pearls by taking in moisture goes back to Pliny in the first century CE:

> When of the genial season the year exercises its influence on the animal, it is said that, yawning as it were, it opens its shell, and so receives a kind of dew, by means of which it becomes impregnated.

These 'pregnant' shells then give birth, and their offspring correspond to the quality of dew they have received. If it was pure, the pearls shine brilliantly, but if it was cloudy, they are a dirty colour.

The idea that pearls were made from water drops was still popular in the sixteenth century, but by then it had become common to associate them with tears rather than dew. In John Webster's *Duchess of Malfi* (1623), Act III scene 5, when the duchess dreams that the diamonds in her crown are changed to pearls, Antonio tells her, 'My interpretation is, you'll weep shortly; for to me, the pearls do signify your tears.'

That 'pearls mean tears' is a common proverb, and many people still think the jewels are unlucky to wear, inviting the mourning they symbolise. A bride who wore them at her wedding in 1984 is reported in Iona Opie and Moira Tatem's *Dictionary of Superstitions* (1989) as saying that 'she knew of the superstition about pearls, but had decided to have them all the same'. A dream of pearls, however, although once thought a bad omen, is said in nineteenth- and twentieth-century dream books to signify money on the way.

ROSSALL POINT, LANCASHIRE

The Penny Stone

In 1554, it was said, the flooding sea swept away a whole village named Singleton Thorp from near Rossall Point. Fleeing from the rising waters, the inhabitants withdrew from the coast and found a new home 'at a place called Singleton to this day'. Writing of this catastrophe in his *History of Blackpool* (1837), the Reverend William Thornber draws a moral:

> Let us beware, lest, in the midst of our eating and drinking, of our revelry and profane oaths, we should be swept into eternity, and have our eternal portion with the unbelieving mockers of the old world.

Thornber's account is a mixture of fact and legend. Although Singleton

Thorp did once exist, its villagers cannot have founded Singleton and named it after their drowned home, since Singleton is recorded as a separate place (some way inland, a little east of Poulton-le-Fylde) long before the sixteenth century. It is, however, true that the tide has made great inroads on the coast around Blackpool. Excavations carried out in about 1886 along the shallow waters of the coast found the remains of an ancient forest stretching from opposite Norbreck to beyond Cleveleys, and the wooden remains of cottages were also dug up.

A landmark pointed out until at least the mid nineteenth century was the Penny Stone, a rock then visible at low tide, though now completely submerged. It was said to have an iron ring fixed to it, perhaps for mooring boats, but traditionally supposed to have been a tethering place for horses outside an inn where travellers could get a pot of ale for a penny (hence the rock's name). Even in the 1930s, although the stone could no longer be seen, it was said that 'on stormy nights when the wind comes howling from the sea the ghostly laughter and jesting of those who held revel within its vanished walls, can still be heard.'

SPANISH HEAD, ISLE OF MAN

Three legs

The crest of the Isle of Man, a curious motif of three legs in armour, derives from an ancient sun symbol, but has also been explained as representing Man's position at the centre of the three kingdoms of England, Scotland and Ireland, while a more picturesque origin for the design is given in Joseph Train's *Account of the Isle of Man* (1845). Many centuries ago, Train was told, Man had been inhabited entirely by fairies. A blue mist hung over it at all times, so that no passing sailor suspected there was any land there, until one day a group of fishermen were blown ashore:

> As they were preparing to kindle a fire on the beach, they were astounded by a fearful noise issuing from the dark cloud which concealed the Island from their view. When the first spark of fire fell into their tinder-box, the fog began to move up the side of the mountain, closely followed by a revolving object, closely resembling three legs of men joined together at the upper part of the thighs, and spread out so as to resemble the spokes of a wheel. Hence the arms of the Island.

The fairy island disenchanted by fire is a theme that occurs elsewhere, for example in the tale of RATHLIN O'BIRNE ISLAND (Northern Eire & Northern

Kings Can't Drown

It was sometimes held that a ruler could not drown, an extreme example of the divine right of kings that has been disproved by history, but the legend could be used to boost morale. Julius Caesar is said to have cheered a nervous boatman with the words '*Quid metuas? Caesarem veluis*' ('Why do you fear? You carry Caesar') and in 1099, when William Rufus wished to sail for France to relieve the siege of Mans:

Although it was almost a storm, and the wind contrary, he insisted upon embarking immediately; and when the sailors pointed out the danger of putting to sea, and entreated him to wait for more favourable weather, he exclaimed: 'I have never heard of a king who was shipwrecked; weigh anchor, and you will see that the winds will be with us.'

Reporting this in *Credulities Past and Present* (1880), William Jones adds that the death of Rufus's nephew (Prince William, drowned in 1120) was understood by contemporary chroniclers as a punishment for the king's boast.

Although the eleventh-century king Canute went to some trouble to convince his courtiers that he could expect no preferential treatment from the waves (*see* WALLASEY, North-West England & Isle of Man), the belief that a monarch was under special providence seems to have continued into the late nineteenth century, when the German emperor Wilhelm I offered his boat to a young couple who wanted to cross Lake Constance in stormy weather. 'Do not be alarmed,' he told them, 'the steamer bears my name, and that ought to reassure you.'

Not only kings were proverbially safe from the sea, but certain people of evil reputation were said to be immune from drowning because they were 'born to be hanged'. In *The Tempest* (1611), Act I scene 1, Gonzalo slanders the Boatswain, saying, 'I have great comfort from this fellow: methinks he hath no drowning mark upon him; his complexion is perfect gallows,' and an early nineteenth-century poem refers the proverb to Napoleon:

Napoleon Bonaparte was considered by his English enemies to be immune from the perils of the sea, since he was innately wicked and thus 'born to be hanged'. He might equally have supposed himself safe in his capacity as emperor, since it was traditionally said that rulers could not drown.

> Buonaparte, the bully, resolves to come over,
> With flat-bottomed wherries from Calais to Dover;
> No perils to him in the billows are found,
> For 'if born to be hanged, he can never be drowned.'

Certain individuals and families were supposed to have a less happy relationship with the waves and weather. Sir William Tracy, one of the four knights who murdered Thomas à Becket in 1170, was said to have intended a pilgrimage to Jerusalem in penance, but was thwarted by adverse gales. This tale developed into the legend that whenever his descendants took a sea journey, they would find the wind against them, harmoniously expressed as the saying that 'The Tracies have always the wind in their faces' – which, some people remarked, would be rather pleasant in hot weather. An eighteenth-century Shetland minister, John Mill, is said to have been threatened with the same curse by the Devil. 'Well, let him do his worst: the wind aye in my face will not hurt me,' he declared, but after that he was unable ever again to leave Shetland.

Ireland), but the three-legged apparition is unique to Man. Legends that the island had once been infested by such beings continued to circulate in the early twentieth century, when the folklorist Walter Gill was told by a retired fisherman that a three-legged race used to make raids on the Manx, from their own island, south-west of Man, which was later submerged.

King of the three-legged tribe's island, said the fisherman, was 'Mannan mac y Lear'. This, more often spelt Manannan mac Lir, is the name of a sea-god or great magician, 'mac Lir' meaning 'son of the Sea'. 'Manannan' comes from *Manaw*, a Welsh term for the Isle of Man (which in turn may come from the name of a tribal deity, so that it is an open question whether Man is named for Manannan or vice versa). He was said to have been Man's first ruler, and to have hidden his realm from passing ships in a magic mist, until at last St Patrick brought Christianity to the island and drove out the heathen sorcerer.

Ancient accounts of Manannan never suggest that he had more than the usual number of legs, but Gill's informant had clearly blended old tales of the sea-god with nineteenth-century explanations of the island symbol, since he wrapped up his account with a description of how, when Manannan was expelled by St Patrick, he took a flying leap off Spanish Head on the southernmost point of the island, and went away over the top of the waves, 'rolling like a wheel'.

WALLASEY, MERSEYSIDE

King Canute

The wide tidal estuary of Wallasey disputes with other places, including BOSHAM (South-East England), the claim to be where King Canute showed that he could not command the waves. Canute, or Knútr as he should properly be spelt, reigned as king of Denmark and England from 1014 to 1035, and is best remembered for the legendary tale, rendered by countless historians and poets. In *The History of Britain* (1670), Milton gives his version:

> He caus'd his Royal Seat to be set on the shoar, while the Tide was coming in; and with all the state that Royalty could put into his countnance, said thus to the Sea: Thou Sea belongst to me, and the Land wheron I sit is mine; nor hath anyone unpunish't resisted my commands: I charge thee come no furder upon my Land, neither presume to wet the Feet of thy Sovran Lord. But the Sea, as before, came rowling on, and without reverence both wet and dash'd him.

Having demonstrated that the sea will not obey him, King Canute reproves his courtiers for their flattery and superstitious veneration of royalty.

Canute quickly got up, and desired his courtiers to consider the 'weak and frivolous' power of royalty. The fervent republican Milton makes him declare that none but God deserved the name of king, a truth 'so evident of it self . . . that unless to shame his Court Flatterers who would not else be convinc't, Canute needed not to have gone wet-shod home'.

This moral and magnificent story has sometimes been misinterpreted as showing that the king genuinely believed the tide would turn at his command. While some people did suppose that the sea respected royalty (*see* KINGS CAN'T DROWN, p. 212), Canute's point was to demonstrate exactly the opposite. Somewhat similar, but told as a near-miracle rather than a parable, was the tale of how a prayer for wind saved Edinburgh from John Paul Jones of ARBIGLAND (Scottish Lowlands).

WALNEY ISLAND, CUMBRIA

A wrecking vicar

In Walney Island, it was said that the locals were given not only to smuggling but to the more malign practice of deliberate wrecking, shining lights to lure boats on to the rocks during bad weather. The islanders used to hang a lantern around the neck of a hobbled donkey so that when it moved, the jerky rays would look like the lights of a bobbing boat, leading unwary crews to approach too near the coast and strand their ships.

The whole community was reportedly involved, even the most respectable. Included in William Rollinson's *Cumbrian Dictionary* (1997) is the story of how a ship was wrecked one day when service was in progress at the church. When the news was brought, the congregation rushed out to see what they could get in the way of loot, followed by the vicar, crying, 'Hod still a lile bit, theer, let's o' hev a fair start!' ('Hold still a little bit, there, let's all have a fair start!') This is an anecdote related of several coastal parishes, with minimal variations, mostly of dialect. Obviously the idea of a minister or vicar involved in wrecking had wide appeal (*see also* HOVE, South-East England; MORWENSTOW, South-West England & Channel Islands).

WORKINGTON, CUMBRIA

Traditional football

Football and rugby as played today are based on archaic sports known as 'mass football' or 'old-style football', games which involved whole communities. A form of this ancient, dramatic, and fairly brutal sport is played at Easter in Workington by two teams, the 'Downies', traditionally sailors and dock workers, and the 'Uppies', miners and steelworkers. The aim is not to score multiple goals but to capture the ball and get it to an appointed base, and when one side has done so the game is won. At Workington, the Downies have to take the ball to the harbour, while the Uppies have to fling it over the wall of Workington Hall. Kicking, throwing, and carrying are all allowed, and a tale is told of how a woman once secured victory by hiding the ball under her skirts until she got it to the goal.

There is no limit to the number of players, and there are reports from some places of games involving thousands, although at Workington they have generally not exceeded a few hundred. Even so, it is a chaotic event, involving frequent injuries, and occasional fatalities.

Another place where the old-style game is played is Kirkwall in Orkney, where the 'Uppies' are the farmers and the 'Doonies' are the fishermen. It may be that such wholesale contests are more likely to thrive where there is a sea community to compete with the landsmen, and in Kirkwall there is an acknowledged ritual element to the game, a victory by the Doonies being a good sign for the fishing, while if the Uppies win, next season's crops will prosper.

In both Workington and Kirkwall, the annual events can only be certainly dated from around the end of the eighteenth century, but games of mass football were played in Britain from at least the twelfth century onwards, making the sport of these towns a revival, if not a survival, of a very old practice.

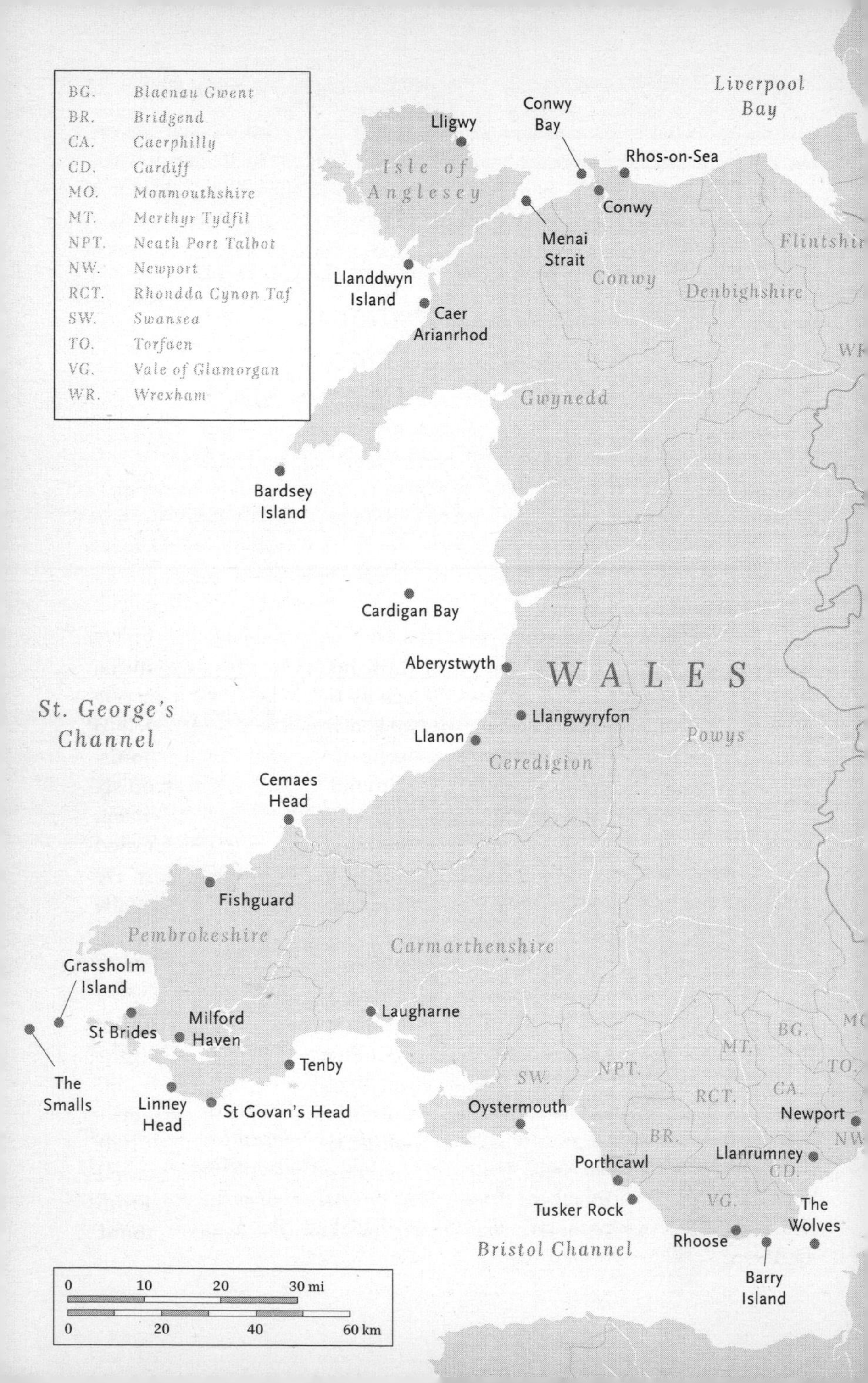
BG. Blaenau Gwent
BR. Bridgend
CA. Caerphilly
CD. Cardiff
MO. Monmouthshire
MT. Merthyr Tydfil
NPT. Neath Port Talbot
NW. Newport
RCT. Rhondda Cynon Taf
SW. Swansea
TO. Torfaen
VG. Vale of Glamorgan
WR. Wrexham
Liverpool Bay
Conwy Bay
Lligwy
Rhos-on-Sea
Isle of Anglesey
Conwy
Menai Strait
Flintshi
Conwy
Denbighshire
Llanddwyn Island
Caer Arianrhod
Gwynedd
Bardsey Island
Cardigan Bay
Aberystwyth
WALES
St. George's Channel
Llangwyryfon
Llanon
Powys
Ceredigion
Cemaes Head
Fishguard
Pembrokeshire
Carmarthenshire
Grassholm Island
Laugharne
St Brides
Milford Haven
The Smalls
Tenby
Linney Head
St Govan's Head
Oystermouth
SW.
NPT.
MT.
BG.
RCT.
CA.
BR.
Newport
Porthcawl
Llanrumney
CD.
VG.
Tusker Rock
The Wolves
Rhoose
Bristol Channel
Barry Island
0 10 20 30 mi
0 20 40 60 km

WALES

Isle of Anglesey, Bridgend, Cardiff, Carmarthenshire, Ceredigion, Conwy, Gwynedd, Newport, Pembrokeshire, Swansea, Vale of Glamorgan

ABERYSTWYTH, CEREDIGION

Watching a mermaid

In July 1826, a farmer near Aberystwyth, whose house stood almost on the seashore, happened to walk down to a rock by the water, very early one morning. There he saw, as he thought, a naked woman with her back to him, washing herself in the waves. Blushing, he turned around to go home, but then it occurred to him that the water around the rock was a good six feet deep, and that surely no woman would go so far into the sea to wash. With that, he realised that she must be a mermaid, and at once he turned back towards the sea, and crept forward until he could see better. He watched her for some time, and said afterwards that she was exceptionally handsome.

After about half an hour, he went back home and called his family to see her too. Some of his children and servants, who crawled forward cautiously as the farmer had done himself, watched for about ten minutes, but when the farmer's wife came down, she walked upright in plain view. Immediately the creature dived into the water and swam off, but not so far that they could not follow. The farmer's family and household, twelve people in all, ran along the shore for about half a mile, keeping her in sight where her head and shoulders appeared above the waves, and all of them agreed on her description. From the waist up she was clearly visible, and looked like a girl of about eighteen:

> Her hair was short, and of a dark colour; her face rather handsome, her neck and arms were like those of any ordinary woman, her breast blameless and her skin whiter than that of any person they had ever seen before. Her face was towards the shore. She bent herself down frequently, as if taking up water, and then holding her hand before her face for about half-a-minute. When she was thus bending herself, there was to be seen some black thing as if there was a tail turning up behind her. She often made some noise like sneezing, which caused the rock to echo.

These details are translated by Jonathan Ceredig Davies in *Folk-Lore of West and Mid-Wales* (1911), and originally appeared, he tells us, in the magazine *Seren Gomer* in June 1823 (there is clearly some mistake, since according to this, the article appeared three years *before* the events it relates). The writer added that 'All the family, the youngest of whom is now eleven years old, are now alive, and we obtained this account, word for word, as it is given here, from them themselves within the last month.'

The creature seen some thirty years earlier off LINNEY HEAD was also described as exceptionally white, and that too spent much time sluicing itself with water. Both reports might relate to a real sea-beast, enhanced with a little imagination.

In 1826, an Aberystwyth farmer was convinced that he had seen a mermaid. Here, a later nineteenth-century illustrator imagines a similar sighting.

BARDSEY ISLAND, GWYNEDD

Twenty thousand saints

A couple of miles off the tip of the Llŷn peninsula lies the small island of Bardsey, about a mile and a half long by half a mile at its widest point, rising to a steep peak at one end and declining at the other to a low plateau that appears almost to dissolve into the waves. Now a nature reserve, with a population (as at 2010) of only seven people, from the sixth century onwards Bardsey was a religious centre, and in its medieval heyday it was one of the most important British sites of pilgrimage.

It continued to attract reverent attention long after the Reformation, when its monastery was demolished. In the late eighteenth century, the naturalist Thomas Pennant travelled to Bardsey from Aberdaron on the Welsh mainland, and wrote that his boatmen 'seemed tinctured with the piety of the place; for they had not rowed far, but they made a full stop, pulled off their hats, and offered up a short prayer'. Sailors paid similar respects to LIHOU (South-West England & Channel Islands) and other islands and rocks, sometimes without being able to give a reason why they did so, but in the case of Bardsey, the 'piety of the place' was specifically attributed not only to its ruined abbey, but to the fabulous quantity of saints said to be buried on the island – no fewer than twenty thousand, it was said. That figure, traditionally used to signify the fact that anonymous martyrs of the early Church had been buried together in a single grave, was never meant to be taken literally, but gave rise to occasional cynical comment. In his *Worthies of England* (1662), Thomas Fuller doubts the possibility of housing so many thousand bodies in so tiny an islet, but adds that it would be 'more facile to find Graves in Berdsey for so many Saints, than Saints for so many Graves'.

The notion of Bardsey as a necropolis, however, seems to have had some foundation, since in the mid nineteenth century, when the island had a population of over one hundred and was intensively farmed, so many bones emerged from its soil that they were used to make fences. Perhaps the island was once used as a general burial ground in line with the Celtic custom of ferrying bodies across to islands for interment, a practice that contributed to ideas of an Isle of the Dead where departed souls lingered (*see* for example THE BULL, Southern Eire).

BARRY ISLAND, VALE OF GLAMORGAN

A prayer book found in a salmon

The sixth-century saint Cadog once sailed to Barry with two of his disciples, Barruc or Barrwg and Gualees, and on landing asked them for his prayer book, which, they confessed, they had left behind on Flat Holm. Cadog furiously commanded them to go and fetch it and not to come back, an apparently contradictory instruction fulfilled when he saw their boat capsize. The two junior saints were drowned, Gualees's body being cast up on the shores of Flat Holm and Barruc's upon Barry.

Cadog was apparently unmoved by this disaster. Feeling hungry, he asked his attendants to get him a fish for dinner, and the salmon they procured was cut open. Inside, they found the book unharmed, which Cadog declared a proof that nothing was impossible to God – but no more was said of poor Barruc and Gualees.

To find a book inside a fish was surprising but not unique, as demonstrated by the celebrated cod caught at KING'S LYNN (East Anglia). So often, indeed, was it reported that treasures might be recovered from fish that in the mid nineteenth century a parodic legend was told, included in T. Gwynn Jones's *Welsh Folklore and Folk-Custom* (1930). One day a woman was gathering shells on the beach, and as she bent down a ring slipped off her finger. At that moment a wave broke near the woman, and she jumped back to avoid it. When the wave had receded, there was no sign of the ring, and the woman went home sad and disconsolate.

> The next day, a man came round to sell fresh herrings. She bought some from him, and went to fry some for dinner. She cut off the head of one and opened it. And what do you think she found inside? 'The ring!' 'No. Guts.'

CAER ARIANRHOD, GWYNEDD

'Go up the hill to see a wonder'

A little way off the coast of Caernarfonshire is a reef, covered by several feet of water at high tide but appearing a little way above the sea at the ebb. It is a natural outcrop of stone, but was popularly said to be a man-made structure, sometimes identified as a Druids' circle or a Roman fort, but most often named as the castle of Aranrhod or Arianrhod.

The princess Aranrhod was a character in the medieval Welsh tales known collectively as the *Mabinogion*, a magical figure who bore two children to an unnamed father, and later helped bring a curse on one of her sons. Her home was in a castle near to or surrounded by the sea, and the Caernarfonshire rock was identified as the remains of her palace at least from the sixteenth century, when a map named the feature as Caer ('Castle') Aranrhod. In the early nineteenth century the scholar William Owen Pughe, in conversation with a friend from Anglesey, learned that

> there was a remarkable ruin in the sea, nearly midway between Llandwyn Point and the church of Clynog, in Carnarvonshire, which sailors in passing over can see in the water, and which is dangerous to vessels, and called by them Caer Arianrod.

The name still appears on modern maps, but the legendary associations of the rock had changed by the later nineteenth century, when the folklorist John Rhys collected some different tales of its origins. Several informants told him that it had once been the home of three women, Gwennan, Elan, and Maelan, who came to the mainland one day and found when they looked back that their island had been completely covered by the sea. They fled respectively to Bedd Gwennan, Tyddan Elan and Rhos Maelan, local place-names said to preserve the memory of the women, although we can be certain that this explanation is back to front, and the heroines of the story were named after the places. Another account, from an old clergyman who said that the story had been passed down in his family for several generations, was that the reef was once inhabited by a family of robbers and murderers. There was a woman living there, however, who was innocent of their crimes, and one evening as she went to fetch water she heard a voice crying out to her, saying, 'Go up the hill to see a wonder.' She climbed Dinas Dinlle, an Iron Age fort nearby, and from the top she saw her home sinking in the sea.

Rhys heard the rock called by various names that he interprets as corruptions of the older form Caer Aranrhod, and he suggests that the princess of the *Mabinogion* was the person originally said to have been punished for her sins when the sea overflowed her home. The folk tales he heard, however, do not bear this out, since none mentions any elements of her story as it appears in the ancient legend. They seem to represent an entirely different strand of tradition adapted from stories like that of Tyno Helig in CONWY BAY, drowned for the wickedness of its inhabitants.

The Ninth Wave

An old belief among sailors was that certain waves of a sequence were more powerful than the rest. Usually the potent number was nine, as in a seventeenth-century passenger's account of a captain who muttered prayers while making the sign of the cross over the sea, and said he was 'breaking the force of a fatal wave'. The traveller asked how he knew which was the fatal one, and was told that only the ninth could sink a ship. The captain pointed out the ninth wave, and told him to count from there:

Strange it is that every ninth wave was much greater than any of the others, and threatened the ship with immediate destruction. This wave, however, whenever it approached, the captain, by his muttering and signing of the cross, seemed to break, and the danger was averted.

The poet Tennyson must have heard the tradition concerning the majestic ninth wave. In his *Idylls of the King* (1869) he makes the legendary king Arthur a godlike child of the ocean itself, found on the shore by the wizard Merlin, who watches the sea roll in:

An illustration from Fletcher Bassett's Legends and Superstitions of the Sea *(1885) shows a captain pointing out the 'ninth wave' to his passenger, warning him that this is the one that could sink the ship.*

> Wave after wave, each mightier than the last,
> Till last, a ninth one, gathering half the deep
> And full of voices, slowly rose and plunged
> Roaring, and all the wave was in a flame:
> And down the wave and in the flame was borne
> A naked babe, and rode to Merlin's feet.

Ancient Welsh poetry describes foaming waves as the sheep of Gwenhidwy (a mermaid) and the ninth as her ram. Other sources, however, specify the tenth wave, or the seventh, or the third, as the greatest. In Scotland, the third wave was held to be *less* powerful than the others, but in Brittany it was supposed to be more dangerous than the first two, and in Shetland the third wave was credited with healing powers, brine from its crest being thought to cure toothache. Sailing around the northern coast of Ireland in the mid nineteenth century, Mr and Mrs S. C. Hall were told that every sixth or seventh wave was called a 'dead wave', which proceeded silently until it touched the shore, when it dashed suddenly into a huge foaming breaker.

It is not surprising that there is no universal consensus, since no wave is regularly bigger than others, but the temptation is strong to imagine a pattern in the sea's chaos. On his remarkable journey from BRANDON CREEK (Southern Eire), Tim Severin found himself looking for the seventh wave, although aware that this was a superstition. Each massive wave in the grey Atlantic vista could have destroyed his small, frail coracle:

> So always, at that brief moment before the boat drops into the next trough, the eye seeks to pick out the seventh waves, real or imaginary, the monsters lifting their heads in menace above their companions . . . before they too then sink down to hide in ambush.

CARDIGAN BAY, CEREDIGION

The Lowland Hundred

Beneath the waters of Cardigan Bay, it is said, lie the remains of fine cities and fertile meadows, drowned and destroyed in the fifth or sixth century when the sea engulfed the land. A nineteenth-century visitor to the area recalled that as a little girl, 'my childish imagination was much excited by the legend of the city beneath the sea, and the bells which I was told might be heard at night. I used to lie awake trying, but in vain, to catch the echoes of the chime.' In 1859, the magazine *The Monthly Packet* described the bustling towns supposed to have stood on the banks of the river Dovey, in an article extolling their 'marble wharfs, busy factories, and churches whose towers resounded with beautiful peals and chimes of bells'.

These details are very much of their time. The 'busy factories' sound more Dickens than Dark Ages, while the chiming of the submerged bells is a feature borrowed from countless other local traditions (*see* for example ST OUEN'S BAY, South-West England & Channel Islands). It was impossible, it seems, for a Victorian to imagine a town without churches, and no drowned city could be complete without a ghostly peal.

The underlying legend, however, is an old one. In the thirteenth-century *Black Book of Carmarthen* is a passage that describes the flooding of a land called Maes Gwyddno ('the Plain of Gwyddno'), under the control of King Gwyddno, a possibly historical ruler of the late fifth to early sixth century. Another prince or king, Seithennin, is summoned to witness the destruction:

> Seithennin, stand thou forth,
> And look upon the fury of the sea,
> Gwyddno's plain it has covered.

The verses go on to call down a curse upon 'the maiden . . . The well-servant' who let the waters overflow. Celtic myth has many tales of springs that overflow to become great lakes, the culprit often identified as a female attendant who neglects her duty to keep the spring covered. Lough Neagh in Ireland, for example, was said to have been formed when the woman guardian of a well forgot to keep proper watch over it (*see* LARNE WATER, Northern Eire & Northern Ireland), and the legend of Maes Gwyddno's destruction seems originally to have been one of this class. It relates, however, not to a lake but to the sea, and another old tale has contributed to its development, the medieval Breton story of Ker-Is (*see* LAND'S END, South-West England & Channel Islands), said to have been drowned in the sixth century when its sluices were opened and the sea was allowed to enter. In later versions of the Cardigan Bay tale, the well-guardian is replaced as the villain of the piece by Seithennin, now said to have been a notorious drunkard who neglected the flood defences of the country and allowed it to be inundated.

Pointed out as 'proof' of this were the lines of debris known as *sarnau* (causeways) stretching out into Cardigan Bay and visible at very low tides. Like the supposed castle of Helig in the rival 'drowned land' tradition of CONWY BAY, these were in fact natural features, but from accounts that reached Britain in the sixteenth and seventeenth centuries from the Netherlands of dykes and dams used there to reclaim land from the sea, the idea arose that the *sarnau* had been constructed for a similar purpose, and later breached by the waves.

All around the Welsh coast, too, are indisputable signs that what is now covered by sea was once dry land. Such indications were noted as long ago as the twelfth century, when Gerald of Wales recorded that after an unusually violent storm, around St Brides Bay in the south 'the surface of the earth, which had been covered for many ages, re-appeared, and discovered the trunks of trees cut off, standing in the very sea itself'. Similarly, at Abergele in the north, in the late eighteenth century, the naturalist Thomas Pennant saw at low tide a long stretch of mud 'filled with the bodies of oak trees, tolerably entire'.

Not only fossil relics but historical events fed into the power of the legend. In 1607, devastating floods occurred around the Bristol Channel, as recorded in contemporary pamphlets giving vivid details of refugees running on to higher ground and lamenting the destruction of their homes. Around two thousand people were drowned, and whole villages were devoured by waves so swift that, according to one chronicler, 'no greyhound could have escaped by running before them'.

All these different factors helped to form the full-fledged tale from the eighteenth century onwards that Wales had once stretched much, much further west. Around 1,200 square miles, embracing almost the whole of Cardigan Bay, was said to have been dry land, and to have contained no fewer than sixteen flourishing cities. The vanished country came to be known as Cantre'r Gwaelod, 'The Lowland Hundred', a 'cantred' or 'hundred' being an area supposed to contain a hundred families, but the word used here to stand for a far greater and more populous area. The calamitous inundation was said to have happened in or shortly before the sixth century, the date attributed in earlier tradition to the events of the Ker-Is legend, and also associated with the original tale of Gwyddno.

It has been proposed that a real flood occurred somewhere in the Celtic world in the fifth or sixth century that helped give rise to the stories. That might be so, but certainly it could not have been on any such scale as that suggested by the Cardigan Bay tale in its developed form. To find such a rise in sea level, one would have to go back to the melting of the glaciers after the end of the last Ice Age. An adventurous idea put forward by F. J. North in his comprehensive survey *Sunken Cities* (1957) is that race memory – inherited subconscious recollection of events lost in the mists of time – has contributed to the legends of devastating floods that feature so widely in mythologies of the Western world. According to this seductive though unprovable theory, all tales of drowned lands really do hold echoes of a remote past, when our prehistoric ancestors retreated from the rising waters several thousand years ago.

CEMAES HEAD, PEMBROKESHIRE

Fairies at market

A tradition current in the nineteenth century told of an invisible land in CARDIGAN BAY. The location links this to the tale of a drowned kingdom beneath the sea west of Wales, but a variant legend developed of a fairy island, ruled long ago by a king called Rhys Ddwfn, a name which could be translated as 'Rhys the Deep' or 'Rhys the Wise'. His people, known as the *Plant Rhys Ddwfn* ('Children of Rhys Ddwfn'), were said to be handsome but very small, and they kept themselves safe from human observation with the help of magical herbs that grew only on Cemaes Head. If someone stood by chance on one particular spot there and looked towards Aberdaron on the Llŷn peninsula, they might glimpse the enchanted country, but as soon as they moved they would lose sight of it, and the place where they had been standing would be nearly impossible to find again.

A receptive viewer may still glimpse the island from the village of Llanon, some way north-east of Cemaes Head but with an equally good view across the bay. The travel writer Jan Morris helpfully pinpoints the exact place to stand, outside the post office. If you wait there, she writes, on a misty morning:

> presently you may just make out, not far off shore, the dim green form of some other country, half away in the tide, and rippled with white waves. Some people claim to hear bells and music, when the wind is right: others can see and hear nothing at all.

According to a story from 1858, the island fairies were regular shoppers at Cardigan market, and raised the prices enormously whenever they came. That was fine by the farmers who had corn to sell, but not so good for the other buyers, who, whenever they found goods too expensive, said simply, 'Oh! They were there to-day' – and everyone knew who *they* were. Although they were often seen at market, they were never observed to arrive or to leave.

The *Plant Rhys Ddwfn* made a friend of one trader, Gruffydd ap Einon, who sold them more corn than anyone else, and even visited them in their home, which he found full of treasures from many lands, since they were great travellers and traders. He asked what might happen if one of their number should betray them, and was told, 'Just as Ireland has been blessed with a soil on which venomous reptiles cannot live, so with our land; no traitor can live here.' Gruffydd's relationship with the fairies remained friendly as long as he lived, but after his death the traders of Cardigan became greedy, and raised their price for corn so high that the *Plant Rhys* abandoned the town. It was said that they went to Fishguard market instead, 'as very strange people were wont to be seen there'.

CONWY, CONWY

The ghostly mermaid

The town of Conwy lies under a mermaid's curse. One of the creatures was stranded here during a storm, and begged the local fishermen to return her to the sea, but whether from fear or from spite they refused to move her even so far as to let her tail touch the water. Morris Hughes, the old man who told this story to the folklore collector John Rhys at the turn of the nineteenth century, added a short verse describing her death. Rhys calls it a 'very crude rhyme', and translates it as follows:

> The stranded mermaid on the beach
> Did sorely cry and sorely screech,
> Afraid to bide the morrow's breeze:
> The cold it came, and she did freeze.

You might expect her to have 'drowned', as it were, out of her element, but it was clearly the cold that killed her.

Before she breathed her last, the mermaid condemned Conwy to eternal poverty, and it was said in Rhys's day that 'when a stranger happens to bring a sovereign there, the Conway folk, if silver is required, have to send across the water to Llansanfraid for change.'

More recently, in *Rumours and Oddities from North Wales* (1986), Meirion Hughes and Wayne Evans write that the old town hall, which stood on the site 'where the mermaid is supposed to have died', was destroyed by fire in 1966, and when it was rebuilt as a library and civic centre it almost immediately burned down again. On both occasions, they add, some local people reported hearing the ghostly laughter of the mermaid.

CONWY BAY, CONWY

'Vengeance is coming'

According to a legend that came to be widely accepted as at least partly history, a prince named Helig ruled on the northern coast of Caernarfonshire in the fifth or sixth century. A flood rose to drown his realm, so suddenly that Helig and his family only just escaped, and the land has remained underwater in Conwy Bay ever since.

This tale first appears in a seventeenth-century manuscript, and has been greatly elaborated in more recent versions. An example that appeared in the Welsh magazine *Y Brython* in 1863, later translated by John Rhys, tells of Helig's sinful ancestors, and of Helig's own evil ways. For generations it had been prophesied that the family's crimes would be punished, a fearful warning echoing throughout the land – 'Vengeance is coming, coming' – and when Helig demanded of the unseen oracle when this might happen, he was told, 'In the time of thy grandchildren, great-grandchildren, and their children.' Thinking himself safe, he continued in his wickedness, but he was a long-lived man, and one evening he sat feasting with four generations of his descendants. A servant went down to the cellar for more wine, saw that water was streaming in, and had only time to warn the musician who was playing the harp for the guests before the sea rose to fill the rooms. No one escaped but the servant and the harper, and a kingdom stretching from Llandudno to Puffin Island was forever drowned. All that remains of Helig's castle is a jumble of stones, submerged except at low tide, between Penmaenmawr and the Great Orme's Head, a structure known as *Tyno Helig* ('Helig's Hollow').

On one level, this is a moral fable, showing that divine judgement may be delayed but cannot be escaped. On another, however, it tells a geographical truth of sorts. The coastline of North Wales has moved quite substantially inland, but the change took place much longer ago than the fifth century CE. The boulders pointed out as the relics of Helig's palace were deposited by melting glaciers at the end of the last Ice Age, and the sea level rose to cover them between the second and first millennia BCE.

Although these are now established facts, an expedition to the 'palace' in 1864 convinced at least one man that he had seen underwater the ruins of 'a grand old Hall of magnificent dimensions', and in the early twentieth century it was still maintained in the *Cambridge County Geographies* that in the fifth century the sea had engulfed a large area of North Wales. 'When the present coast-line is compared with that of the fifth century' – inferred entirely from the legend – 'it is seen that the ancient Flintshire must have been nearly twice as large as at present,' it is stated.

Such reports, explaining natural features as human artefacts, and the results of slow prehistoric evolution of the landscape as those of sudden catastrophe, resemble the tale of cities drowned in CARDIGAN BAY, and indeed the earliest references to Helig associate him with Ceredigion rather than Conwy, indicating that his legend was imported from further south. The distinct tradition concerning Helig flourished, however, and in the early twentieth century, according to the folklore collector Marie Trevelyan, the prince himself acted as a weather warning for the sailors of North Wales, his voice being heard around the coast before rough weather and shipwreck.

FISHGUARD, PEMBROKESHIRE

Whistling for a wind

Walking one day along the pier at Fishguard with his friend Captain Owen, the writer Ferrar Fenton began to whistle. As he reports in *Pembrokeshire Antiquities* (1897), his companion 'started as if he had trod on a nail barefoot', and said after a while, 'I wish you would not whistle here!' Fenton asked him what harm it could do.

> 'Well, you know,' he said slowly, as if shy at his words, 'we Welshmen and sailors are superstitious over some things, and whistling, as you now do, is one of them.'

When he was a boy, his mother and all the old people he knew had told him that whistling was the way magicians talked with the Devil, and sailors, he said, believed something similar, so that 'it always makes my heart start to hear it, especially on the seashore'. He pointed out the clouds gathering behind Pencaer, and said that Fenton would bring on a storm, at which his companion promised never again to whistle near the sea.

The idea that whistling was somehow infernal was reported elsewhere, as in the case of Mrs Leakey's whistling ghost at MINEHEAD (South-West

Pirates and Privateers

Interrogating some prisoners, the seventeenth-century pirate François L'Olonnais, 'being possessed of a devil's fury', ripped open one of the men with his cutlass, 'tore the living heart out of his body, gnawed at it, and then hurled it in the face of one of the others', threatening to do the same to him unless he came up with the required information.

This piece of Grand Guignol comes from *The Buccaneers of America* (1678) by Alexander Exquemelin, who had himself served in the crews of both L'Olonnais and Sir Henry Morgan of LLANRUMNEY (Wales), but whose tales of pirate life were almost certainly exaggerated to titillate his audience. The book became a bestseller in many languages, and helped inspire the even more influential *General History of the Pyrates* (1724), by 'Captain Charles Johnson', perhaps a pseudonym for Daniel Defoe.

Whatever his true identity, 'Johnson' took the pirate legend a step further. His portrayal of Edward Teach, alias Blackbeard, was a horrid caricature, wreathed in pistols, the trademark beard covering his whole face and twisted in dreadlock-like tails. With little more than his fierce eyes visible, he was 'such a Figure, that Imagination cannot form an Idea of a Fury, from Hell, to look more frightful'. Nobody who met the real Captain Teach confirms this striking appearance, but the description has coloured the buccaneer image ever since. As for Blackbeard's burial of his loot, declaring that 'no Body but himself, and the Devil, knew where it was, and the longest Liver should take all', this too is fantasy – a pirate's wealth was generally spent as soon as he came on shore – but tales of hidden treasure had been told of historical characters such as Captain Kidd of GREENOCK (Scottish Lowlands) before Robert Louis Stevenson's *Treasure Island* (1883) perfected the myth with Long John Silver's coded map.

In reality, pirates were simply sea-robbers, and atrocities were no commoner among them than among their land-based counterparts. Many had legal sanction, operating under privateers' commissions or 'letters of marque' that allowed them to plunder enemy vessels in wartime, and even the famous Caribbean buccaneers tended not to attack shipping of their own nation. They looked much like other

The buccaneer Edward Teach, also known as Blackbeard.

sailors, such picturesque accoutrements as earrings, eye patches, and parrots on the shoulder being additions from later literature and film, and even the Jolly Roger by no means always featured a skull and crossbones, other pirate flags showing skeletons, hourglasses, cutlasses, or other symbols.

In folklore and fiction, however, their unique customs and costume set them apart. While seventeenth- and eighteenth-century accounts generally depict monsters of cruel depravity, nineteenth-century Romantic taste favoured the noble outcast, a prototype being Byron's mysterious *Corsair* (1814). Later still, Gilbert and Sullivan's *Pirates of Penzance* (1879) and J. M. Barry's Captain Hook in *Peter Pan* (1904) made fun of the melodramatic legacy, as, more recently, does Johnny Depp's charming villain in *Pirates of the Caribbean*. How the legend will develop now that genuine piracy is in the news again, around the coast of Somalia and elsewhere, remains to be seen.

England & Channel Islands), and in the mid sixteenth century, when the Protestant bishop of St David's, Pembrokeshire, was burnt for heresy, the charges brought against him included an accusation that he had taught his baby son to communicate with the Devil by whistling. Far more common, however, and more intuitively reasonable, was the idea that whistling, by imitating the sound of the wind, brought a breeze or a storm without any satanic interference. In *Scenes and Legends of the North of Scotland* (1835), Hugh Miller neatly describes the practice and the thinking behind it:

> People acquainted with seafaring men, and who occasionally accompany them in their voyages, cannot miss seeing them, when the sails are drooping against the mast, and the vessel lagging in her course, earnestly invoking the wind in a shrill tremulous whistling, – calling on it, in fact, in its own language.

In a storm, the last thing you should do is whistle. When Miller, in his youth, thoughtlessly began to do so aboard a Cromarty fishing boat labouring in a brisk gale, one of his companions told him to shut up: 'whisht, whisht, boy, we have more than wind enough already'.

See also CEMAES HEAD.

GRASSHOLM ISLAND, PEMBROKESHIRE

A floating island

One summer morning, Captain John Evans had a strange experience, reported in the *Pembroke County Guardian* on 1 November 1896. While sailing up the Channel past Grassholm Island, in what he knew to be deep water, Evans was surprised to see a beautiful green meadow. 'It was not, however, *above water,* but just a few feet *below,* say two or three, so that the grass waved and swam about as the ripple flowed over it.' Evans had heard old people say that there was a floating island near there, which sometimes rose to the surface or nearly to the surface, and then sank down again fathoms deep. Nobody saw it for years, but when they were not expecting it, up it would come again for a while.

This need not have been an illusion. On Redes Mere in Cheshire and Derwent Water in Cumbria, large bundles of peat or matted vegetation have been known to float and give the appearance of solid land. Something like this may account for what Captain Evans saw, but there are other explanations for intermittently visible islands. Some reefs are only rarely exposed by the waves, and mirages, too, can play their part. William

Wordsworth's *Guide through the District of the Lakes* (1853) records a strange experience the poet had at Grasmere in Cumbria, when he and a friend were 'alarmed by the sight of a newly-created Island', larger, higher, and clearer to view than the island they knew to be in the lake. 'At length the appearance underwent a gradual transmutation; it lost its prominence and passed into a glimmering and dim *inversion*, and then totally disappeared . . .' They saw that a patch of ice had produced the illusion, by reflecting and refracting a section of the opposite mountain.

Wordsworth had thought for a while that the new island was the result of 'an earthquake or some other convulsion of nature', a credible hypothesis. There have been well-documented although rare instances of islands produced overnight by volcanic eruptions, such as Surtsey, which reared up from the ocean off southern Iceland in 1963, accompanied by a towering column of steam and flying rocks.

The most famous floating islands of classical legend were the Symplegades, rocks which periodically clashed together to the peril of passing craft, but were successfully evaded by Jason's ship the *Argo*. Larger land masses too were said to have been free-floating. In the seventeenth century, it was believed that the Isle of Man had once travelled unattached around the Irish Sea (*see* KING WILLIAM'S BANKS, North-West England & Isle of Man), and a tale current in Mayo in the early nineteenth century was that Ireland had reached its present position after a long sea journey, having broken off from the far north of America. This is actually true, in a sense, of Antrim, as well as much of Scotland: around four million years ago, their rocks were part of the North American land mass, and joined what would become Britain as a result of tectonic shifts. Islands, in some cases, really do move.

LAUGHARNE, CARMARTHENSHIRE

See MILFORD HAVEN.

LINNEY HEAD, PEMBROKESHIRE

An eighteenth-century mermaid

An 'extraordinary creature' was seen near Milford Haven in about 1782 by Henry Reynolds, a farmer, as recorded a few years later by Mary Morgan. It seems that she only read the account taken down by Dr George Phillips, who interviewed Reynolds, but at second hand she was impressed by the simple and circumstantial narrative.

Reynolds stated that he had gone one morning to the cliffs near Linney Head, where he saw what he thought was a bather. He knew the sea was deep in that place, and was surprised to see the person clear of the water from the waist up, so went along the cliffs to get a better look.

> As he got towards it, it appeared to him like a person sitting in a tub. At last he got within ten or twelve yards of it, and found it then to be a creature much resembling a youth of sixteen or eighteen years of age, with a very white skin, sitting in an erect posture, having, from somewhat about the middle, its body quite above the water; and directly under the water there was a large brown substance, on which it seemed to float.

The 'substance' was a tail like that of an enormous eel, which shifted in circles. The top half of the creature looked human, though with thick short arms and a sharp nose that began high up between its eyes. It was bald, but with what Reynolds (or perhaps Phillips) again describes as a brownish 'substance' rising from its head and going down over its back into the water. 'This substance did not at all resemble hair, but was thin, flat, and compact, not much unlike a ribbon,' making it clear that we are not dealing with the archetypal long-haired siren. Nor was it a crest attached to the spine, because the creature lifted it up from its neck from time to time and washed underneath it, as well as washing under its arms and around its body.

Reynolds stayed watching the creature for about an hour. It swam quickly through the sea and turned smoothly, putting one hand into the waves.

> It never dipped under the water all the time he was looking at it. It looked attentively at him, and the cliffs, and seemed to take great notice of the birds flying over its head. Its looks were wild and fierce; but it made no noise, nor did it grin, or any way distort its face.

Eventually Reynolds left to fetch some other people to look, but when they came back the creature was gone.

Mrs Morgan did not doubt that the farmer himself believed he had seen what he described, and added that 'one would almost be tempted to think there was something more than imagination in it', while Dr Phillips confirmed that Reynolds was an honest man, and said it seemed almost incredible that he should have been 'either able or willing' to make up such a story. Certainly the detailed yet sober description gives a strong impression of a real animal, as does the account, similar in many respects, of the mermaid seen at ABERYSTWYTH in 1826.

LLANDDWYN ISLAND, ISLE OF ANGLESEY

St Beuno and the curlew

A curlew once rescued a saint's book from the sea, and in return was granted divine protection.

In the early seventh century, St Beuno lived at Celynnog on the Caernarfonshire coast, and regularly used to preach on Llanddwyn Island at the south-west tip of Anglesey. The holy man could walk on the water and was always able to cross to Llanddwyn on foot, but one Sunday he dropped his book of sermons into the sea, where it floated out of his reach. A curlew swooped down and picked up the volume, and placed it on a rock above the waves, from which the saint was able to retrieve it. In gratitude, Beuno prayed for a protective blessing on the curlew, and since then, nobody ever knows where that bird makes its nest.

LLANGWYRYFON, CEREDIGION

Eleven thousand virgins

The origin of the name Llangwyryfon is 'somewhat curious', writes Samuel Rush Meyrick in *The History and Antiquities of the County of Cardigan* (1810). In the fourth century, a Welsh prince named Conan Meriadoc was given control of Brittany. Many men of Wales followed the prince, and after their arrival on the Continent they sent home for wives. No fewer than eleven thousand virgins were sent to meet the demand, but their ships ran into stormy weather and were driven on to the German coast. There the maidens were taken captive by a hostile tribe of barbarians, 'who pressed them earnestly to share their beds'. They resisted, and were massacred, but their chastity remained inviolate, and in memory of their martyrdom a church was erected in Wales and called Llan gwryfôn, 'The Church of the Virgins'.

Llangwyryfon's church is in fact dedicated not only to the Virgins but to their legendary leader, St Ursula, and Meyrick's account, derived from Geoffrey of Monmouth's twelfth-century history of Britain, closely follows Ursula's story, although without mentioning the saint herself. She was said to have been martyred in Cologne in the fourth century at the hands of the Huns together with her eleven thousand virgin companions. The extravagant number of the maiden martyrs was first mentioned in the tenth century, and was probably due to a mistaken reading of Latin manuscripts giving XI MM VV as shorthand for 'eleven martyred virgins', misinterpreted as signifying 'eleven thousand [millia] virgins'.

LLANON, PEMBROKESHIRE

See CEMAES HEAD.

LLANRUMNEY, CARDIFF

Henry Morgan the buccaneer

According to the rum bottles that bear his name, Captain Morgan was a fine figure of a sailor. 'A natural born leader, his great charm won him the respect of noblemen and the loyalty of his crew. He and his men became famous throughout the Caribbean for their love of adventure,' declares the label, making them sound a wholesome bunch. The folklorist Horace Beck, on the other hand, calls Morgan 'a miscreant so foul no sink or drain in Hell is vile enough to contain his remains', and 'a homicidal psychotic' who attained knighthood only 'through prudent application of a fortune in gold obtained by torture and extortion'.

The first account of Sir Henry Morgan's life was written in Dutch by Alexander Exquemelin and translated into English in 1684. Exquemelin had served under Morgan, and his account has been taken as gospel by many later writers, but some have maintained that his more lurid tales of sadism and rape are slanders. Morgan himself sued the English publishers for libel, and was awarded £400 – a substantial sum in those days, although nothing like the £20,000 he had claimed.

He was born in about 1635 in Llanrumney, now a Cardiff suburb. He went to sea as a boy, joining a ship bound for Barbados, and once there, according to Exquemelin, was sold into slavery – something hotly denied by Morgan himself, who declared that he 'was never a Servant unto anybody in his life, unless unto his Majesty, the late King of England'. Whatever the truth of his early career in the West Indies, by the 1660s he was in charge of his own ship, and took part in a profitable expedition to Central America to plunder cities held by the Spanish. All his notable campaigns took place on shore, and indeed one of his biographers notes that he was a brilliant general but a terrible seaman: 'His victories on land were remarkable . . . By comparison he rarely went to sea without losing his ship.'

Back in Jamaica, Morgan invested in plantations, made a good marriage, and cultivated the friendship of the island's governor, Thomas Modyford. In 1668, Modyford claimed that the Spanish were planning to invade Jamaica, and ordered Morgan to get together all the English privateers that he could, and to capture Spanish prisoners to get information.

For the next few years, Morgan and his fleet caused havoc across the Caribbean, looting and destroying many Spanish-owned cities. In 1671, having assembled a force of around two thousand men, nearly the whole complement of the Caribbean's buccaneers, Morgan led an expedition against Panama, the largest ever pirate raid in the Americas. Making their way towards the city on foot through the jungle, the men ran out of food and captured a herd of cows, which they ate raw. Once at Panama, writes Exquemelin, they spared 'no Sex, nor Condition whatsoever'. One Spaniard was strung up by the testicles and beaten before having his nose and ears

Alexander Exquemelin's seventeenth-century exposé of pirate life, The Buccaneers of America, *included an engraving of the infamous Sir Henry Morgan, looking every bit as villainous as one might expect.*

cut off, in an effort to make him reveal where treasure was hidden. Insisting to the last that he was only a poor man, he was murdered.

Women generally met with no mercy unless they consented to prostitute themselves to the pirates, but an exception was eventually made in the case of the beautiful young wife of a rich merchant, who held out against both bribery and threats, refusing to gratify Morgan's wish for 'voluptuous Pleasures'. Even when stripped and locked in a dark cellar she remained steadfast, declaring that she would sooner be killed than submit. 'I my self was an Eye-witness unto these things here related, and could never have judged such constancy of Mind, and vertuous Chastity, to be found in the World, if my own Eyes and Ears had not informed me thereof,' writes Exquemelin. She sent two monks to collect a ransom, but when they used the money to redeem their own friends instead, Morgan relented and set her free, keeping the monks as substitute prisoners to punish them for their treachery. His other six hundred or so captives were treated with barbarous cruelty, to encourage them to find money for the pirates, and Morgan, having entirely destroyed the city of Panama, departed with huge amounts of silver, gold, and other treasure.

While many of the buccaneers squandered their takings on riotous living, Morgan was more prudent. He acquired huge estates in Jamaica, and rather touchingly called his home Llanrumney after his Welsh birthplace. Although in 1672 it briefly seemed that his crimes were catching up with him when he was arrested and sent to England, he was never actually imprisoned. In English eyes he was Jamaica's defender against the hated Spaniards, and in 1675 he was knighted and made lieutenant governor of the island. His enemies, however, campaigned for his dismissal, and in 1683 he was removed from office, spending the next few years in heavy drinking and obscene diatribes against the government until his death in 1688.

See also PIRATES AND PRIVATEERS (p. 232).

LLIGWY, ISLE OF ANGLESEY

A witch in the water

A piece of local tradition was attached to a large stone at Lligwy known as Arthur's Quoit. A fisherman on his way down to the sea stopped to shelter beside the stone when a sudden storm blew up, and as the rain and wind died down again he looked out to sea and saw someone struggling in the water. He hurried down to the shore, and found that it was a woman with long dark hair, dressed in white, wearing jewelled bracelets on her arms. With the fisherman's assistance she got to land, and asked him to help her to the great stone, but once she was there she started to laugh.

> 'Ha, ha!' she cried. 'If I had been swimming in my usual raiment, you would have allowed me to sink. I am a witch, and was thrown off a ship in Lligwy Bay; but I disguised myself, and was rescued.'

She was not ungrateful for his kindness, however, and as a reward she gave him a small ball which she told him contained a snakeskin. He was, she said, to keep it in a secret place, where nobody could find it, and once a year he was to take it to the sea and dip it in the water. As long as he kept to his instructions, he would have good luck, but if he ever lost the ball, misfortune would surely follow.

The witch then vanished, but an hour later the fisherman caught sight of her leaping from rock to rock in Lligwy Bay until she reached a boat that was waiting for her and sailed away.

The man hid the lucky ball in a hole beside Arthur's Quoit. As the rock was reputed to be haunted, he knew that people were unlikely to come poking round it, and he thought his treasure would be safe there. So it proved, for a long time. Every year he took out the ball and dipped it in the sea, and then hid it again, and he and his family had wonderful success throughout. Then came a day when he went to fetch the ball and found it gone. For the next seven years, nothing but disaster came to the fisherman, but at the end of that time a neighbour, on his deathbed, confessed to having stolen the ball. When he restored it to its owner, good fortune at once returned.

On the fisherman's death, the ball was left to his eldest son, who preserved it as his father instructed him. He and his brother both made considerable fortunes in Australia in the early nineteenth century, and the charm continued to be passed down in the family, as Marie Trevelyan reports in *Folk-Lore and Folk-Stories of Wales* (1909): 'A descendant in the female line of the old fisherman considered the ball one of her most precious treasures, and carefully preserved it in her far-away home in India. It was last heard of about forty years ago.'

MENAI STRAIT, ISLE OF ANGLESEY

Lucky Hugh Williams

Before the Menai Bridge was built, the only way to get across the strait was by ferry, often a risky journey. In 1783, according to G. N. Wright's *Scenes in North Wales* (1833), the boat overturned with seventy on board, all of whom were drowned but one man named Hugh Williams. In 1820 another disaster took place, with twenty-five passengers losing their lives and, again, only one survivor – whose name was Hugh Williams.

Ken Radford's *Tales of North Wales* (1982) dates the earlier event to 1785, and raises the odds by mentioning a third (or rather first) catastrophe in 1664, when the ferry crossing the strait was swept into deep water and capsized, with the death of all passengers except, yet again, Hugh Williams.

If true, the tale is a surprising coincidence but hardly qualifies as folklore. Radford adds, however, that the 1664 wreck was held to be predestined, since the boat had been built with timbers stolen from Llanidan church on the south coast of Anglesey. To use materials from a sacred structure for everyday purposes was always considered sacrilegious and potentially dangerous, and the building at Llanidan moreover contained a particularly potent object, a stone shaped like a human thigh bone that was the subject of an ancient legend. According to the twelfth-century historian Gerald of Wales, wherever the stone might be taken, it always returned the following night to its original position. This was tested by Hugh, Earl of Shrewsbury, who ordered the stone to be chained to a larger boulder and thrown into the sea. The next morning it was duly found back where it had come from. Miraculous return after removal was a property reported in folklore of other relics such as bells and skulls (*see* for example ST GOVAN'S HEAD, and CLARE ISLAND, Northern Eire & Northern Ireland), but the Llanidan stone had more sinister powers. When a local man, rashly wishing to prove the truth of the tale, tied it to his own thigh, not only did it return to its place, but his leg became gangrenous. It was said too that if any couple made love near the church – which, according to Gerald, they frequently did – great drops of sweat would drip from the stone, and that no child was ever conceived on such occasions. The stone's reputed contraceptive influence must have been the reason why it was a popular place to have sex. It could still be tested, since the stone remains embedded in a wall of the now-ruined church.

MILFORD HAVEN, PEMBROKESHIRE

The Green Islands of the Sea

Magical lands known as the Green Isles (*Gwerddonau Llion*) were reported to exist off the Welsh coast, and to be inhabited by the fairies. In *Cambrian Superstitions* (1831), W. Howells writes that these islands, 'which appeared to be beautifully and tastefully arranged', could be seen from the shores of Carmarthenshire and Pembrokeshire, and were joined to the mainland by subterranean passages through which the fairies regularly came to visit the markets at Milford Haven and Laugharne.

> It is reported, they were particularly fond of purchasing their meat from one butcher, to whom they often came invisible, and after taking the meat deposited the proper payment.

The story about the fairies' favourite market trader was one that travelled (*see* CEMAES HEAD), and their enchanted islands too were located in various places off the Welsh coast. According to the American consul Wirt Sikes, in the late nineteenth century Welsh sailors still spoke of 'the green meadows of the sea', lying west of Pembrokeshire in the Irish Channel, where they were said to be clearly visible from Milford Haven:

> There are traditions of sailors who, in the early part of the present century, actually went ashore on the fairy islands - not knowing that they were such, until they returned to their boats, when they were filled with awe at seeing the islands disappear from their sight, neither sinking in the sea, nor floating away upon the waters, but simply vanishing suddenly.

In the fifth century, adds Sikes, King Gavran sailed with his family in search of these green islands, and was never heard of again. This tale stems from the inventive pen of Iolo Morganwg (1747–1826), whose fabrications of 'ancient traditions' were widely credited in the nineteenth century. Iolo's book *The Triads of Britain* is organised (as its title proclaims) in groups of three, following an old Welsh form of literary composition. Among the legends he adapted or invented are what he calls the 'three disappearances by loss in the Isle of Britain', one being Gavran's journey, and the others a voyage made to an unknown destination by Merlin and nine bards in a ship of glass, and the mass exodus of Prince Madoc and his followers, discussed at RHOS-ON-SEA. The poet Robert Southey was beguiled by these mysteries. In his long poem *Madoc* (1805), he concentrates on the prince's supposed journey to America, but pauses to speculate on Gavran and Merlin:

> Where are the sons of Gavran? where his tribe,
> The faithful? following their beloved Chief,
> They the Green Islands of the Ocean sought;
> Nor human tongue hath told, nor human ear,
> Since from the silver shores they went their way,
> Hath heard their fortunes. In his crystal Ark,
> Whither sailed Merlin with his band of Bards,
> Old Merlin, master of the mystic lore?

NEWPORT

Rude mermaids

In 1957, Newport City Council wanted to improve its civic badge by adding 'supporters' (figures on each side of the central escutcheon or shield, traditionally often fabulous beasts such as griffins or unicorns). The city's motto is *Terra Marique*, meaning 'By land and sea', and the College of Arms supplied a design they thought suitable, with results reported in the *Daily Express* on 12 September:

> When the drawings came back there were some red faces in the council chamber. Instead of unicorns there, holding up the shield, were two amply proportioned young mermaids with wings.
>
> 'They represent the land, sea and air,' said the college designers.
>
> 'The rudest things I have ever seen,' said a councillor. To a man the council rejected the design.

In the end they settled for a golden sea lion and a red sea dragon, both winged. Sadly, neither the city council nor the College has preserved the original picture of the rude mermaids.

OYSTERMOUTH, SWANSEA

The nameless saint

Where a lighthouse now stands off the Mumbles Head, south-east of Oystermouth on the Gower peninsula, there was once, according to tradition, a monastic chapel, home to an aged hermit. One evening, looking out over the sea, the monk saw a boat approaching, and from it alighted a solemn man who directed the rowers to carry a dead body to the cave below the chapel.

> The body was bravely dressed, like that of a man of high degree, and his still features were white as chiselled marble. The monk looking on him could not help saying, 'So young and so handsome!'

The corpse was deposited in its rocky tomb, and the monk said Masses for his soul, but 'to this day', according to Mr and Mrs Hall's *Book of South Wales* (1861), 'it is believed that the spirit of the poor murdered man cries out from that cave for Christian burial in consecrated ground.'

No source is given, and the Halls do not explain why the man is said to

The lighthouse on the Mumbles, and in the foreground the rock arch on Oystermouth beach, a few years before it was destroyed in a storm.

have been murdered. Their tale seems to be a version of a very much earlier legend, in a ninth-century collection of manuscripts, that tells how St Iltut or Illtud, worshipping in a cave by the sea, saw a ship sail in bearing the corpse of a holy man with an altar travelling in the air above it. Two sailors told Illtud that they had been instructed to bring the body to him for burial, and that he was not to reveal the name of the deceased to any person. The saint did as he was told, establishing a church near the grave, and the altar continued to float above the place, held up by the divine will.

This report probably originated from a desire for some holy relics to justify the building of a shrine. In the absence of St Illtud's own body, the church founders seem to have come up with a clever solution resting on the anonymity of the dead man in question. Whether or not any actual bones were ever preserved at the site is unknown, but the rumour of their presence would have supplied a reason for worshipping there, and the tale also gave a sacred explanation for a natural wonder, a rock arch on the beach at Oystermouth that must have strongly suggested the idea of a levitating stone. The arch was destroyed in a storm of 1910, but from pictures it looks just the sort of feature to attract a legendary interpretation.

PORTHCAWL, BRIDGEND

The Devil's ship

A story current among Porthcawl sailors in the early nineteenth century was that the Devil once made a great three-masted ship in which he put wicked souls. Whenever he added a fresh crop of sinners to his cargo, he would celebrate noisily, while the wood of his craft, cut from trees in hell, smelt so strongly of sulphur that it plagued the whole coast of Wales. At last St David (or St Donat, according to some) lost patience and attacked the ship with his spear, piercing the hull. The Devil, who was counting souls on board at the time, had to swim for it, and the ship was wrecked on the Gower coast. A local giant took the mast for a toothpick, and made the mainsail into a handkerchief.

A French version of this legend says that the Devil's ship was so sulphurous and pestilent that it glowed. It was sunk by St Elmo, but 'when the night is dark and the air warm the ship burns again', an explanation for St Elmo's fire (*see* CROMER, East Anglia) or perhaps for submarine phosphoresence (*see* HAUXLEY, North-East England).

RHOOSE, VALE OF GLAMORGAN

Swan maidens

A young man from Rhoose once went with a friend to shoot wildfowl at Whitmore Bay on Barry Island, walking across when the tide was out. There they stayed until sunset, when they found that the tide had risen, cutting off their return for the moment, and so to pass the time they walked to Friar's Point, where they saw two white swans between the rocks. As they watched, the swans took off their feathers and became two beautiful maidens. While the girls bathed in the sea, the young men hid the feather cloaks, and then persuaded the swan maidens to marry them. The Rhoose man's marriage lasted for seven happy years, but at the end of that time (probably by mistake) he threw the swan's wings out with the rubbish into the farmyard, where his wife found them. She took to the air and was never seen again. The other man's wife was eventually run over by a wagon. When the bystanders ran to pick her up she changed into a swan, and she too flew away.

'The children of these marriages were said to be somewhat conspicuous by reason of their swan-shaped necks,' writes Marie Trevelyan, who found this story among the manuscript collection made by her family in the nineteenth century. It was, she says, a variant of a tradition common on the Gower coast of a man who saw a swan maiden bathing in the sea, stole her wings, and married her, only to lose her when she found her feathers. It is the same story as that told of seals or mermaids in many places including BARRA (Scottish Highlands & Islands), and the Gower tradition may have been influenced by the Irish legend of the Children of Lir (*see* INISHGLORA, Northern Eire & Northern Ireland).

RHOS-ON-SEA, CONWY

Prince Madoc's voyage to America

In the garden of Odstone, a house at Rhos-on-Sea adjoining the golf course, stands a flint-and-stone structure with a plaque attached that reads 'Prince Madoc sailed from here to Mobile, Alabama'. This (so it is claimed) is the very quay from which, in the twelfth century, Madoc's ships set sail on an epic voyage of discovery, after the prince dreamed of a land to be found far beyond the western horizon.

Madoc is a mythical figure, said to have been a son of the twelfth-century king of North Wales Owain Gwynedd (who had, in fact, no son of that name). According to the tale, after the king's death, Madoc's brothers struggled for the crown, but Madoc himself left them to fight it out and sailed westwards, returning to collect more ships and followers and then departing again, never more to be seen in Wales. It was said that he reached the New World nearly 350 years before Columbus and spent the rest of his life among the Native Americans.

The legend became popular in the sixteenth century, when it was used to bolster English claims to dominance in the New World, through the assertion that sovereignty in the territories won by Madoc had passed to

An illustrator of Robert Southey's Madoc *(1805) imagines the prince's ship on its legendary journey to America. The vessel, however, looks more eighteenth century than twelfth century.*

the Welsh Tudor dynasty. As Welsh-speakers began to explore the continent, they returned with tales of meeting natives who spoke their language. In 1669, the Reverend Morgan Jones told how on a missionary trip in North Carolina he and his companions had been captured by Indians (as they were then known). When Jones muttered a prayer in Welsh, the Indians understood his language and greeted the men as cousins. Other reports told of a light-skinned tribe called the Mandans, whose language shared many words with Welsh, and who buried their dead in Celtic mounds.

The Mandans soon died out, victims of the Sioux and of smallpox imported from the Old World, but the story of Madoc's discovery refused to die. It was cited again in the eighteenth century during struggles between the English and the Spanish for control of the upper Missouri valley, and in 1791 the minister John Williams wrote *An Enquiry into the Truth of the Tradition, concerning the Discovery of America, by Prince Madog ab Owen Gwynedd*, concluding that 'there are stronger Reasons for admitting the Truth of Prince Madog's landing on the American Shores, than for the contrary'. Among his reasons was Native American nomenclature. He had heard that certain chieftains were called Mad Dog, and commented: 'No Man, however fond of Titles, would take a name which conveys so alarming and vile an idea; therefore I am disposed to believe that this name is a corruption of Madawg or Madog.' A young preacher named John Evans set off in 1792 to search along the Missouri for the Welsh-speakers, but having followed the river for 1,600 miles without success, he died of fever in 1799.

Before his death Evans had concluded that the Welsh Indians did not exist, but the myth was a persistent one. Even in modern times, researchers have reported similarities between Welsh usage and that of some North American tribes, such as the Kutenai of British Columbia, whose numerals are identical with those of Wales except for the suffixes. A book published as recently as 1986 states as fact that settlers advancing westwards across North America in the seventeenth and eighteenth centuries met with a tribe of pale-skinned Indians and found that 'remarkably they spoke a language that was almost identical to Welsh'. The same source gives the information that in Mobile Bay, Alabama, stands a memorial with the legend:

> In memory of Prince Madoc, a Welsh explorer
> who landed on the shores of Mobile bay in 1170
> and left behind, with the Indians,
> the Welsh language.

In purely practical terms there is no reason why a Welsh sailor could not have reached America in the twelfth century. It has been claimed that some six hundred years before that, St Brendan crossed the Atlantic from BRANDON CREEK (Southern Eire) in a leather coracle, an adventure reproduced in detail by Tim Severin in the 1970s to prove at least the possibility of such a journey.

ST BRIDES, PEMBROKESHIRE

Water-horses

The water-horse of Wales, the Ceffyl-dwr, most often haunted lakes but was sometimes described as a denizen of the seashore or even of the sea itself. One of them, a hulking beast, was seen in St Brides Bay after a storm and captured by a farmer, who put it to the plough. For several weeks it worked biddably, but one day it was seized with the urge to return to its native element. Setting off across the field, it dragged plough and ploughman after it to the shore, where it plunged into the waves and was seen no more. Something very similar occurred in Carmarthen Bay, where a Ceffyl-dwr was caught, bridled, and used as a carthorse, until one day the bridle came undone, and the animal ran into the sea with its cart and driver.

Until about the mid nineteenth century, according to the folklorist Marie Trevelyan, older people in rural Wales firmly believed in the Ceffyl-dwr. He could be identified by his hoofs being turned the wrong way. Sometimes he could be seen plunging up and down in the sea, white like the foam or leaden grey like the clouds, before a thunderstorm or gale of wind.

Accounts like these make the Ceffyl-dwr more of an elemental beast and less of a demon than the Scottish water-horse, of whom tales were told at LOCHBOISDALE (Scottish Highlands & Islands) among other places. The Ceffyl-dwr was also said to mate with ordinary horses (something never reported of the Scottish creature), and a breed of small Welsh mountain ponies known as merlyns were traditionally supposed to be of this mixed stock.

See also INISHMAAN (Northern Eire & Northern Ireland).

ST GOVAN'S HEAD, PEMBROKESHIRE

The silver bell

St Govan's chapel, nestling among the rocks.

St Govan's chapel is reached from Bosherston by a lane leading to the clifftop, from which about fifty broken stone steps lead down a cleft in the rock to the tiny stone building, and a further flight to a covered well, now dry. In the chapel belfry there was once, so tradition states, a beautiful silver bell, but many hundreds of years ago it was taken down by a crew of pirates, who planned to sail away with it. On their way down to the sea, however, they rested it for a moment on the brink of the well, where it was miraculously swallowed up by a stone, inside which it still remains. 'This stone is still shown, and emits a metallic sound when struck.'

This account was printed in *Notes and Queries* in 1852. A slightly later account relates that the pirates got the bell as far as their ship, but as they left the shore a violent storm blew up out of nowhere, wrecking the boat and drowning the crew. The bell was magically transported back to land and enclosed in the stone, from inside which, again, it can be heard giving its silvery chime.

To hear the bell was no good omen, according to a rhyme which also refers to Bosherston Mere, near St Govan's Head, a spot said to be haunted:

> There is nothing to hope and nothing to fear
> When the wind sounds low on Bosherston Meer;
> There is much to fear and little to hope
> When unseen hands pull St Gowan's rope;

> And the magic stones, as the wise know well,
> Promise sorrow and death like St Gowan's bell.

THE SMALLS, PEMBROKESHIRE

Death at the lighthouse

The Smalls, a set of rocks about twenty-two miles off the coast of Pembrokeshire, were a notorious danger to ships until the late eighteenth century, when a lighthouse was built there. The original structure was a curious one, balanced on pillars that allowed the waves to dash between, and entered by a rope ladder through a trapdoor. The builder, Henry Whitesides, spent some time in the lighthouse to test his own work, and he and his companions suffered considerable hardship, being reduced in February 1777 to sending a message in a corked bottle to beg for help from Thomas Williams, the lighthouse agent:

> Being now in a most dangerous and distressed condition upon the Smalls, do hereby trust Providence will bring to your hand this, which prayeth for your immediate assistance to fetch us off the Smalls before the next spring, or we fear we shall all perish; our water near all gone, our fire quite gone, and our house in a most melancholy manner.

They took the precaution of dispatching three identical notes, and amazingly enough one of them reached its mark within a few days. The men were rescued, and Whitesides was able to make some vital adjustments to his design.

From the early nineteenth century the lighthouse was manned by three keepers at any one time, but this had not been so when it came into operation, as reported in Richard Fenton's *Historical Tour through Pembrokeshire* (1811):

> I believe at first in this light-house . . . there were but two men employed, one of whom died, and the other was afraid to make away with the corpse; nor was he relieved from the painfulness of his situation till he had suffered much from the stench of the unburied body.

Later accounts elaborate on this episode with macabre relish. A booklet of 1858 records that not long after the lighthouse was completed a severe storm blew up. A distress flag was hoisted on the Smalls, but because the weather was so bad, the closest any boat could get was just near enough

This picture of the Smalls lighthouse from 1875 makes its storm-swept desolation graphically clear. In the late eighteenth century, a man was confined here for three weeks with only a corpse for company.

to see 'the dim outline of one of the men' standing on the gallery that ran around the lighthouse. Only three weeks later did the wind drop, and then a horrid situation was discovered. One of the two keepers had died, and the survivor had not wanted to throw the body into the sea in case he was later accused of murder. Instead he had made a coffin for the corpse, which he had lashed to the railing of the platform.

The 'dim outline' was a detail seized on by more recent writers such as Kenneth Langmaid, who enjoys himself with the story in *The Sea, Thine Enemy* (1966). During the great storm, he writes, a vessel sailed close enough to the lighthouse for her crew to glimpse something odd:

> The men stated that they had seen a solitary figure standing on the lantern gallery and, on their approach, it had appeared that this person waved a hand but otherwise had made no move. It seemed very strange that anyone should remain in this exposed position for any length of time.

The explanation was that the home-made coffin had come apart, allowing the body to flop out. 'It was a corpse's arm that had waved a greeting to the oncoming boatmen.'

Whitchurch parish register provides some confirmation of the story, recording for 28 October 1780 that 'Joseph Harry, died and lay dead on the Smalls 2 mths'. The 1858 booklet, however, gives the date of the tragedy as 1802, and the name of the dead man as Howell, errors reproduced in many subsequent versions.

TENBY, PEMBROKESHIRE

Disappearing shoals of fish

The late arrival or sudden disappearance of shoals of fish was normally ascribed by fishing communities to some paranormal cause. One example was in the MORAY FIRTH (Scottish Lowlands), and another is described by William Jones in *Credulities Past and Present* (1880):

> At Tenby, in Pembrokeshire, there is a tradition of some extraordinary bank, or rock, at sea, called Will's Mark, on which cod-fish in great abundance were formerly taken. This spot is no longer to be found, and the loss is said to have been occasioned as a judgment for some enormity, formerly committed by the inhabitants of Tenby.

The 'enormity' is elsewhere specified as the citizens' barbarous usage of a deaf and dumb man who had come begging to Tenby. In 1858, Jones continues, rumours spread that the submerged bank had been found, perhaps because the people of the town had done penance or some act of expiation, but its rediscovery seems to have been short-lived, or the cod had abandoned it, since the fishing ground was never in use again.

TUSKER ROCK, VALE OF GLAMORGAN

The tragedy of Dunraven

In about 1575, four young people were stranded on the Tusker Rock when their boat drifted away, and drowned in the rising tide. From this simple tragedy grew a dark legend of mutilation, revenge and murder. The tale seems to have developed in the nineteenth century, and was well known by 1926, when W. H. Wyndham-Quin, Earl of Dunraven, wrote in his history of Dunraven Castle that 'there is a story long current in South Wales' that one of the Vaughans, the family who once owned the castle, was guilty of wrecking ships by luring them on shore with false lights, and that as a consequence of his crimes his children were drowned. Overwhelmed by remorse, he sold the estate to the Wyndham family, left the country and was never heard of again.

By reference to legal records, the earl was able to demonstrate that this was untrue as far as the Vaughans were concerned. Turning to earlier owners, the Butlers, he found from seventeenth-century manuscripts that the children of Arnold Butler had been drowned as described, and commented with some justice that 'Surely it is extremely unlikely that a calamity of this kind should happen to two families.' His documents made no mention of wrecking, and he concluded that 'the exact truth of the story will ever remain in the lap of the gods'.

Not all writers have been so historically scrupulous. In *Annals of South Glamorgan* (1913), Marianne Robertson Spencer gives a highly coloured account which she claims to have found in a work of 1769, *England Displayed* – a book that in fact makes no mention at all of the legend. Spencer did not, however, simply invent the story, which was well enough known to be the subject of a ballad written in 1894. Whatever its original source, the tale is a gripping one. Vaughan of Dunraven was a clever but vain man, who once swam out to save the crew of a ship wrecked near his castle. Following this success, he devised some apparatus (not further described) for rescuing shipwrecked sailors, but his invention was ignored by the authorities, and Vaughan was so embittered that he began to lose his reason, and started to squander his money recklessly. His wife died of a broken heart, presumably because of his insane prodigality, and his eldest son went abroad. Left with three young children, Vaughan was determined to recoup the family fortunes. At about this time another ship ran aground near the castle, and as lord of the manor he was able to claim the wreck. He was inspired with the notion of engineering more shipwrecks for his own benefit, and consulted 'a man of desperate habits living in the neighbourhood'. This unsavoury character had been a pirate captain, and years earlier had been arrested on the orders of Vaughan himself, then a magistrate. In the struggle the pirate had lost a hand, later replaced with a metal hook, and was thereafter known as 'Mat of the Iron Hand'. In his unhinged mental state, Vaughan was unaware that Mat still bore a grudge, and went into partnership with him. They conspired to hang out false lights, tempting unwary ships on to the dangerous rocks.

One afternoon, Vaughan was looking out to sea, and saw two of his sons rowing out to some rocks which were dry at low water, but covered at high tide. They moored their boat and began to swim, but then a storm blew up. Vaughan saw his sons' boat adrift, and realised that the children were doomed. Nobody could reach them in time to save their lives, and the boys were drowned before their father's very eyes. At the same moment, the youngest child, who had been left alone while Vaughan's servants rushed to the shore in vain, tumbled into a barrel of whey and also drowned.

Devastated and repentant, Vaughan mended his ways, but his confederate Mat of the Iron Hand continued his criminal career. Vaughan, now an old man, spent his days watching the horizon for a ship in which his eldest son might return to him, and one evening he saw a sail approaching in a gathering storm. From the cliffs Mat's false lights shone out. The listeners on the shore heard a terrible crash, and cries for help followed by a more terrible silence. Mat came to Vaughan, and announced that of the ship's crew, all had drowned but the captain, a native of Dunraven.

> 'Was he saved?' broke out Mr Vaughan. A fiendish laugh was the answer as the wrecker thrust a death-cold hand into that of the questioner. A gleam of light from the fire in the cave lit up a ring on the dead hand, and the father saw that it was the hand of his long-lost and only son! The one-handed man had had his revenge!

Vaughan sold his estate to Humphrey Wyndham, left the country, and was never heard of again. Some said that Mat of the Iron Hand was hanged a few months later for another murder, some that he was pursued and shot by Vaughan's faithful harper.

It came to be said that Tusker Rock was haunted by the ghosts of Mat, Vaughan, and Vaughan's sons. In *Sea Phantoms* (1963), Warren Armstrong writes that 'strange shapes' are seen there, usually when bad weather is on its way, including two boys trying to climb out of reach of the sea, a youth 'dressed in captain's rig of three centuries ago', a figure with a clawlike right hand, and a demented man running along the shore. 'All these phantoms,' writes Armstrong, 'have been seen and reported by the crews of passing ships and by local residents around Glamorgan.' An earlier account reports more simply that a phantom light is seen hovering around the Tusker Rock before a storm.

THE WOLVES, CARDIFF

Wolves turned to stone

The sixth-century saint Cadog, who sometimes resided on BARRY ISLAND, also spent time on Flat Holm in the Bristol Channel, and put his sheep out to pasture there. On one occasion, two wolves from England swam across to the island and caused havoc among the flock, slaying and tearing. Having devoured their fill, they set off again, swimming towards Wales, but judgement came upon them. For their crime in killing the saint's sheep they were turned into two rocks, known to this day as the Wolves.

In a footnote to this tale in W. J. Rees's *Lives of the Cambro British Saints* (1853), it is observed that Cadog conferred small benefit upon posterity by placing the Wolves in position, since they are notoriously dangerous rocks that have caused the wrecks of many ships – but, adds the sceptical editor, 'we may safely acquit him of having had anything to do with them.'

CL.	Clackmannanshire
EA.	East Ayrshire
ED.	East Dunbartonshire
EDIN.	City of Edinburgh
EL.	East Lothian
ER.	East Renfrewshire
FK.	Falkirk
GL.	City of Glasgow
IN.	Inverclyde
ML.	Midlothian
NA.	North Ayrshire
NL.	North Lanarkshire
RN.	Renfrewshire
SA.	South Ayrshire
SL.	South Lanarkshire
WD.	West Dunbartonshire
WL.	West Lothian

Orkney Islands

St Combs
Portessie
Broadsea
Moray Firth
Rosehearty
Buckie
Macduff
Burghead
Portgordon
Rattray Head
Highland
Moray
Peterhead
Bullers of Buchan
Collieston
Aberdeen
Aberdeenshire
River Dee

SCOTLAND

Coll
Croig
Tobermory
Tiree
Duart Castle
Angus
North Sea
Perth and Kinross
Lismore
Treshnish Isles
Staffa
Lady's Rock
Iona
Lower Largo
St Monans
Elachnave
Inveraray
Crail
Stirling
Buckhaven
Fife
Corryvreckan
Argyll and Bute
CL.
Colonsay
Dunfermline
Inchkeith
Firth of Forth
North Berwick
Greenock
WD.
ED.
FK.
Coldingham
IN.
EDIN.
RN.
GL.
NL.
WL.
EL.
Rothesay
ML.
ER.
Eyemouth
Islay
Gigha
Edinburgh
Prestonpans
NA.
Irvine
SL.
Scottish Borders
EA.
Firth of Clyde
Ailsa Craig
SA.
North Channel
Knockdolian
Dumfries and Galloway
Collin
Lochar Moss
Northumberland
Arbigland
Little Ross
Solway Firth
NORTHERN IRELAND
ENGLAND
Cumbria
Isle of Man
North Yorkshire

0 20 40 60 80 mi
0 40 80 120 km

SCOTTISH LOWLANDS

Aberdeenshire, Argyll & Bute, Ayrshire, Dumfries & Galloway, East Lothian, Fife, Inverclyde, Moray, Scottish Borders

ABERDEEN, ABERDEENSHIRE

Sea ceremonies

The launch of a ship is an important event, whatever her size. Until at least the nineteenth century it was the custom in many places for fishing boats to be sprinkled with liquor or grain when they first went to sea, and larger vessels are still 'christened' by having a bottle of wine broken over the prow (*see* COWES, South-East England).

A more violent ritual used to be carried out in Aberdeen. Captain Andrew Shewan, who witnessed the proceedings as a boy in the 1860s when he was about to take his own first voyage, describes how young apprentices and other lads were ducked in the sea by the older, stronger workers:

> As the ship left the ways two or more stalwart shipwrights took up positions by the waterside, and, at the critical moment, seized any victims within reach and plunged them head over heels into the recoiling wave thrown up by the vessel's passage into the water. They were ducked not once but three times, unless they were agile enough to escape; and I have been told by an old Aberdonian that he has seen unfortunate youths chased dripping wet half-way up the street, and, when caught, brought back and ducked until it was deemed their baptism had been sufficiently thorough.

Shewan saw the same thing done at Peterhead, and it may be relevant that both Peterhead and Aberdeen, like HULL (North-East England), were whaling ports. The whalers were a rough bunch, their life at sea a constant

This picture from Robert Chambers's Book of Days *(1863–4) shows the ceremony of 'crossing the line'. On the left, a reluctant sailor is about to be shaved by his shipmates, while on the right, Neptune brandishes his trident, and other figures in long wigs impersonate mermaids and mermen.*

battle with nature on the grandest scale, and perhaps old and brutal rites like this were kept up longer in the whaling towns.

Shewan suggests that the ceremony represented human sacrifice to the ocean gods, holding echoes of ancient times when a victim might really have been drowned. Alternatively, perhaps the immersion of the lads symbolised the new ship entering the water, and the fact that they did *not* drown was thought to bode well for her safe return. Although Shewan refers to the boys' 'baptism', the point of the proceedings was not to initiate them personally, as is clear from the fact that it was not only shipwrights' apprentices who were ducked, but any bystanders hanging around by the water's edge.

There was, however, a ritual performed well into the twentieth century, expressly to induct new sailors into the brotherhood of the sea. When a ship crossed the equator, an elaborate pageant marked the event of 'crossing the line', involving two of the crew dressing as the sea-god Neptune and the sea-goddess Amphitrite, the latter a sort of pantomime dame in a woman's nightcap with ribbons of seaweed, carrying in her lap one of the smaller members of crew to represent her 'baby', and attended by sailors as sea-nymphs. Any man who had never crossed the line before then had his face roughly shaved and was dunked in a tub of sea water. This was the general outline of the ceremony as performed in the nineteenth century, but earlier accounts describe a more savage initiation which could maim or even kill a particularly unpopular member of crew.

AILSA CRAIG, SOUTH AYRSHIRE

Tickling a giantess

Ailsa Craig is a towering volcanic rock in the Firth of Clyde, rising 1,114 foot above sea level. Like other isolated sea-crags, it was said to have been dropped in place by the Devil, or, in a more detailed and dirtier legend, by the Cailleach, an ancient Gaelic goddess. The Cailleach of Arran liked to eat sailors, and to catch them she would stand astride Loch Fyne, legs wide apart, her great left foot planted in Carrick on Bute, and her great right foot more than thirty miles away on Kilmorey, Knapdale. Whenever a ship tried to pass she would drop stones on them to sink them, and then pick the men out of the water and pop them in her mouth.

One day a French captain wanted to sail that way, and saw her waiting with an immense rock to drop on his ship. Being an artful fellow, 'by a very adroit handling of his mizzen-mast he tickled her in the obvious place that you would expect a Frenchman to tickle a woman.' The Cailleach screeched and dropped the rock, and that was Ailsa Craig.

The story was recorded in 1962 from Alexander Archibald, a Prestwick policeman, who in turn had heard it from a Ballantrae fisherman. It was not a unique idea that somebody could escape from a giantess by tickling her, although tales including this motif are not generally so precise about which bit of the giantess gets tickled.

ARBIGLAND, DUMFRIES AND GALLOWAY

John Paul Jones the pirate

Was John Paul Jones a villain or a hero? The sources describe him so differently that at times it is hard to believe they mean the same man.

John MacTaggart of Galloway, writing in 1824, drew his information 'from the lips of many who personally knew him, and all about his singular ways', and his account makes Jones a menacing figure:

> The late celebrated sea robber; a Gallovidian, I am rather sorry to say, but he was a clever devil, had strong talents of the infernal stamp; he was a short thick little fellow, about five feet eight in height, of a dark swarthy complexion.

John Paul (he added the Jones later) was born in Arbigland in 1747. He was a skilled sailor, but generally disliked for his violence and ambition. When his ship was boarded by a press gang in Liverpool, he killed three of the men and threw the rest into the Salthouse Dock, and later, while in command of a cargo ship, he threw a ladle of hot pitch at a young sailor, causing such pain that the boy jumped overboard and was drowned. Hearing of this, the Earl of Selkirk threatened Jones with dire punishment if he ever returned to his homeland. Exiled, Jones turned to piracy, 'bent on bloody deeds', and found his true vocation. MacTaggart concludes with reluctant admiration that 'his undaunted courage, his penetrating judgment, and his savage temper, befitted him in an extraordinary manner for the damnable trade.'

So much for the Scottish point of view. To the Americans, on the other hand, Jones was a champion, the 'Father of the US Navy'. In the 1770s he took service with America, then allied with the French against the British, and made himself the terror of the North Sea in his coastal attacks. His most memorable encounter took place near Flamborough Head, Yorkshire, with the English frigate *Serapis*. The battle with Jones's ship *Bonhomme Richard* was fierce and bloody, and after heavy losses on both sides the English captain shouted to Jones to ask if he would surrender. The answer was, 'I have not yet begun to fight!' Faced with such dogged courage, the *Serapis* was finally forced to yield, and the conquest of the English in their own waters was a tremendous boost to American morale. Later, Jones became a rear admiral in the Russian navy. He died in Paris in 1792, and in 1913 his remains were taken to the US Naval Academy in Annapolis.

Legend was bound to gather around such a character. Popular ballads of the day celebrated, or deplored, Jones's exploits, and in Galloway, an off-shore crag became known as the 'Frenchman's Rock', because it was said that a local vigilante force had mistaken it for Jones's ship, fired on it, and rejoiced when the tide covered the reef, thinking they had sunk the pirate. He was equally dreaded on the other side of Scotland, and a tale current into the mid twentieth century told how his threatened attack on Leith harbour was turned away. As told in Horace Beck's *Folklore and the Sea* (1973), a bishop took a chair down to the shore at low tide and prayed:

> 'Lord, I have served you well over forty years. Either send a gale to drive away the pirate who is coming to attack our town or I will sit here until I drown. Amen.'

For a while it seemed that his prayer would be no more effectual than King Canute's challenge to the waves at WALLASEY (North-West England & Isle of Man). The water rose to the bishop's ankles, and then to his chest, and still he sat there, grimly confronting the tide. Then finally divine assistance came. The wind turned, and blew strongly from the west, driving the attackers out to sea, and the old man got up and took his chair ashore. After waiting another twenty-four hours for a favourable breeze, Jones gave up and sailed away, leaving Edinburgh in peace.

BROADSEA, ABERDEENSHIRE

See ROSEHEARTY.

BUCKHAVEN, FIFE

A fiery devil

The people of Buckhaven, a fishing town on the Firth of Forth, were said to be of foreign origin, descended either from the crew of a Dutch ship wrecked on the coast in the fifteenth century, or, more colourfully, from a gang of buccaneers who took shelter at Berwick-on-Tweed when pursued by the authorities:

> After a smart battle, among themselves, they divided, and 'tis said, the party who gained this Bucky-battle, fearing the English law to take place, set northward and took up their residence at this Buck-haven, so called not only from the great quantity of Buckies [periwinkles or whelks] that are found in and about that place, but on account of the battle they had with their neighbours at Berwick when they divided, which was then called bucking one another, but is now named boxing or fighting.

This cheerful etymological hotchpotch appears in a pamphlet by Dougal Graham (1724–79) devoted to the peculiarities of Buckhaven and its inhabitants. Graham reports that they believed in witches, who sailed over the sea in cockleshells, and in 'Ghosts, Willy with the Wisp, and the Kelpy, Fairies and Maukens, and boggles of all sorts'. The Kelpy, a water-spirit which often appeared as a horse, was said to make a dreadful roaring before any boat was lost at sea, and 'maukens' – hares – were dreaded here, as they were all along the east coast of Scotland and in many other places too, from a widespread belief that witches turned themselves into those animals. 'Willy with the Wisp', also known as 'Spunkie', was:

> a fiery devil, and leads people off their road in order to drown them, for he sparks sometimes at our feet, and then turns before us with his candle, as if he were twa or three miles before us, many a good boat has Spunkie drown'd; the boats coming to land in the night-time, they observe a light off the land, and set in upon it and drown.

In England, the will-o'-the-wisp is the pale flame of marsh gas flickering above boggy ground, but here some sort of marine phosphorescence seems to be the light that leads the sailors astray.

BUCKIE, MORAY

The Burry-Man

The ceremony of the 'Burry-Man', as performed at several ports on the east coast of Scotland, involved a procession around the town led by a man whose clothes were thickly covered in burrs from thistles or burdocks. In the *Banff Journal*, it was reported in the mid 1850s that:

> The herring fishing being very backward, some of the fishermen of Buckie, on Wednesday last, dressed a cooper in a flannel shirt, with burs stuck all over it, and in this condition he was carried in procession through the town in a hand-barrow. This was done to 'bring better luck' to the fishing.

Writing of 'primitive beliefs' in 1929, J. M. McPherson interprets the ritual as a remnant of human sacrifice, the Burry-Man representing a scapegoat offered to the sea-gods. Neither secular nor religious education had stamped out such pagan survivals. As the *Banff Journal* writer commented with evident disapproval, the Buckie procession occurred 'in a village where there are no fewer than nine churches and chapels of various kinds, and thirteen schools'.

A similar event took place in Fraserburgh, forty miles or so to the east of Buckie, in 1864, when the burr-covered man rode through the town on horseback. He was followed by a second rider playing the bagpipes, and a third man, wearing 'a high-crowned hat, which was hung round with herrings by the tails', brought up the rear with a flag, while a cheering crowd followed the trio.

In Fraserburgh and Buckie this seems to have been a one-off procedure, but at South Queensferry on the Firth of Forth it was, and is, an annual event. As described in 1885, the leader went on foot, 'encased as if he were in armour in his suit of close-sticking burrs, grasping staves adorned with flowers', and the townspeople greeted him with shouts and gifts. Old locals said at that time that this had happened since at least the mid eighteenth century, and even now the Burry-Man parades through the town during the August fair, though nowadays he is collecting money for charity, his association with the fishing industry having been largely forgotten.

BULLERS OF BUCHAN, ABERDEENSHIRE

The mermaid of the cave

The Bullers of Buchan is a sea cavern on the north-east coast of Scotland, a feature 'worth going six hundred miles to see', according to one nineteenth-century admirer. A 'buller' is a whirlpool, and also refers to the noise it makes: a bellowing roar. In this great oval basin between the rocks, entered by boat through a vaulted arch, not only the waves but the screams of sea-birds echo with deafening effect. Dr Johnson regarded the place with awe, remarking that 'If I had any malice against a walking spirit, instead of laying him in the Red Sea' – to which demons and ghosts were traditionally banished by exorcism – 'I would condemn him to reside in the Buller of Buchan.'

According to Hugh Reid's *Past and Present* (1871), a man once met a mermaid here. He took out his boat one evening from Springhaven, and under a great cliff by the Bullers he saw 'the unwelcome monster, half-woman half-fish'. She was combing her long brown hair, and holding 'a looking-glass made from some sea-weed', an intriguing invention. Below the waist she had 'a sort of double tail', and beautifully expanded fins, which she constantly waved in the water. The upper half of her body was 'like that of a full-grown female, though much smaller in size; the breast full and plain', and she had a small round head, a slender neck, flat features, white teeth and small clear eyes. All this detail suggests that the man must have got very close to her, as indeed appears from what follows.

> The mermaid beckoned the fisherman to approach; he obeyed; she whispered some message and suddenly disappeared. The fisherman hastened towards the shore; and whatever that message was, he never ventured out to sea again.

He described his encounter when he got home, but he never mentioned it again. That night a terrible storm swept along the coast, resulting in the destruction of many boats, and ever afterwards the man refused to talk about what he had seen. If anyone tried to discuss it he would silently walk away into his house, and not emerge again for days or weeks.

Mother Carey's Chickens

As suggested by its common name, the storm petrel was associated with tempests, and although weather predictions could be made from watching other gulls (*see* THE POWER OF SEAGULLS, p. 304), this small swallow-like bird was the most significant to a sailor. Captain William Dampier wrote in his *Voyage to New Holland* (1703):

They are not so often seen in fair Weather; being Foul-weather Birds, as our Seamen call them, and presaging a Storm when they come about a Ship; who for that reason don't love to see them.

In rough seas, he continues, they hover under the ship's stern in the smoother water of its wake, and 'pat the Water alternately with their Feet, as if they walkt upon it; tho' still upon the Wing'. From this trait, according to Dampier, the birds get the name 'petrel', alluding to St Peter's walking upon the water (Matthew 14:29). The etymology may not be correct, since the name appears first in 1676 as 'pitteral', less suggestive of a derivation from Peter or Petrus, but the birds do drop their legs when feeding close to the sea's surface, as if attempting to walk on the waves.

A storm petrel 'paddling' with its feet, as if walking on the surface of the waves.

The petrel goes by many other names, some referring to its strong musky smell. Among these, according to the *Encyclopaedia Britannica* (1910–11), 'that of "Stink-pot" is not the most opprobrious', although the worse ones are not given. The stink arises from the fact that when attacked, the petrels discharge a jet of foul-smelling stomach oil. Their stench may have led to their association with the Devil, indicated in the names 'Witch' and 'Waterwitch', '*oiseau du diable*' and '*satanique*'. Like other birds they were believed to embody the souls of drowned mariners (*see* PORLOCK, South-West England & Channel Islands). Breton fishermen explained petrels specifically as the spirits of cruel captains who were condemned after death to flutter forever over the sea, and another French name for the birds was '*ames damnées*', literally 'damned souls'. A friendlier epithet was 'Mother Carey's chickens', a name in common use among English sailors, perhaps from the Latin *mater cara*, 'mother dear', an endearment applied to the Virgin Mary.

From Captain Marryat's seafaring tale *Poor Jack* (1841), traditions concerning the petrel are clear. The captain predicts dirty weather, pointing out 'some Mother Carey's chickens dipping in the water astern'. A passenger is eager to shoot one and has to be restrained, since the captain says he would as soon murder one of his own companions:

'I believe that they *were* every one sailors like ourselves in former times; they are now the sailors' friends, come to warn us of the approaching storm.'

He adds an incident from the experience of one of his crew, when three storm petrels were shot and soon afterwards three sailors died, because 'if one of these birds is killed, it is certain that some one of the crew must die and be thrown overboard to become a Mother Carey chicken, and replace the one that has been destroyed.'

BURGHEAD, MORAY

Burning the clavie

In 1689, a Burghead man appeared in court accused of 'paying a superstitious worship, and blessing the boats after the old heathenish custom'. His crime was that he had made a huge torch called a clavie, and carried it around the harbour to purify the fishing fleet from witchcraft.

The ceremony of 'burning the clavie' still takes place in Burghead, though its links with the sea and the fishing industry have weakened. On New Year's Eve, half a barrel is fixed to a long pole and filled with tar and wood, which is then set alight, carried in procession around the town and finally taken to burn on the cliffs. As practised today, this is a non-specific though picturesque fire festival, but in earlier centuries the important part of the proceedings was to visit the boats, throwing grain and spirits over them to symbolise gifts of food and drink, and making the sign of the cross with flame – a hybrid pagan/Christian rite that was repeatedly condemned by the Presbyterian Church, but continued to be performed until the mid nineteenth century, the last recorded instance being in 1875.

COLDINGHAM, SCOTTISH BORDERS

St Cuthbert and the otters

To show their devotion and mortify their flesh, the ancient saints practised immersion, standing in the cold sea while they recited psalms and prayers. St Cuthbert was one of these ascetics, and while he was staying at Coldingham he used to slip away from the monastery in the evening, go down to the shore and walk out into the rough waves, where he would stand all night. He told nobody what he was doing, and one of the Coldingham monks, who wondered what could be keeping the holy man out from dusk till dawn, followed him one night and watched as Cuthbert marched into the dark and freezing water. There he stayed, singing songs of worship, until just before sunrise, when he returned to land and continued to pray.

> While he was doing this, there came forth from the depths of the sea two four-footed creatures which are commonly called otters. These, prostrate before him on the sand, began to warm his feet with their breath and sought to dry him with their fur, and when they had finished their ministrations they received his blessing and slipped away into their native waters.

The spying monk, who had seen all this from his hiding place among the rocks, was terrified at the thought that Cuthbert might know he had been watching, and begged the saint's pardon. This was freely granted on condition that he kept silent about what he had observed while Cuthbert was alive, a promise he kept, 'but after the saint's death he took care to tell it to many'.

Bede's eighth-century account quoted above identifies the animals as otters, although an earlier anonymous life of the saint calls them only 'two little sea animals'. Another suggestion is that they were seals. Whatever they were, their care of the saint and his blessing them in return are typical of Cuthbert, who was an animal-lover. The eider ducks that breed on the FARNE ISLANDS (North-East England), where he lived for a while, were known at least until the mid twentieth century as 'St Cuthbert's birds' because he was so fond of them, and an Irish life of the saint tells how when he once dropped a book of psalms into the sea, it was retrieved for him by a seal.

COLL, ARGYLL AND BUTE

Sea-serpents of the Isles

The early nineteenth century saw a rash of sea serpent reports around Britain. The famous Animal of STRONSAY (Scottish Highlands & Islands) was washed ashore in November 1808, and a few months earlier the Reverend Donald Maclean, minister of Eigg, saw off the coast of Coll a giant beast that he was able to describe in detail. While rowing around

A sea-monster as seen by a Mr C. Renard, from A. C. Oudemans's The Great Sea Serpent *(1892). Like this one, the beast spotted by several observers between the islands of Coll, Canna, and Rhum in 1808 was reported to have an enormous head and eyes 'as large as a plate'.*

the island, he wrote, he had seen something about half a mile off which looked like a rock. Knowing that there was no rock at that point of the ocean, he peered more closely:

> Then I saw it elevated considerably above the level of the sea, and after a slow movement, distinctly perceived one of its eyes. Alarmed at the unusual appearance and magnitude of the animal, I steered so as to be at no great distance from the shore. When nearly in a line betwixt it and the shore, the monster, directing its head (which still continued above water) towards us, plunged violently under water.

Convinced that it was chasing them, Maclean and his crew rowed for the rocks, and just as they had reached a place where they could jump out, they saw the monster speeding towards them. Finding the water too shallow, it writhed away, and swam off with its head once more above the waves.

In shape it bulged below the neck and then tapered away at the tail, and Maclean estimated its length to be between seventy and eighty foot. It seemed to have no fins, but moved by undulating up and down, and in this way it raced through the water at great speed. The same beast, Maclean reported, was seen near Canna by the crews of thirteen fishing boats, who were so frightened that they fled to shore, and between Canna and Rhum it advanced towards another vessel, its head high above water:

> One of the crew pronounced the head as large as a little boat, and each of its eyes as large as a plate. The men were much terrified, but the monster offered them no molestation.

The Glenelg minister John Macrae reported that Maclean was 'quite a man to be believed', and thought that this creature was probably the same, or at least the same breed, as the one he himself watched for two days, on and off, in the SOUND OF SLEAT (Scottish Highlands & Islands).

COLLIESTON, ABERDEENSHIRE

The sea's prey

At least until the mid twentieth century, among fishermen on the east coast of Scotland and probably in other places, it was taboo to use a boat if any of its crew had drowned. Not only would other men refuse to put to sea in the craft, they would not even use the timbers for firewood. According to Peter Anson's *Fisher Folk Lore* (1965), when a fishing boat ran ashore at Collieston with the loss of all on board, even though the boat was scarcely damaged, the men's friends would neither sail the vessel nor sell it, 'for they were afraid to handle the money paid for a boat which had demanded the sacrifice of human lives'.

Anson notes the sailors' conviction 'that the spirits of the waves and the sea gods must have their prey', a belief expressed more succinctly in a fishermen's proverb quoted in twentieth-century folklore collections that 'the sea takes its own share'. This idea underlies the widespread belief that to save someone from the water was dangerous, since doing so would put you at risk yourself. The sea was thought to demand a life for a life, so that when any man was rescued, another would have to drown in his place.

COLLIN, DUMFRIES AND GALLOWAY

A death coach drives across the waves

On land, the 'death coach' was often said to fetch away souls, and sometimes these black vehicles were reported to drive across the sea. They took only the wicked, either libertines like the lord of Tirawley, as reported at CASTLE GORE (Northern Eire & Northern Ireland), or cruel oppressors such as Grierson, laird of Lagg, a government official who pursued the Presbyterian Covenanters in late seventeenth-century Scotland and was one of the most cordially hated men of his time. Tales of his arbitrary cruelty, including summary execution not only of the rebels themselves but of anyone who had offered them help, circulated for a long time. His name was still used as a threat to scare children in the mid nineteenth century, when it was the custom in Dumfries homes for someone to dress up as the laird at Hallowe'en and chase people round the house.

According to John MacTaggart's *Scottish Gallovidian Encycylopedia* (1824), after Grierson's death in 1700 'a ship at sea met with a singular *sail*, a chariot drawn by six horses, and conducted by three drivers, all of the Pandemonium stamp'. Black clouds swirled above it, vomiting thunder and lightning. As the carriage plunged past across the waves, the sailors hailed it, asking where it was going, and got the answer, 'from Hell to Colinn'. Colinn, or Collin as it is spelt now, was Grierson's home, and this was the vehicle sent to transport the evil laird to the land of demons.

COLONSAY, ARGYLL AND BUTE

Macphie and the mermaid

John Leyden's ballad 'The Mermaid' tells the story of the laird of Colonsay, who was captured by an entrancing though dangerous sea-nymph:

> That sea-maid's form, of pearly light,
> Was whiter than the downy spray,
> And round her bosom, heaving bright,
> Her glossy, yellow ringlets play.
>
> Born on a foamy-crested wave,
> She reach'd amain the bounding prow,
> Then, clasping fast the chieftain brave,
> She, plunging, sought the deep below.

She kept him prisoner for some time, and pleaded for his love, but he was faithful to the memory of his sweetheart at home on Colonsay. Eventually the mermaid released him, but continued to adore him from afar, singing her sorrow from the depths:

> And ever as the year returns,
> The charm-bound sailors know the day;
> For sadly still the mermaid mourns
> The lovely chief of Colonsay.

Leyden based his poem on a story he heard in his travels around the Highlands in 1801, a darker traditional tale than his romance, which told how Macphie of Colonsay was kidnapped by a predatory mermaid who confined him in a cave on the shore. She showered gifts upon him, but the only thing he wanted was his freedom, and finally he escaped, to the mermaid's fury. She pursued him viciously, and was only overcome when Macphie threw his dog into the water to fight her. The noble animal killed the siren, but met its own death in the battle.

Macphie of Colonsay features in several local legends, and so does his black dog, said to have saved its master's life in another supernatural contest set on ISLAY. The Macphies or Macduffies were lairds of Colonsay until the middle of the seventeenth century, and it was rumoured that they were descended from a marriage between a man and a seal. The hero of the tales may be based on Malcolm Macphie, who led a rebellion against the Campbells in 1615, and although we do not know whether he really had a notable dog, the beast is such a constant feature of the stories that it is tempting to suppose so.

CORRYVRECKAN, ARGYLL AND BUTE

The devouring whirlpool

A whirlpool is a circular current produced by opposing tides or winds. In rivers, eddies (tiny whirlpools) are fairly common, but large-scale sea whirlpools are rare. The best known examples from ancient geography and myth are Charybdis, one of the perils encountered by Odysseus, usually identified as Garafalo off the Sicilian coast, and the Maelstrom, near the Norwegian Lofoten Islands, which was said to be an entrance to the underworld (*see* SWILKIE, Scottish Highlands & Islands). In fact, neither of these is a true whirlpool, since they form no vortex, although they both present very real perils to shipping.

Corryvreckan, however, is a genuine vortical whirlpool, one of the greatest in the world. Its roar is heard from twenty miles away, and the waves can reach twenty feet in height. To pass it is fraught with risk, and should never be attempted without good local knowledge. Even a vessel with a strong engine can be dragged towards the whirlpool, and to a sailboat or rowing boat it is almost inevitably fatal. Martin Martin's *Description of the Western Isles of Scotland* (1703) paints a vivid picture:

> Between the North end of Jura, and the Isle Scarba, lies the Famous and Dangerous Gulph call'd Cory Vrekan, about a Mile in breadth, it yields an impetuous Current, not to be matched any where about the Isle of Britain. The Sea begins to boil and ferment with the Tide of Flood, and resembles the boiling of a Pot, and then increases gradually, until it appear in many Whirlpools, which form themselves in sort of Pyramids, and immediately after spout up as high as the Mast of a little Vessel . . .

The whirlpool was sometimes known as the Cailleach, a Gaelic word meaning 'hag' or 'witch', and usually the name given to an ancient elemental spirit or goddess of the Highlands. It was supposed to be the place where she cleaned her clothes, a sort of vast washing machine. One legend says that when she was scrubbing them there, all the tartan in Scotland turned white, an image inspired by the way the foam whitens the dark water as it spins. A local saying has it that 'When the Cailleach wears her white cap' – when the whirlpool foams – 'a gale will surely come.'

The name Corryvreckan or *Coire Bhreacáin*, meaning either 'the speckled cauldron' or 'the cauldron of Breccán', was earlier applied to the turbulent sea channel south of RATHLIN ISLAND (Northern Eire & Northern Ireland), and the old legend of a prince named Breccán, drowned with his fifty ships, was originally an Irish one, but by the fourteenth century, both legend and name were associated with the Scottish whirlpool. Irish settlers in Scotland, confronted with these ferocious waters, must have been reminded of Prince Breccán's fate and begun to tell the story of his death in this setting, and by the late seventeenth century the tale was so firmly located in Scotland that a site on Jura was identified as the prince's grave.

A local version of the tale, current in the mid twentieth century, relates that the prince loved a woman of Crinan, west of Corryvreckan across the Sound of Jura, but was told she had taken a vow to marry only a man who had spent three nights in a boat on the whirlpool. The prince consulted a witch, who advised him to procure three ropes, one of horsehair, one of sealskin and one made from the hair of a virgin. He obtained the first two without trouble, and for the third used the locks of his beloved. One night he rode the whirlpool in safety, anchored with the horsehair rope. A second night he was held by the sealskin rope, although the sea was rougher. On the third night the hair cable held for a while as the whirlpool raged, but just after midnight it gave way – as his sweetheart's virtue must have done – and the prince drowned.

In the 1970s, a Crinan fisherman told the folklore collector Horace Beck what happened to a 450-ton boat taking cattle to the market at Oban. The captain had collected over a hundred beasts, together with their owners, and was about to set off from Port Askeig on Islay.

> Now just as he was leaving the minister came running down to the pier to go with him and the crew saw him and they said, 'Quick, take in the gangplank and cast off!' for they didn't want to take him but they weren't quick enough and the minister came aboard.

It was notoriously unlucky for a boat to carry a clergyman of any kind aboard, and at first it seemed that the minister would earn his bad reputation, for as they came round to the west of Jura the weather got worse, the vessel was rolling, and the owners of the cattle became alarmed. They sent a delegation to the captain to ask him to take the eastern route round Jura, and insisted that they would take the responsibility if he came through Corryvreckan. Reluctantly he set out that way, but sure enough, 'she [the whirlpool] took charge of them'. The boat started to go down and the cattle all slid to one side, so that the deck was almost vertical.

> Now they had this minister aboard, the one they didn't want to take and it was this minister that managed to open the gangway and managed to get all of the cattle out and the minute the cattle went out it choked the whirlpool and the vessel righted and floated out. Those cattle bloated and they washed ashore for weeks after that.

Corryvreckan can be seen in full and fearful action in Powell and Pressburger's lyrical film *I Know Where I'm Going!* (1945), the nearest you can safely get to feeling what it would be like to cross the Scarba channel in a storm.

CRAIL, FIFE

A phenomenal skate

Travelling in Scotland in the early nineteenth century, the Reverend James Hall visited Crail and heard that the local fishermen had recently been much alarmed by a remarkable skate caught there – a *lusus naturae* (freak of nature), he says. When first landed the fish lay quiet, but when someone began to cut it up in preparation for the market, 'it leaped from the table, bit and wounded many of them, and the pieces they had cut off leaped from place to place into the street'.

Amazed and terrified, everyone ran away, but one elder of the Kirk, braver than the rest, dared to return, and the others followed cautiously.

> At length they collected the pieces, which, by being put together, seemed to collect new life; and, having provided a decent coffin, they buried the fish, though not in the church-yard, yet as near the church-yard-wall as possible. As it was enormously large, they all supposed that it had fed upon some human body at the bottom of the sea, and had, with the flesh, imbibed some part of the nature and feelings of man.

Skate can grow to immense sizes, weighing more than an average man. Even less monstrous specimens were held to be potent, their flesh widely recommended as an aphrodisiac, and in eighteenth-century Yorkshire the fertility of fishermen around Robin Hood's Bay was attributed to their eating dried skate, known as 'merry meat', while the liquid in which the fish had been boiled (skate broth, or 'skate-bree') was supposed to awaken the sexual instincts. The belief has persisted until recent times. 'Awa' an' sup skate bree!' is recorded in the early twentieth century as a taunt to a childless woman, and in 1960 an old Scottish fisherman said that he had given a dried skate to a newly married couple, to help them have a large family.

CROIG, MULL, ARGYLL AND BUTE

Sailors' yarns

It was long thought that a certain bird, the Barnacle or Bernicle Goose, was born from barnacles which were themselves spontaneously generated from rotten wood (*see* BANNOW BAY, Southern Eire). The legend was sometimes associated with Scotland, as appears from John Marston's *The Malcontent* (1604), Act III scene 1, where a character refers to 'your Scotch barnacle, now a block, instantly a worm, and presently a great goose'.

By the nineteenth century, such reports were no longer generally believed, but had passed into the sailor's repertoire of tall tales. A famous yarn-spinner of the 1860s or 1870s, mentioned by John Gregorson Campbell in *Superstitions of the Highlands & Islands of Scotland* (1900), was Calum of Croig, 'the delight of youngsters by his extraordinary tales of personal adventures and of wonders he had seen'. He liked to tell how in the Indian Ocean he and a companion had jumped overboard to swim to land:

> They swam for a week before reaching shore, but the water was so warm they felt no inconvenience. The loveliest music Calum ever heard was that made by Bernicle Geese as they emerged from barnacles that grew on the soles of his feet!

Sailors' yarns became a branch of folklore in themselves. A story often told in one form or another is of a mariner who is not believed on a rare occasion when he is telling the truth, as in Stanley Rogers' *Sea-Lore* (1929):

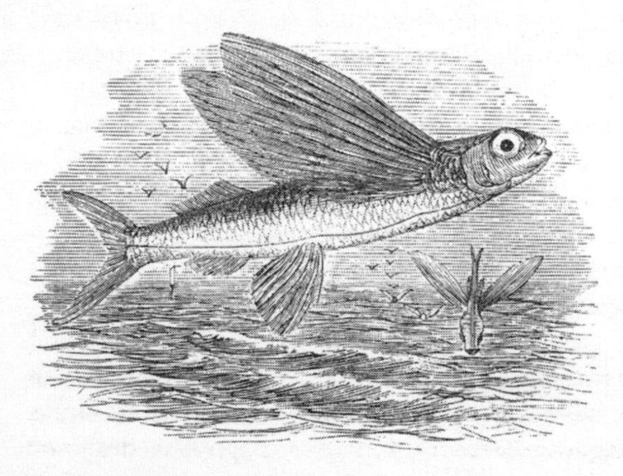

Flying fish really exist, but sailors who mentioned them were often accused of lying.

> Sailor Jack returns home to his mother and spins her a tale about the wonders he has seen, and, to make it good, he tells of the marvels of Jamaica, where there are rivers of rum and mountains of sugar, and of the Red Sea, where the anchor brought up a wheel of Pharaoh's chariot. Then, his imagination failing him for extravagant illustration, he told of other wonders, flying fish. Here his mother interrupted with, 'Nay, nay, Jack! I'll believe your mountains of sugar and rivers of rum, but flying fish? Never! That's too much for any old woman to swallow.'

This was familiar from over a century earlier, when the poet Southey wrote that 'I have somewhere seen an anecdote of a sailor's mother, who believed all the strange lies which he told her for his amusement, but never could be persuaded to believe there could be in existence such a thing as a flying fish.'

A similar tale is held to account for an old naval saying, when King Charles (unstated which, but more convincingly Charles II) was told by one of his sailors of fish that flew. The king, incredulous, turned to a marine officer and asked if he had ever seen such a thing. The officer confirmed what the sailor said, and the king commented that if he ever doubted anything again, he would first 'tell it to the marines'.

RIVER DEE, ABERDEENSHIRE

Patriotic mermaids

> Near the place where the famous DEE payeth his Tribut to the German Ocean, *if curious Observers of wonderfull things in Nature*, will be pleased thither to resort, the 1, 13, and 29 of May; and on diverse other days in the ensuing Summer; as also in the Harvest tyme, to the 7 and 14 October, *they will undoubtedly see a pretty Company of* MAR-MAIDS, *creatures of admirable beauty*, and likewise hear their charming sweet Melodious Voices
>
> 'In well-tun'd measures and harmonious Lay's,
> Extoll their Maker, and his Bounty Praise;
> That Godly, Honest Men, in every thing,
> In quiet peace may live, GOD SAVE THE KING.'

This notice, culled from a seventeenth-century almanac and reprinted in *Notes and Queries* in 1858, has caught the attention of many writers since. It is certainly an intriguing advertisement, perhaps a simple hoax designed to disappoint the credulous visitor, but the fact that the mermaids were meant to show themselves more than once – booked for a whole season, as it were – suggests that some enterprising Aberdeen showman had girls dressed as mermaids ready to make their entrance.

Robert Chambers writes in his *Book of Days* (1864) that 'The piety and loyalty of these predicted mermaids are certainly remarkable characteristics.' He seems to assume that the mermaids will actually sing 'God save the King', but the announcement probably includes the verse at the end as an illustrative quote, rather than as the sirens' lyrics. It is possible, however, that the extract inspired Robert Hawker, who really did conclude his celebrated impersonation of a mermaid at BUDE (South-West England & Channel Islands) with a rendition of the national anthem.

DUART CASTLE, MULL, ARGYLL AND BUTE

Herring live on foam

When herring are gutted, there is often very little evidence of food in their stomachs. Accordingly it was sometimes said that they lived on the foam they made themselves by thrashing their tails in the water, explaining the fact that they could not be caught with a baited hook. It became proverbial that 'A herring was never caught by its belly', and the line was used in several narratives as a warning that someone should beware of eating in certain company. In John Gregorson Campbell's *Superstitions of the Highlands & Islands of Scotland* (1900) is a story that the laird of Duart Castle, Mull, once laid a plot against M'Kinnon of Skye, and invited him to a banquet, intending to kill or capture him. When the unsuspecting M'Kinnon arrived, the doorkeeper, who knew of the plan and wished to save him, asked casually if they were getting any herring in the north at present, and then praised the herring as a royal fish 'that never was caught by its mouthful of food or drink'. The prospective victim took the hint and left at once.

DUNFERMLINE, FIFE

Sir Patrick Spens

> The King sits in Dunfermline town,
> Drinking the blude-red wine;
> 'O whare will I get a skeely [skilful] skipper,
> To sail this new ship of mine?'

O up and spake an eldern knight,
Sat at the King's right knee, –
'Sir Patrick Spens is the best sailor
That ever sail'd the sea.'

The king is Alexander III (1249–86). The ballad, included in Walter Scott's *Minstrelsy of the Scottish Border* (1801–3), is loosely based in history, dealing with an expedition to collect little Queen Margaret, 'the Maid of Norway', who inherited the crown of Scotland after her grandfather Alexander's death. It was planned to marry her to the eldest son of Edward I of England, but long before the match could take place she died in Orkney, en route from Norway to Scotland, aged only seven.

The messengers who went to fetch Margaret in 1290 did not include Sir Patrick Spens, and indeed were not dispatched by Alexander (who was dead by that time), but 'this is no conclusive argument against the truth of the tradition', writes Scott. He suggests that there might have been an earlier excursion, while the king was still alive, whose envoys failed even to bring the Maid away from Norway.

In the song, Sir Patrick foresees disaster as soon as he gets the king's orders to make a winter voyage across the North Sea. He and his men reach the Norwegian court safely, but are then accused of wasting money, and without Margaret they set out in a hurry on their return journey, although one of the crew warns that 'I fear a deadly storm':

'I saw the new moon, late yestreen,
Wi' the auld moon in her arm;
And, if we gang to sea, master,
I fear we'll come to harm.'

This refers to authentic sailors' lore. When the pale disc of the 'old moon' could be seen within the brighter crescent of the new, it was taken as an omen of foul weather, and in this case rightly so, since the ship has barely sailed three leagues when it is overtaken by the tempest. Sir Patrick climbs the topmast to see if he can spy land, but even as he does so the timbers are broached and the sea pours in. All aboard are drowned, and the ladies of Scotland are left to mourn their lost sons and lovers.

O forty miles off Aberdeen,
'Tis fifty fathoms deep,
And there lies gude Sir Patrick Spens,
Wi' the Scots lords at his feet.

EDINBURGH

A portent of whales

A terrible battle of whales took place in the sea near Edinburgh on 25 April 1707. The mighty contest was described in a contemporary Scottish broadsheet:

> At the foresaid time there was heard in the Firth a most Dreadful Noise of Sea Monsters, Roaring, plunging and threshing upon one another to the great Terror of those who heard the same. At length about 35 of them were run a Shore upon the Sands of *Kirkaldy*, where they made yet a more dreadful Roaring and Tossing.

Having remarked that 'what the unusual appearance of so great a number of them at this juncture may portend, shall not be our business to enquire into', the author cannot resist doing so, and connects the sea-fight with an event that took place a few days later on 1 May, the Act of Union between England and Scotland. Some people, he writes, suggest that the appearance of so many unusual creatures might be a bad sign, particularly the fact that so many of them beached themselves 'out of their own Eliment', a clear comment on the notion that either the Scots or the English, or both, might be figuratively out of their element in the treaty. The omen could, however, be interpreted patriotically:

> Others, will have the said Animals to be of two several Tribes the one *Scots* and the other *English*, who meeting together and contending for Seniority, came to decide the matter by Blows, the *Scots* Monsters having Surrounded those of *England*, drove them up the Firth, and brought them in Prisoners to *Kirkaldy* Sands.

One of the main bones of contention between the countries was the matter of imports and exports, Scottish trade with England having been threatened if the treaty was not complied with, and a further imaginative suggestion about the whales' behaviour was that in order to obtain 'an Ease of Duties', the whales seized the opportunity to run ashore before the new regulations came into force. The broadsheet writer concludes in verse:

> Some Days before the British Nations two
> Into the band of Wedlock Strick did go,
> In *Edinburgh* Firth a dreadful noise was heard,
> Of Scaled Armies that anone appear'd . . .

How dreadful Furious must the batle be,
Where Squadrons of great Leviathans Flee?

Unusual natural phenomena were widely interpreted in a political sense, or more simply as omens of momentous events. The enormous fish that was beached in 1167 in CARLINGFORD (Northern Eire & Northern Ireland) was thought to foretell an English invasion, and a whale that swam up the Thames as far as Greenwich in 1658 did so, it was said, to predict the death of Cromwell.

ELACHNAVE, ARGYLL AND BUTE

Fastitocalon and other monsters

Before he set out on his epic voyage across the ocean from BRANDON CREEK (Southern Eire), St Brendan spent some time in Scotland. Standing one day on a high cliff on Elachnave, one of the Garvellach islands, he saw two monsters emerge from the depths of the sea, fighting desperately together. One of the beasts, which seemed to be getting the worst of the fight, called with a human voice on the saints of Ireland. First it invoked St Patrick, then St Brendan, and finally St Brighid. When she was named, the attack stopped.

Brendan, marvelling at this, returned to Ireland and asked St Brighid why the monsters of the deep should respect her name more than his own. She replied that while he, Brendan, sometimes considered worldly matters, she never for a second allowed her attention to be diverted from God.

The 'monsters' may well have been whales, which were reasonably common around the British coast until at least the fourteenth century, while occasional battles between them were reported much later, for instance near EDINBURGH. St Brendan had another encounter with a whale on his great sea journey, when his companions landed on an island, as they thought, and made a fire to cook their dinner. When the fire was hot, however,

> then this island began to move, whereof the monks were afeard, and fled anon to ship and left the fire and meat behind them, and marvelled sore of the moving. And S. Brandon comforted them and said that it was a great fish named Jasconye, which laboureth night and day to put his tail in his mouth, but for greatness he may not.

Although vast, Jasconye was not hostile, and he helpfully gave the travellers a lift across the ocean. They remained on good terms with the creature and visited him every Easter, a hint at the Resurrection story prefigured in the biblical tale of Jonah, swallowed but later released by a whale.

Other fables were told of huge sea-beasts mistaken for islands, but the monsters were rarely so benevolent as Jasconye. The eleventh-century Exeter Book tells of Fastitocalon, who lies in the ocean like a great weed-covered bank. Sailors fasten their ships to him and are glad to rest and kindle a fire, but as soon as he feels them settled on his back, he suddenly dives down with his prey into the watery abyss, delivering men and ships to the hall of death.

The name 'Fastitocalon' is derived from a Greek word meaning 'tortoise shield', and the legend was originally told of a giant turtle, although it later became associated with the whale. The story belongs to a branch of medieval literature, the Bestiary, in which accounts of birds and beasts both real and fabulous were used as parables for the human condition. Such tales are found in many languages, and provided Christian art and literature with a colourful repertoire of animal symbolism. John Milton refers to the whale island in *Paradise Lost* (1667), comparing the fallen Satan to Leviathan, a name given in the Bible to the greatest beast of the sea:

St Brendan's companions alight on the back of a whale, and begin to prepare their dinner. Supposing themselves to be on an island, the monks are quite serene, but the whale looks uneasy.

> Him haply slumbering on the Norway foam
> The pilot of some small night-foundered skiff,
> Deeming some island, oft, as seamen tell,
> With fixèd anchor in his scaly rind
> Moors by his side under the lee, while night
> Invests the sea, and wishèd morn delays . . .

Milton's imagery combines legends of Fastitocalon and of the Norwegian Kraken (*see* BURRA, Scottish Highlands & Islands) with an allegorical tradition that used the whale to represent the Devil. One idea underlying that metaphor is a medieval belief that the whale had breath so fragrant that little fish could not resist it, and swam in to be eaten, as described in the Exeter Book: 'They enter there in a thoughtless throng, till the wide jaw is filled. Then suddenly the fierce jaws snap together, enclosing the plunder.' This notion may have been inspired by the sweet smell of ambergris, a waxy substance found in the intestine of the sperm whale and long valued as an ingredient of perfume, but whatever the explanation, the analogy is obvious. The whale represents the temptation offered by the Devil to lure the sinner to his doom, and thus its gaping mouth becomes an image of the entrance to hell.

EYEMOUTH, SCOTTISH BORDERS

'Is it salt ye want?'

Salt was once unmentionable among sailors and fishermen in some places. Illustrating both the taboo and its local nature is an anecdote of 1905, when an Eyemouth boat ran short of salt, and hailed a Yarmouth vessel. The Scottish captain called out, 'We need something that we dinna want tae speak aboot.' The English skipper, who evidently knew of the prohibition but did not share it, shouted back, 'Is it salt ye want?' and handed some over, but noticed that the crew of the Eyemouth boat had disappeared below, to prevent themselves hearing 'the terrible word'.

Whether the superstition was particular to Eyemouth or shared by Scottish fishermen in general is not clear from this story, but it was probably fairly widespread. The fatal significance of salt, to a seaman, must have stemmed from the idea that if you asked for it, you might soon get more than you wanted and meet a briny death.

FIRTH OF FORTH

St Mungo commands the tides

Where deep-sea tides enter shallower coastal areas, there is sometimes a 'double tide', when both the flow and the ebb are halted or reversed for a while before carrying on to reach the full high or low tide. Among the places where this happens is the Firth of Forth, between Culross and Alloa, and the feature, known locally as 'lakies' or 'leakie' tides, was explained in folk tradition as due to a miracle performed by St Mungo, patron saint of Glasgow. When the holy man was sailing up the river Forth to Stirling, his ship went aground on the ebb, but he caused the tide to return and refloat the vessel so that he and his companions could continue their journey, and ever since then there has been a double tide in this part of the Forth.

GIGHA, ARGYLL AND BUTE

The Great Well of the Winds

On the island of Gigha, between Islay and the Mull of Kintyre, a sacred well was carefully tended in the seventeenth century. A structure of stones and clay was built above it, since it was thought that if left uncovered the spring would flood the island, a belief that echoes ancient tales of overflowing wells such as that told at CARDIGAN BAY (Wales). Powers of healing were attributed to the water of Gigha's spring, and as reported in Martin Martin's *Description of the Western Isles of Scotland* (1703), it possessed another miraculous quality:

> That when any Foreign Boats are Wind-bound here (which often happens) the Master of the Boat ordinarily gives the Native that lets the Water run a piece of Money, and they say that immediately afterwards the Wind changes in favour of those that are thus detain'd by contrary Winds.

The 'Native that lets the Water run', in other words the local well-attendant, thus shared the ability of certain witches to change the wind, the gift in this case directly associated with a sacred spring. Another well with power over the weather was on INISHMURRAY (Northern Eire & Northern Ireland), but that had no attendant, the spell being workable by anyone who needed it.

There are a few ancient wells on Gigha, and Martin does not give specific directions, but websites suggest a location on the north-west side of Cnoc Largie that is sometimes known as the 'Great Well of the Winds'.

GREENOCK, INVERCLYDE

Captain Kidd

One of the late seventeenth century's most notorious pirates was William Kidd of Greenock. He entered folklore as a bloodthirsty murderer and robber, and tales of his terrible deeds were told among English and American sailors, the latter, for some reason, remembering him as 'Robert Kydd'. As the anti-hero of a popular song, he declares that:

> My name is Robert Kydd, God's Laws I did forbid
> and much wickedness I did, as I sailed.

The last verse has a moral: 'Take warning now from me and shun bad company,' the pirate concludes, 'Or you'll go to hell with me, as I sailed.'

In fact, although Kidd's fearsome reputation has clung to him for over three hundred years, he was less a fiend than an unlucky victim of poor advice, false friends, and an unruly crew, and you could say that it was indeed 'bad company' that was responsible for his downfall. In the 1690s he was a respectable merchant sea captain, with a family in New York, and was offered a commission by the American authorities to deal with piracy in the Indian Ocean. He set off in 1696 in his ship the *Adventure Galley*, but had no luck against the pirates. After a year in which he had made no captures, and therefore no money, his men were getting impatient, and some of them thought the best way of turning a penny would be to turn pirates themselves.

Kidd anchored near the entrance of the Red Sea, where pilgrim ships from Mecca were due to pass on their way to India, and according to *A General History of the Pyrates* (1724) encouraged his men with these words: 'We have been unsuccessful hitherto, but Courage, my Boys, we'll make our Fortunes out of this Fleet.' This promise was not quite fulfilled, but they did loot one vessel, and successfully fought off two Portuguese warships.

A little later, the *Adventure Galley* took the *Quedah Merchant*, a French ship under lease to the Indian government. This was Kidd's only substantial prize, but probably his greatest mistake, since the Indian emperor, enraged by the attack, threatened to expel all European traders – including the powerful East India Company, the worst enemies Kidd could have made. To keep the Indians sweet, Kidd was disowned by his patrons and declared a pirate, and a warrant was issued for his arrest.

In 1699, he was taken prisoner and sent to England, where he spent more than a year in Newgate prison before he was brought to trial in 1701. 'My Lord, I am innocentest of them all, but I have been sworn against by perjured persons,' he protested, arguing that the ships he had taken were French, and were therefore legal prizes since France was then at war with Britain. The French passes from the ships, however, had mysteriously vanished while in the possession of the Admiralty, and Kidd was found guilty. He was hanged at Execution Dock in Wapping, where his ghost is supposed still to walk.

What continued to fascinate later generations was what had become of Kidd's loot. It was generally thought that he had buried a vast fortune, and possible hideouts were ransacked all along the New York coast. A cache was supposedly discovered at the eastern end of Long Island in the nineteenth century, and in the 1930s interest in the case was reawakened when the missing French passes were finally unearthed in London's Public Records Office. Shortly afterwards Kidd's desk and chests 'turned up', writes Horace Beck in *Folklore and the Sea* (1973), although he does not say where they were found:

> Each and every one contained a secret compartment, and in each compartment was discovered an 'authentic' map of what was thought to be a treasure island. Unfortunately, the latitude and longitude of the island had been improperly worked, but the captain stated that it was located in the China Sea.

'The China Sea' or South China Sea, which stretches between southern China, Vietnam, Borneo and the Philippines, is a wide area containing a multitude of islands, and the whereabouts of Kidd's hidden wealth remains a mystery.

See also PIRATES AND PRIVATEERS (p. 232)

INCHKEITH, FIFE

Four-and-twenty mermaids

On Inchkeith 'in very olden times' there lived a hermit, who was the victim of local sea-nymphs, according to a rhyme given in an 1882 *Hand-book to Fife*:

> Four-and-twenty mermaids, who left the port of Leith,
> To tempt the fine auld hermit, who dwelt upon Inchkeith;
> No boat, nor waft, nor crayer, nor craft had they, nor oars nor sails;
> Their lily hands were oars enough, their tillers were their tails.

This seems to pre-date the famously filthy 'Ball of Kirriemuir', with its four-and-twenty virgins who went to Inverness, but did not return as they set out. A waft, incidentally, is a flag, and a crayer is a small sailing boat.

INVERARAY, ARGYLL AND BUTE

White stones

George Gomme, addressing the Folk-Lore Society in 1893, recorded a custom among the Inveraray fishermen of placing white pebbles on the graves of their friends, a tradition he traced back to the Stone Age. When a prehistoric cairn at Achnacrie had been excavated, 'the first thing that struck the eye' on entering the innermost chamber was a row of quartz pebbles, arranged on a ledge above the burial slab. At Burghead, likewise, smooth white pebbles were found on graves, although these must be of much later date as the stones are arranged to form crosses, and several more examples have been recorded of white stones found on or in graves of both the Christian era and pre-Christian times.

White stones or pebbles seem to have had a wide though enigmatic significance. In Orkney and Shetland it was said that they symbolised white water, the foaming waves seen in storms, and were for that reason unlucky. In the Isle of Man too they were to be avoided, particularly as part of a boat's ballast. There is nothing to say when this taboo originated, but the folklorist John Rhys, writing in 1901, suggested that it might be connected with the white pebbles on graves, the stones considered ominous from their association with death. The son of a clergyman went out a few times with the Manx fishermen, who caught little or nothing on each occasion, and considered the boy to be the cause of their ill fortune. They were probably prejudiced against his father more than against him, ministers being generally unpopular passengers on a boat, but the interesting point here is that because of his reputation as a jinx the boy was nicknamed *Clagh Vane*, 'white stone'.

In Ireland, too, boatmen would carefully remove white stones from their ballast. In *The Islands of Ireland* (1936) Thomas Mason records being told that the sailors of Clare Island 'raised many objections' when a new priest asked them to carry some white pebbles from the mainland to decorate his churchyard:

> 'They were too heavy' – 'The tide was against them' – and the matter was finally and definitely closed by 'Them's prohibited.'

In some contexts, however, a white stone was preferred. In the mid nineteenth century, at Macduff and Rosehearty, two fishing communities on the north coast of Aberdeenshire, witches or 'wise women' used to throw stones into the sea as a protective charm, and white stones were regarded as the best for this purpose.

IONA, ARGYLL AND BUTE

St Columba and the whale

Exploring Scotland in 1773, Dr Johnson famously remarked that the man was little to be envied 'whose piety would not grow warmer among the ruins of Iona', and his further thoughts on the occasion, as reported by Boswell, are a fine example of the sort of imaginative response to history *in situ* that inspires local legends of heroes and saints:

> Whatever withdraws us from the power of our senses, whatever makes the past, the distant, or the future, predominate over the present, advances us in the dignity of thinking beings. Far from me, and from my friends, be such frigid philosophy as may conduct us indifferent and unmoved over any ground which has been dignified by wisdom, bravery, or virtue.

When Johnson visited, the medieval abbey and its cloisters had been neglected for centuries. Now they have been restored, as has the twelfth-century chapel of St Oran, the oldest building on the island – but ruinous or not, Iona still casts a spell. A view from the central hill at evening, across the monastery and the silver sea beyond, evokes visions of 'the past, the distant'.

St Columba or Colm Cillé founded the religious community here in 563, and it became the heart of a religious network that stretched from Ireland to the Western Isles of Scotland. Columba's mission was just one, though one of the most notable, of many voyages by Celtic seafarers. The sea was spiritually significant to the Celts, from pagan beliefs in an Otherworld land of the dead across the western ocean (*see* THE BULL, Southern Eire), and in Christian times those seeking the 'white martyrdom' of exile from their homelands (as opposed to the 'red martyrdom' of death) often took up residence on small islands. The word *cill* (a monk's or hermit's cell), which appears as part of place-names throughout the Hebrides, may perhaps, as suggested by John Marsden in his excellent book *Sea-Road of the Saints* (1995), have been an early equivalent of a

These spouting, tusked, and spiny beasts are supposed to be whales, but were probably drawn by somebody who had never seen one. The creature that met a monk's ship travelling from Iona to Tiree sounds similar, its mouth 'bristling with teeth'.

bed-and-breakfast sign, indicating to travelling monks where they could find food and lodging.

Those so intimate with the sea as the Celtic saints could, it was thought, control it. Columba's mastery was reputed to include the ability to raise storms and still them, and other holy men who visited him on Iona were often assisted on their way by his conducting of the winds. According to Adamnan's seventh-century life of Columba, he could even provide consecutive breezes in opposite directions, to oblige two monks who wanted to travel north and south respectively.

He was also able to advise another traveller, about to cross from Iona to Tiree, to take a roundabout route in order to avoid meeting a 'huge monster'. The monk foolishly ignored Columba's advice, and in mid-channel, 'he and the sailors who were with him looked out, and lo, a whale, of huge and amazing size, raised itself like a mountain, and as it floated on the surface, it opened its mouth, which, as it gaped, was bristling with teeth.' The crew were terrified, and nearly swamped by the waves. Next morning another holy man, Baithene, made the same journey, receiving the same warning from Columba: 'Last night, at midnight, a great whale rose from the depths of the sea, and it will float this day on the surface of the ocean between the Iouan and Ethican islands [Iona and Tiree].' Baithene was undisturbed, answering that the beast and he were both under the power of God, and when he and his companions met the vast creature, he calmly blessed it. The whale dived under the waves and vanished from sight.

It has been suggested by naturalists that the ancient mariners' frequent encounters with whales were no accident. Whales (and dolphins too) use echo-location, projecting and receiving patterns of sound through the water in order to find food and society and escape danger. Timber or metal vessels would produce alien echoes, and whales would therefore tend to avoid them, but the boats used by the early navigators were hide-covered curachs, and it is thought that their reverberations might be similar to those of another whale. The beasts might therefore actively seek out the craft, under the impression that they were going to meet one of their own kind. The same idea occurred to Tim Severin, whose hazardous voyage in imitation of St Brendan is described at BRANDON CREEK (Southern Eire). His boat *Brendan*, built in the manner of a traditional Irish craft with a wooden frame covered in oxhide, was visited day after day by whales, which would stay alongside for half an hour or more at a time, although as soon as they entered areas of general shipping they would dive and disappear.

'There was no doubt,' writes Severin, 'that a leather boat, becalmed on these northern waters, held some sort of attraction for whales. It was not an exaggeration to say that it drew them from the depths.' If this was so in the twentieth century, when whales had become scarce, it must have been extraordinarily noticeable in the sixth and seventh centuries, and it is hardly surprising that the lives of the ancient seafaring saints feature so many meetings with 'sea-monsters'.

IRVINE, NORTH AYRSHIRE

Witches and the winds

'I'll give thee a wind,' promises the Second Witch in Act I scene 3 of *Macbeth* (1606). 'And I another,' responds the Third Witch. The First Witch then boasts:

> I myself have all the other,
> And the very ports they blow,
> All the quarters that they know
> I' th' shipman's card.

At the beginning of the seventeenth century, witches were still being condemned to death in Scotland for raising storms. In 1618, when Margaret Barclay of Irvine had been falsely accused by her brother-in-law of theft, she cursed his ship, on which the provost of Irvine was a passenger, praying that the sea might not bear it up and that crabs might devour all on board. The ship sank, and news of the disaster was brought to Irvine by John Stewart, a vagabond who was arrested on the grounds that he was a soothsayer. To get himself off the hook, he testified that one night he had gone to Margaret Barclay's house and found her and other women making clay images of the provost and the ship in order to work their spells on them. While they did so, he said, the Devil appeared in the shape of a little black dog, and went down to the sea with the women. There the clay figures were thrown into the water, which turned red, and a storm arose. On this evidence Barclay was tried, put to the torture and admitted her guilt. Although she later declared that she had confessed only 'in agony of torture', she was executed for witchcraft and her remains were burned.

That it is within human power to control the weather is a very old idea, but concepts about the source of that power changed with time. In the sixth century, Columba's ability to direct the winds (*see* IONA) was adjudged a gift from God, but around a thousand years later, in 1555, the Swedish historian Olaus Magnus ascribed the remarkable skills of Eric, king of Sweden, to devilish influence. He was 'so familiar with the Evil Spirits, which he exceedingly adored, that which way soever he turned his Cap, the Wind would presently blow that way' – for which reason he was known as Windy Cap. People from the far north of Scandinavia were generally talented in this way, according to Thomas Fuller in *The Profane State* (1648):

> There be many Witches at this day in Lapland, who sell winds to Mariners for money, (and must they not needs go whom the devil drives?)

Belief in the special wind-raising powers of the Lapps and Finns persisted into the nineteenth century and even later, as reported at EDAY (Scottish Highlands & Islands).

ISLAY, ARGYLL AND BUTE

Attack by a banshee

Macphie, laird of COLONSAY, was once hunting on Islay and went to rest in a hut, but his sleep was disturbed when a dark thing or banshee came to the door, was attacked by the laird's dog and set up a hideous screaming. The dog's hair began to smoke, and seeing this, Macphie ran down to his boat. Soon the dog came running after him like a beast possessed, with green flames streaming from its jaws, and leapt to savage its master. Macphie had had the forethought to load his double-barrelled gun with crooked silver sixpences (silver being the only metal thought capable of destroying a demon), and fired on the dog, killing it. The banshee had been terribly wounded by the dog's teeth, and dragged herself to the shore, leaving a track between the rocks, which, John Gregorson Campbell reported in the late nineteenth century, was still known in his day as *Sgrìob na Caillich*, the Furrow of the Cailleach. ('Cailleach' is used as a general term for a hag, and is also the name given to a specific and very powerful witch-goddess of the Highlands.)

Arriving at the edge of the sea, the Cailleach threw a ball of thread after Macphie's departing vessel, a time-honoured spell to catch a boat. In the eighth-century Irish saga *The Voyage of Bran* (*see* WONDER VOYAGES AND LOST LANDS, p. 458), when the hero arrives at the Land of Women and is reluctant to get too near the shore, the chief of the women throws a ball of thread:

> Bran put his hand on the ball, which clave to his palm. The thread of the ball was in the woman's hand, and she pulled the coracle towards the port.

A very similar charm features in the story of RATHLIN O'BIRNE ISLAND (Northern Eire & Northern Ireland). As used on Islay by the Cailleach, however, it failed for once, and Macphie escaped with his life.

KNOCKDOLIAN, SOUTH AYRSHIRE
A mermaid kills a baby

A legend first told in 1810, set at the mouth of the Dalbeattie Burn in Kirkcudbright, told of a woman who destroyed a mermaid's favourite seat, a boulder in the stream, and later found her own baby dead in its cradle. By 1870, the tale had been transferred to Knockdolian in Ayrshire, a change that seems to have been influenced by an interesting account of a curse given in James Paterson's *History of the Counties of Ayr and Wigton* (1863–6). In the late seventeenth century, according to this work, the head of the McCubbin family, who lived at Knockdolian, refused help to the Presbyterian preacher Alexander Peden, who was trying to escape from government persecution. Peden is said to have cursed McCubbin, declaring that Knockdolian would not stay in his family's possession, and soon afterwards both the laird's sons were killed, one drowned in Ballantrae Bay, the other falling from a tree. The property then passed by Fergus's daughter Margaret to the Cathcarts.

More than one family had already died out while in possession of Knockdolian – first the Grahames, at the beginning of the seventeenth century, then the Kirkmichaels, and the final extinction of the McCubbins (in terms of male heirs) seems to have inspired the rumour that a curse rested on the house. In the 1870 edition of Robert Chambers's *Popular Rhymes of Scotland*, the Dalbeattie tale was relocated at Knockdolian, and the mermaid was said to have declared, after her stone seat was broken up, that 'there'll never be an heir to Knockdolian again'.

The story was still current in 1962, when a version was recorded from Alexander Archibald, a fifty-year-old policeman who had lived most of his life in Ayrshire. In the Waters of Moyle, the sea between Scotland and Ireland, there were once many mermaids, said Archibald. One in particular was most beautiful, with 'long yellow hair, fish tail, and all that sort of thing', and every night she used to swim up the river Stinchar to a pool next to Knockdolian, then owned by the Grahame family. In the pool was a large boulder known as the False Craig, because Scottish sailors steering home from Ireland sometimes used to mistake it for Ailsa Craig in the Firth of Clyde and were wrecked as a result. On this rock the mermaid loved to sit and sing, but a woman of the Grahames, who was trying to get her baby to sleep, became fed up with the noise and told her servants to smash up the boulder. They did so, and next night they heard the mermaid sing these words:

> Ye may keep your cradle, while I'll ne'er have my stane,
> But there'll never be an heir to Knockdolian again.

The Grahame baby died, 'under some queer, mysterious circumstance', and never again was there a direct male heir in the family.

'Now I'm telling you a fact,' Archibald concluded. 'Because I got this from old Jimmie Hannah, the skipper of the sloop Annabella . . . say forty-five years ago. Jimmy was then, say, eighty. His father would tell him it. So in his lifetime, and in mine, there never was a male heir. Never.'

LADY'S ROCK, ARGYLL AND BUTE

MacLean of Duart's wife

In the Firth of Lorn, between Lismore and Mull, is a reef known as the Lady's Rock or Lady Isle, once a significant danger to shipping, and now marked with a beacon. In the mid twentieth century, Horace Beck heard a local legend accounting for the name of the rock. Long ago, the Duke of

Argyll's daughter married MacLean of Duart, but when he found that she was barren he marooned her on the crag, assuming that the tide would drown her, and then wrote to the Duke of Argyll, giving him the sad news that his daughter had met with an accident while fishing and died. The duke replied, asking MacLean to come to dinner and give him the details. When MacLean arrived, who should be sitting next to the duke but MacLean's wife, who had been rescued from the rock by some fishermen.

'Apparently the girl's father didn't care much for her,' continued Beck's informant, 'for he didn't do anything about it at the time,' and she returned to Duart Castle, where 'things seemed to go on much as before' until ten years later, when the wife's brother finally avenged her by murdering MacLean. Lady Rock, the storyteller added, is actually always above water, and nobody could drown on it. He thought the name must have been transferred from another smaller rock nearby.

For the far longer and more complex saga of 'MacLean of Duart and the Spanish Princess', from which this tale derives, *see* TOBERMORY.

LISMORE, ARGYLL AND BUTE

The severed hand

The Campbells and the MacDonalds were old rivals, and one day, according to a tale repeated to the folklore collector Horace Beck in the 1960s, there was a race between the two clans to see who should take possession of the island of Lismore by landing on it first.

> Of course the Duke of Argyll, tough old codger that he was, he was settin' in the bow of the boat – eggin' his men on, eggin' his men on and he was ready to jump from the bow of the boat and he seen he was goin' to get beat . . .

With an axe, the duke cut his own hand off and threw it on to the shores of the island, and so Lismore became his own.

Of course this tradition is not remotely historical, and perhaps it hardly matters which Duke of Argyll might be meant, but one could guess at Archibald Campbell, the first duke (d. 1703), traditionally blamed for the Glencoe Massacre of 1692 when thirty-eight members of the MacDonald clan were butchered by Argyll's troops.

The tale of the severed hand appears in many different contexts, most notably to account for the Red Hand used as a heraldic symbol. It was originally the badge of the O'Neils of Ulster, of whom it was said that their ancestor had gained their territory by cutting off his hand and throwing it on to the coast. The legend was also specifically attached to Lismore in another version, which tells how St Moluag and St Columba were racing for the island, hoping to found their missions there, and St Moluag won when he severed his little finger and cast it ahead of him to shore. All parts of a person's body were felt to be symbolically indivisible, whether or not they actually remained attached, so that even something as small as a finger could represent the whole and be used to make a claim to a piece of land.

Certainly, however, the hand or finger would have been carefully retrieved and preserved. When there was widespread belief in bodily resurrection at the Last Judgement, it was vital that no corpse was buried incomplete, a belief underlying stories of ghosts searching for their lost bones, and accounting for the Spanish king's anger when he found that his daughter's body had been returned minus one of her toes (*see* TOBERMORY).

LITTLE ROSS, DUMFRIES AND GALLOWAY

Janet Richardson

Off the headland known as Big Ross or Meikle Ross is a small island, Little or Wee Ross, and between the two is a rock called Janet Richardson. At least, that was how it was known in the early nineteenth century, named after a woman who went there at low tide to gather mussels and was overtaken by the waves. She drew up her skirts and plunged in, making for the shore, but was swept off her feet and eventually came safe but exhausted to land, according to John MacTaggart's *Scottish Gallovidian Encyclopedia* (1824), at 'the Milton Lands'. If he means Milton in Luce Bay, more than thirty miles away from Little Ross by sea, it is not surprising that the journey passed into local legend.

The rock is still sometimes known as Richardson's Rock, so that Janet's adventure has its memorial yet.

LOCHAR MOSS, DUMFRIES AND GALLOWAY

Land drowned by a witch

In the sixteenth century, the young Prince James (later James V of Scotland) was entertained with stories of the Gyre Carline, a celebrated witch whose fame lasted into the nineteenth century. She used to lead her coven in wild gallops known as the Hallowmass Rades (Rides), and a rhyme supposedly sung by her followers when they gathered was preserved in local tradition:

The Naming of Ships

In February 1885, it was reported in the *Telegraph* that the ship *Ianthe* had sunk, and not for the first time. She had been launched as the *Daphne*, and under that name had capsized, drowning 124 workmen:

She sank in the Clyde as the *Daphne*; she was raised, and then sank in Portrush harbour as the *Rose*; she was raised again, and, still as the *Rose*, she ran ashore on Big Cumbrae. Then she was got off and lost sight of for a little, and now reappears as the *Ianthe*, comfortably lodged on the mud which she seems to love so well . . .

'She is evidently an unlucky ship,' the writer concluded. Her changes of name may have been intended to deflect ill fortune, since re-christening a ship was sometimes a desperate attempt to change her luck for the better.

It is much more often said, however, that to change a ship's name is to invite disaster. In Robert Louis Stevenson's *Treasure Island* (1881),

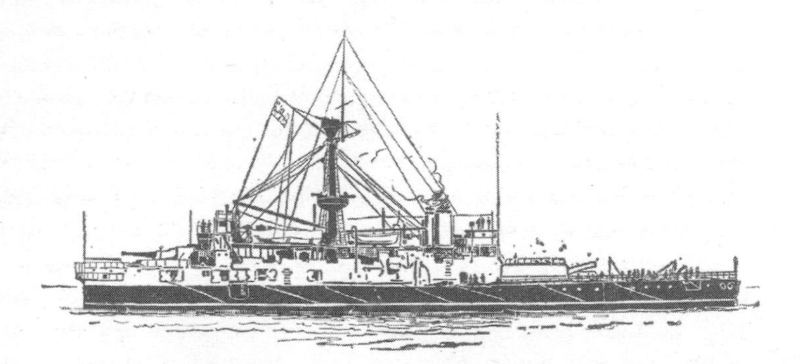

HMS Victoria *sank in 1893, after an ill-advised manoeuvre ordered by her captain, Sir George Tryon. It was sometimes said that her name, ending in an 'a', had made her an unlucky vessel.*

Long John Silver attributes a man's death to changing the name of his ship: 'Now, what a ship was christened, so let her stay, I says.' The same opinion was given in 1980 by a captain's wife, who said, 'When we buy a boat we never change her name – it's very unlucky.'

In the 1930s, a Folkestone fisherman told the story of a wrecked boat on the beach. 'It were like this,' he said:

> 'Bill's first wife were named Bertha, and this 'ere boat were named *Bertha* – and quite right, too. But she died last year, and Bill married agin – quite right, too – but damned if 'e don't go and re-christen the boat *Beatrice*, 'cause that were the name of 'is fresh wife. It's all wrong – agin Providence – for ain't it true that you must never rename a boat or a ship?'

Bill's boat had been smashed, and what was more, his new wife was 'something tragic, a she-devil'. Get a new boat by all means, said the fisherman, but never change the name of an old one.

Some names were unlucky in themselves. Ships called after 'things that sting' were thought likely to sink: the navy has lost five *Serpents* and three *Vipers*, a *Cobra* and a *Rattlesnake*. Words ending in 'a' were also ill-omened, a superstition that probably began after the torpedoing of the *Lusitania* in 1915, although the *Victoria*, which sank in 1893 (*see* WEYMOUTH, South-West England & Channel Islands), is also cited.

There is occasionally controversy about gender. In France, where ships are masculine, the *Normandie* was firmly called '*Le*' *Normandie* by dockyard workers, although the French Academy insisted that it should be '*La*', since '*Normandie*' is a feminine word. In Germany there have been similar arguments. English-speakers, on the other hand, seem happy to call even the *Royal George*, for example, 'she'. Almost uniquely among inanimate objects, ships retain a sex in standard English, although railway engines and cars sometimes get the same treatment, probably by extension, since these vehicles too may be invested with a lot of personality.

When the gray Howlet has three times hoo'd,
When the grinning cat has three times mewed,
When the Tod has yowled three times i' the wode,
At the red moon cowering ahint the clud;

When the stars hae cruppen deep i' the drift,
Lest cantrips had pyked them out o' the lift,
Up Horsies a', but mair adowe.
Ryde, ryde, for Locher-brigg knowe!

An English version would run roughly:

When the grey owl has three times hooted,
When the grinning cat has three times mewed,
When the fox has yowled three times in the wood,
At the red moon cowering behind the cloud;

When the stars have crept deep in the mist,
Lest spells had picked them out of the air,
Up horses all, without more ado,
Ride, ride, for Locherbrigg hill!

The stretch of ground between the Solway Firth and Locharbriggs (as it is now spelt) is still known as Lochar Moss, although it was drained in the nineteenth century and is no longer a 'moss' or peat bog. Legend relates that this is not the first change it has undergone. It was once open sea, providing a good harbour for shipping, but was converted to a quagmire by the Gyre Carline after one of her rides was disrupted by a high tide that swept away several horses. In fury, she stretched out her magic staff and had her revenge on the sea, turning it to marshland.

The notion that this patch used to be entirely below the sea is supported by the fact that in the early nineteenth century, remains of boats were sometimes dug up from the bog by peat-gatherers. An old rhyme said of Lochar Moss:

First a wudd, and syne a sea; [First a wood, and then a sea]
Now a moss, and aye will be.

Quoting this in 1826, Robert Chambers remarks on the fact that 'the modern naturalist' accounts for the production of peat bogs in the same way, by water rising to flood a wooded area, then sinking again to leave it waterlogged but above sea level. The accuracy of the verse did not, of course, go so far as to predict the draining of the marsh.

LOWER LARGO, FIFE

Alexander Selkirk and Robinson Crusoe

Daniel Defoe's *Robinson Crusoe* (1719) was inspired very largely by the life and surprising adventures of Alexander Selkirk or Selcraig, born in Lower Largo in 1676. He was remembered as a youth of high spirit and uncontrollable temper, traits which helped lead to his career as the king of all castaways.

Selkirk was, like a true fairy-tale hero, the seventh son of his family. At the age of nineteen he ran away to sea, and a few years later joined the explorer, navigator and buccaneer William Dampier on an expedition to the South Seas. Arriving in 1704 at Más a Tierra, the largest of the Juan Fernández islands, some four hundred miles west of Chile, Selkirk fell out with his captain, Thomas Stradling, and asked to be put ashore. Before the ship left he had changed his mind and asked to be taken back on board, but Stradling refused, and Selkirk was marooned on the island, which was uninhabited except for a colony of wild goats. Here he remained alone for the next four years.

For the first few months he lived a melancholy life, terrified by his surroundings and so forlorn that he was tempted to suicide, but as time went on he became accustomed to his solitude, built himself two huts (one for a kitchen, the other for sleeping), and learned to kill and skin goats. He was relatively well supplied, having a gun, shot and powder, some tobacco, a knife, a kettle, a Bible, and some other books. What else he needed the island could provide: fresh water, fruit, turtles and crayfish. When his clothes fell to pieces, he made new ones of goatskin, sewing them with a piece of goat sinew threaded through a nail, and when his knife wore out, he beat and ground a piece of iron on the rocks until it was a passable blade.

In 1709, Captain Woodes Rogers put in at Más a Tierra and found, as he wrote himself, 'a Man clothed in Goat-Skins, who look'd wilder than the first Owners of them'. Rogers took Selkirk back to Britain, where he returned briefly to Largo, but he had become so unaccustomed to human company that he spent much of his time in a hillside cave. Richard Steele, who met him at this time, recorded that 'the Man frequently bewail'd his return to the World, which could not, he said, with all its Enjoyments, restore him to the Tranquillity of his Solitude.'

Robinson Crusoe cast up on his desert island, in an illustration by George Cruikshank from 1890.

A romance followed with a young girl named Sophia Bruce whom Selkirk saw tending her parents' cow, and after the couple eloped to London, Selkirk never came back to his home town. He left behind his sea chest and a cup he made on his island, which are preserved in the Museum of Scotland, Edinburgh. Sophia died not long afterwards, and although Selkirk married again (his widow, Frances Candia, turned up a few years later to claim his estate), he seems soon to have returned to sea, and died in 1721 while acting as a lieutenant of the Royal Navy.

In the meantime, his experiences had become the subject of several published accounts, but the great work was *Robinson Crusoe*. It is possible that Defoe interviewed Selkirk in Bristol, but even if they never met, the initial inspiration for Defoe's novel came from Selkirk's life, although he embroidered it considerably. Crusoe, shipwrecked rather than marooned, tells us how he survives with the help of a few tools and stores saved from the wreck, builds himself a house, domesticates the goats and constructs a boat, all alone until he meets a friendly native whom he calls Friday. The pair are finally rescued, not just four but twenty-eight years after Crusoe's wreck.

Although the book was rejected by several publishers, when it appeared it was a smash hit, with four editions appearing in as many months. The spell of the masterpiece lies in its matter-of-fact narration of extraordinary events. To generations of children, Crusoe's island has remained immortal – not a paradise, but a far more interesting place where real problems must be encountered and overcome by unaided human courage and resourcefulness.

MACDUFF, ABERDEENSHIRE

See INVERARAY; ROSEHEARTY.

MORAY FIRTH

Herring dislike quarrels

The dangerous dependency of fishing communities on the HERRING (p. 130) is reflected in local tales and traditions. Herring shoals have long been known to be fickle. 'The herring is one of the most eccentric little fishes that frequents our seas,' wrote Hugh Miller in 1835. 'For many years together it visits regularly in its season some particular firth or bay; – fishing villages spring up on the shores, harbours are built for the reception of vessels.' At length, however, the season comes but not the herring, 'and, ere the shoal returns to its accustomed haunts, the harbour has become a ruin; and the village a heap of green mounds.'

This happened in the Moray Firth, Miller tells us, in the early eighteenth century. Although the previous season had been busy and successful, one year no herring came, and for more than fifty years afterwards hardly a fish was taken. Two stories survived in his day to account for this sudden and inexplicable departure. According to the first, huge quantities of herring had been caught and brought ashore on a Saturday, and the packing continued during the night. By morning, much remained to be done, so, as the weather was sultry and they did not want the fish to go off, the packers worked through the Sabbath day. Towards evening, the minister came to preach to them. The packers soon became impatient, some going back to work, others pelting him with filth. The minister abruptly concluded by praying that 'the besom of judgement' would come and sweep every herring out of the firth. On the following Monday, the boats went to sea as usual, but returned empty. Season followed season, but still no herring were caught. None of the packers who had insulted the minister lived to witness the return of the shoal.

The other story was that on the last day of the successful fishing, two nets became tangled together, and one boat's crew took a knife to their neighbours' net. A fight followed, with the two sets of men leaning over their gunwales to attack each other. 'Blood was spilt, unfortunately spilt in the sea; the affronted herrings took their departure.' Miller says that one of the men who took part in the affray was known ever after by the nickname 'Bloody', and there were men still living in Miller's own time who remembered him.

The tradition that herring hated human blood and human quarrels was reported earlier by Martin Martin in his *Description of the Western Isles of Scotland* (1703):

> It is a General Observation all *Scotland* over, that if a Quarrel happen on the Coast where Herring is caught, and that Blood be drawn violently, then the Herring go away from the Coast without returning, during that Season.

Today we prefer to attribute the sudden departure of herring – from Great Yarmouth in Norfolk, for example, in 1951 – to global warming or pollution, but the same sense of guilt is there: we have only ourselves to blame.

NORTH BERWICK, EAST LOTHIAN

Drowning a cat to raise a wind

On a journey from Denmark in 1590, James VI of Scotland encountered a storm that he believed was raised by witchcraft, and around three hundred men and women of North Berwick were subsequently charged with seeking to destroy their king.

One of the accused, Agnes Sampson, confessed that she and her coven had travelled to meet the Devil in North Berwick kirk, taking their sea voyage in sieves, traditionally witches' favourite boats (since only by magic could anyone sail in a sieve). As reported in the contemporary pamphlet *Newes from Scotland,* they made a party of it, 'with flaggons of wine, making merrie and drinking by the way', and dancing to the music of a Jew's harp after they landed.

Under Satan's instruction, the witches had thrown a cat into the sea, a method of conjuring the weather that remained in popular report, if not necessarily in practice, for hundreds of years. Henry Fielding writes in *The Journal of a Voyage to Lisbon* (1755) that sailors on his ship asserted 'that the drowning a cat was the very surest way of raising a favourable wind', and in *Notes and Queries* (1854) it is stated likewise that 'An infallible recipe for raising a storm is to throw a cat overboard.'

Agnes Sampson, whose confession was extracted under torture, was burned to death, along with many others. Protests were made by a few, including Reginald Scot, whose *Discoverie of Witchcraft* (1584) was a diatribe against 'the tyrannicall crueltie of witchmongers and inquisitors', and the absurdity of supposing that anyone could command a storm, 'when she being but an old doting woman, casteth a flint stone over hir left shoulder, towards the west', or by any other traditional technique. Confessions were not to be relied on, he insisted, being the result of 'melancholie' and 'merveilous imaginations' on the part of lonely old women. Such voices of reason spoke in vain, however, against the prevalent fear of witches as the Devil's instruments, and the Scottish witch-hunts continued for over a century, involving nearly four thousand trials and around two thousand executions.

See also IRVINE.

PETERHEAD, ABERDEENSHIRE

Prophetic dreams

Jimmy Buchan was a sailor of the east coast, renowned for strength and courage. He ascribed the beginning of his good fortune to a dream he once had, which he remembered vividly many years later in the 1960s.

He started by explaining that in his part of Scotland nobody ever sets sail on a Sunday (*see* PRESTONPANS), and so he was waiting for midnight in order for his journey to start on Monday. He went to sleep and dreamed that his boat was already steaming down the river towards the sea, with the dawn ahead of him. Glancing aft, he saw his deceased sister-in-law pursuing him in a little rowing boat. For eight hours, he dreamed, he went onwards towards the rising sun, while behind him the gap narrowed as the rower caught up. At last she came alongside, stood up in her skiff and offered him a slice of golden cake she had just baked. He reached for the cake, but it fell into the sea between the ship and the rowing boat. Another piece was cut and offered, and that too fell in the water. Then she held out the whole of the rest of the cake, and this time Buchan took it and bit into it, and found that it was the sweetest, most delicious cake he had ever eaten, 'and yellow as gold'.

Waking, Buchan found that it was time to set out. As soon as his ship was clear of the harbour, he told his helmsman to steer east at full throttle, and kept her on that tack for eight hours. After the time had passed he instructed the crew to stop the ship and cast the nets. This was done, but when the nets were pulled in they were empty. Buchan told his men not to be discouraged, and to cast the nets again. Again they came back empty, and this time he said, 'Now, lads, go below and I would have you eat hearty and have a wee dram, for we shall shoot again and there will be nae rest aboard this ship this day.'

The third time the nets were drawn in, they were full to bursting. The ship came back to Peterhead so heavily laden that it was lucky there was no strong wind, as any storm would have capsized her. ''Twas the greatest catch of fish ever to be landed in Peterhead and for mony the day after,' said Buchan.

Seafarers generally paid attention to dreams and premonitions, which were more often warnings of disaster than promises of good fortune. In *Legends and Superstitions of the County of Durham* (1886), William Brockie reports the case of a sailor prosecuted for deserting his ship, who told the magistrate that although he was satisfied with his officers, his food and his working conditions, 'he had had a dream that the ship would be lost, and would not go to sea in her for any amount of money'. Any sailor who had a similar dream, Brockie wrote, would much sooner face imprisonment than ignore the warning.

PORTESSIE, MORAY

Bloodshed

You should never wish a fisherman good luck (or an actor either). Where luck is all-important, it is not to be lightly named. At Portessie, it was said that the ill fortune thus invited could be averted by hitting the well-wisher so as to draw blood, and the same was done if anyone asked a sailor on his way to his boat the ill-omened question where he was going (since the true answer might be to the bottom of the sea).

The shedding of blood was a powerful counter-charm, often used against witches and also at sea. All along the south side of the Moray Firth, reported Peter Anson in *Fisher Folk-Lore* (1965), it was once the custom to mark the New Year by ceremonial bloodshed:

> If it was too stormy for the boats to put to sea on New Year's Day, men who owned guns went to the beach before dawn, waiting for the sight of any living creature they could wound or kill, so as to have proof of shedding blood.

Otherwise, the blood of the first fish caught would do.

All this is in contrast to the belief reported from the MORAY FIRTH that herring in particular would desert their usual grounds if the waters were stained with human blood, though what they abhorred was presumably blood spilt in anger, as opposed to the rituals mentioned here.

PORTGORDON, MORAY

A merman and mermaid

'As the existence of Mermaids is a point that has long been disputed,' wrote George McKenzie, the schoolmaster of Rathven, 'I send you the following account, which I received from two fishermen residing at Port Gordon.' His letter to the *Aberdeen Chronicle*, printed on 20 August 1814, noted that he believed the men in question, Thomas Johnston and William Gordon, to be truthful and respectable, and went on to give their experiences when returning from fishing on 15 August. It was broad daylight, and the weather was calm. When they were about a quarter of a mile off shore, they saw within a few yards of their boat 'a Creature of a tawney colour, appearing like a man sitting', its body half out of the water, and its back towards them until they came within a few yards of it, when the noise of their approach made it turn round.

> His countenance was swarthy; his hair short and curled, of a colour between a green and a grey; he had small eyes, a flat nose, his mouth was large, and his arms of an extraordinary length. Above the waist he was shaped like a man, but, as the water was clear, my informants could perceive that, from the waist downwards, his body tapered considerably; or, as they expressed it, like a large fish without scales, but they could not see the extremity.

After staring at them for a few seconds, the merman dived, and then came up again further off, this time accompanied by what the men took to be a female of the species, since she had breasts. Her hair was straight and a little more than shoulder-length. By this time the men were frightened, and made for shore as fast as possible, but they could see the two creatures watching them all the while.

As soon as they landed, they went to tell McKenzie, who took down their report, noting that they gave it 'without the smallest variation between them'. Like the mermaids seen at ABERYSTWYTH and LINNEY HEAD (both Wales) in 1826 and 1782 respectively, these creatures are convincingly and straightforwardly described, and one would give much to know what it was that the men really saw.

The Power of Seagulls

Seagulls were said to foretell storm and death, but sometimes they were thought able to save sailors from danger and bring them safely back to shore.

Creatures of air and water, seabirds were thought to be in tune with the weather, and forecasts have been made from their behaviour since ancient times. It was recorded in the first century BCE that when seagulls were unusually active and noisy, that was a sign of storm, and reports persist that gulls sense a rising wind and fly away from it, so that if many birds are seen on land, it means a tempest must be expected at sea. The late eighteenth-century *Statistical Account of Scotland* notes that seagulls were 'considered as ominous', flying to the fields before a gale, and returning to the shore when it was over. In Northern Ireland, a similar belief was recorded in the 1950s, and it is current today in coastal districts of the British Isles. 'The higher the gulls, the harder the gale,' goes an English saying.

In folklore, as in common language, the word 'seagull' is applied indiscriminately to various sorts of gull. It is the specific name for the Common Gull (*Larus canus*), probably the bird whose appearance inland figures in traditional weather signs, since it is less likely to be found far from the sea than, say, the Black-headed Gull (*Larus ridibundus*), which breeds and lives inland in great numbers.

By a natural association of ideas, the harbingers of storm were also regarded as a death omen. Three gulls wheeling overhead were sometimes taken to foretell death, and Breton fishermen believed that when a man died at sea, gulls and curlews would fly round his house and beat their wings against his windows. This is parallel to the landsman's belief that certain birds, notably white doves and robins, would appear at or tap on the window of a dying person.

Sometimes birds were not merely considered portents, but were imagined to influence events. In nineteenth-century Brittany, women would scatter flowers on the waves and ask the gulls to bring their loved ones safely home from the sea, a practice that may have been tantamount to asking for protection from the spirit world, since gulls were often supposed to embody the souls of dead sailors (*see* PORLOCK, South-West England & Channel Islands).

An unusual case was reported in 1857, when a strange grey bird flew on board a ship in mid-Atlantic and attacked the captain. 'I regarded the appearance of the bird as an omen,' he reported, and changed course accordingly, soon afterwards falling in with a wrecked steamer and rescuing forty-nine of her passengers and crew. Belief in the supernatural significance of seabirds continued into the twentieth century. Writing of his experiences in the First World War, Rear Admiral Gordon Campbell records that a member of crew told him that a submarine would be sighted next day:

On inquiring how he knew, he said that a bird had flown into my cabin, and although it had never struck me particularly before, a similar thing had happened on each occasion of engaging a submarine – and sure enough the omen came true.

See also MOTHER CAREY'S CHICKENS (p. 264).

PRESTONPANS, EAST LOTHIAN

Sailing on Sunday

Depending whose accounts you read, Sunday was either a lucky day to put to sea (being the Lord's day) or unlucky (because by sailing then you would break the Sabbath). Writing of his own experiences on a merchant ship in the early nineteenth century, the American author Richard Dana explains the pragmatic considerations behind the practice as he knew it:

> The main reason for sailing on the Sabbath is not, as many people suppose, because Sunday is thought a lucky day, but because it is a leisure day. During the six days, the crew are employed upon the cargo and other ship's works, and the Sabbath, being their only day of rest, whatever additional work can be thrown into Sunday, is so much gain to the owners.

Catholic nations, he continues, never started their journeys on a Sunday, 'but the American has no national religion, and likes to show his independence of priestcraft by doing as he chooses on the Lord's day' – not what the sailors themselves chose, but what would be of most benefit to their employers.

It seems that, like the Catholics Dana writes of, fishermen on the Presbyterian east coast of Scotland generally stayed ashore on the Sabbath, as mentioned above at PETERHEAD. In the mid nineteenth century, however, it was 'a favourite custom' among the fishermen of Prestonpans to start for the fishing grounds on Sunday. They seem to have been a pagan lot, according to a correspondent in *Notes and Queries* (1852):

> A clergyman of the town is said to pray against their sabbath-breaking; and to prevent any injury accruing from his prayers, the fishermen make a small image of rags, and burn it on the top of their chimneys.

If Sunday was a bad day to start a journey, Friday was generally worse, as reported at GREAT YARMOUTH (East Anglia).

RATTRAY HEAD, ABERDEENSHIRE

A sea-monster

In the early 1900s, the steam trawler *Craig-Gowan* was making her way round Rattray Head when her chief engineer, J. Watt, reported to Captain Ballard that a 'large animal' was following the vessel. Ballard, who subsequently wrote a report of the incident, joined the engineer at the rail, where he saw 'a very large animal of a dark colour, which seemed racing with us, but which was about fifty feet to windward'. Ballard had seen whales, and was sure that this was not a whale but 'some sea monster, the like of which I have never seen in my life'. As it rose in the water, several portions of its body could be seen at the same time, like a snake undulating through the sea. Its skin was brown and hairy.

> Securing the furnace rake to a stout line, I threw it at the animal, but it fell short. I again tried; this time the rake landed across the animal's back and we suddenly drew the line. The monster raised its body (the fore part) clean out of the water and made direct for the *Craig-Gowan*. I plainly saw the

A Punch *cartoon from 1890 shows a steersman horror-struck at seeing a monster rising from the waves. Some who have encountered sea-serpents, such as the engineer who reported a 'very large animal' off Rattray Head, have been foolhardy enough to attack them, but have never succeeded in a kill or capture.*

> monster rise up until its head was over our gaff peak, when it lowered itself with a motion as sudden as lightning, carrying away the peak halliards and sending the gaff, sail and all, down on deck. The utmost consternation ensued among the crew and it was a time before we got matters squared up. The animal had then entirely disappeared and we did not see it again.

Ballard was presumably trying to scare off the serpent – he cannot have expected to catch it with a rake – but as Peter Jeans comments in *Seafaring Lore and Legend* (2004), 'One is somewhat surprised at the boldness, not to say rashness, of those folk who take it into their heads to offer violence to what they universally describe as a large "monster" of unknown habits and preoccupations.'

ROSEHEARTY, ABERDEENSHIRE

Fishermen's superstitions

In an article written for the *Folk-Lore Journal* in 1885, the Reverend Walter Gregor listed a range of miscellaneous fishermen's traditions he had heard about on the north coast of Aberdeenshire, in the ports of Broadsea, Macduff, Rosehearty, St Combs and elsewhere.

Some waves or states of the sea were given names. The swell preceding a storm was 'the dog afore his maister' or 'the sheep afore the dog', and the roll when a storm had dropped was 'the dog ahin [behind] his maister'. In ordinary weather, the sound of the sea breaking on the beach was called 'the sang o' the sea', and if the Rosehearty fishermen heard this song coming from the west in the morning, that was regarded as a forecast of fine weather in which they could take their boats further out to sea than on other days. A particular mournful sound known as the 'knell' was made when a body lay drowned near the shore, or before a shipwreck.

To raise a breeze, sailors might softly whistle or scratch the mast with their fingernails, practices almost universal at sea, and mentioned for instance at FISHGUARD (Wales). Other beliefs were possibly unique to the area: stormy weather, it was said in Rosehearty, was brought on by marriages, and it was usual to marry after the end of the herring season, between September and January.

Eggs were not supposed to be carried on board a boat, for fear of raising unfavourable winds, a taboo possibly connected with the widely held belief that witches used eggshells as boats (which is why, when you eat a boiled egg, you should always crack the bottom of the shell to make it unseaworthy). One Rosehearty woman, Jessie Ritchie, fishing at Barra in the Western Isles, was forbidden to take any eggs into her boat, but smuggled a dozen of them home without any subsequent ill fortune.

It was thought that the luck of the fishing would disappear if anything, even the smallest article, were loaned to anyone during the herring season or whenever fishing was being done. Jessie Ritchie (who hid the eggs) was once cooking for a crew fishing on the west coast, and wanted to borrow a washing tub from the housekeeper of another crew. The woman was willing to lend it, but one of the crew, an old man from Broadsea near Fraserburgh, 'in most vigorous words forbade the loan'.

While the lines were being baited indoors, it was a matter of much importance who crossed the doorstep – the 'first fit' (foot). In St Combs, it was the custom to bolt the door at this time, and if anyone unlucky came knocking, they would immediately accept the explanation that baiting was in progress and would go away again. Gregor does not explain here who would be regarded as having 'an ill fit' (unlucky foot) except a red-haired person, traditionally looked on with disfavour in many places, but other probable candidates would be a clergyman or a woman – both notoriously ominous to sailors, as mentioned for instance at WESTON-SUPER-MARE (South-West England & Channel Islands). If someone with an 'ill fit' did come into the house while the fishing lines were being made ready, the end of the line would be drawn through the fire to purify it or remove the bad luck, and if a fishermen met an unlucky person on his way to sea, he lit a match and threw it after the ill-omened one.

When about to go to sea, fishermen much disliked to be asked where they were going. If the question was put, various evasive, misleading or impossible answers might be given: 'A'll speed the better it ye've speert' (I'll get on better now you've asked); 'T' the lånart' (To the landward); or 'Awa t' the back o' the moon.'

As at ST MONANS, the pig was neither to be eaten nor to be mentioned, and the word 'cat' too was banned:

> A man told me that one suggested to him when a boy to go to the door of a certain fisherman's house and call out 'Cat.' He did so. The fisherman was engaged in some work by the side of the fire. No sooner did he hear the word, than he seized the tongs, and threw it at the head of the offender.

Luckily it missed, or it might have killed him, he said.

Other unlucky animals at Broadsea were the hare, the rabbit, the salmon (called 'the beast wi' the scales' to avoid naming it directly) and the rat, but at St Combs the rat was regarded as fortunate, and the arrival of rats in a house was interpreted as meaning money to come. The idea that rats leaving a house or ship meant disaster was, as Gregor points out, quite general, and remains proverbial ('rats leaving a sinking ship').

A skilful midwife could tell at birth someone destined to be drowned from 'the water glance' on their brow, a slight hollow which, it was believed, would gradually fill up. When it was smooth, death came.

Common curses in Macduff and other towns were a wish for Norway to be the first landing of a boat, spoken when the craft was putting to sea (meaning that she would be blown far off course), and 'May yir boat be yir bonnet' (meaning she would turn upside down). A Rosehearty man told Gregor about a time many years earlier, when he was a young lad and he and some friends borrowed a boat for a joyride. The owner spotted them pushing off, hurried down to the shore and shouted to them to come back. They paid no attention. He continued to shout until one of the boys cried out, 'May she be yir bonnet the neest time ye gyang oot wi' ir' ['May she be your bonnet the next time you go out with her']. The owner then called that they were to take the boat and be off.

See also INVERARAY; LYBSTER (Scottish Highlands & Islands); SUPERSTITIONS (p. 360).

ROTHESAY, BUTE, ARGYLL AND BUTE

Giant Squid

'As there are still some people who doubt the existence of the Kraken,' wrote John Blair in 1775, 'I enclose you an authentic Deposition upon the subject.' It was a document which, he was sure, would convince the most incredulous. The testimony was that of Robert Jamieson of Rothesay, who swore before Blair, the commissary of Bute, that in 1774, while sailing off the west coast of Ross, one of his men called him on deck to look at 'an Island which had just made its appearance in the sea'. Jamieson and his crew of nine men observed the island and agreed that it was about a mile and a half long, rising to a height of about thirty foot, and tapering towards each end. They could not estimate how broad it was. It stayed motionless for five minutes or so, then slowly sank, but reappeared after a few minutes, and sank and rose twice more before finally vanishing. Immediately after it disappeared for the last time, Jamieson saw on the surface of the water an even odder apparition. Closer than the island had been, was what looked like 'a Regiment or a considerable Assemblage of men all in white', which soon went off to the south-east 'in a confused fighting-like manner'.

What Jamieson and his crew saw may have been a fog bank, or the refracted image of a piece of land further off than appeared. Optical illusions of this kind have been reported from several places, including YOUGHAL (Southern Eire), and there are well-attested accounts of groups of soldiers, for example, being observed at exercise apparently in the air, when in fact they were securely on land somewhere beyond the horizon, and could only be seen due to an odd reflection. Jamieson himself had no explanation, maintaining only 'that the appearance excited the Idea at the time that they were really Islands', but John Blair, who took his statement, was firmly persuaded that this was the 'fish call'd the Kraken', and he even explains the final 'Assemblage of men' as being the 'points or tentacula of the fish' poking out of the water.

Jamieson's deposition and Blair's comments, in elegant faded handwriting, are contained in a letter dated August 1775. The manuscript is bound into

In his Histoire Naturelle des Mollusques *(1802), Pierre Denys de Montfort included a picture of a* 'poulpe colossal', *reproduced in many later works. No octopus or squid grows anything like as large as this, and the French naturalist must have been partly influenced by ancient tales of the gigantic Kraken.*

an English translation of Erich Pontoppidan's *Natural History of Norway* (held in the British Library, London, under shelfmark 459.c.3), in the chapter dealing with the Kraken, which Pontoppidan calls 'incontestibly the largest Sea-monster in the world'. He had discussed the subject with many Norwegian fishermen, who agreed that they often saw a part of the enormous creature appear above the waves. Its back, all that ever emerged, was 'about an English mile and a half in circumference', and was surrounded by things that looked like floating weeds. These were the creature's arms or tentacles, which sometimes stood up 'as high and as large as the masts of middle-siz'd vessels', and it was said that with them it could pull down the largest ships, although in fact it never deliberately caused harm to humans, and was only dangerous from the whirlpool that it caused by sinking in the ocean.

Tales of the Kraken belong to an ancient tradition. Homer's *Odyssey*, composed in the eighth century BCE, features the monster Scylla, whose lower parts swarmed with serpents, and reports of many-tentacled beasts were rife in the Mediterranean by the time of Pliny, whose *Natural History* (77 CE) reports them as highly dangerous, prone to embracing and drowning humans with their many feelers and suckers. He gives an account of a monster 'polypus' with massive arms that emerged from the sea each night to steal from fish tanks at Carteia (Rocadillo in Spain). A fence was built to keep it out, but this 'the polypus managed to get over by the aid of a tree', and it was only dispatched at last with the aid of several three-pronged fish spears and a pack of trained dogs. It weighed seven hundred pounds, and its arms were thirty foot long, with suckers as large as an urn.

These days it is generally agreed that accounts like this were partly based on sightings of the Giant Squid. Until quite recently, few distinctions were made between the larger cephalopods - squids, cuttlefish and octopuses - and Pliny's monster seems to combine the octopus's ability to travel short distances on land with the Giant Squid's dimensions. The octopus attracted its own legends, described at BRECQHOU (South-West England & Channel Islands), but the real Giant Squid (*Architeuthis princeps* or *Architeuthis harveyi*) is perhaps still more extraordinary. It has a tubular body ten to fifteen foot long and a hooked beak surrounded by ten tentacles. Two of these are longer than the rest and may measure as much as fifty foot. Its eyes are the size of dinner plates.

Popular legend and sensational films have continued to portray cephalopods as capable of devouring whales, dragging down the largest ship or squeezing it until its timbers burst, and picking off a crew one by one with its suckers - as according to one theory a Giant Squid took the crew of THE MARY CELESTE (p. 334).

Until very recently it was thought that the Giant Squid was the largest animal without a backbone on land or sea, but in the early twenty-first century the astounding discovery was made that there exists an animal even more enormous: the Colossal Squid or Antarctic Squid (*Mesonychoteuthis hamiltoni*), which may measure as much as sixty foot from top to tentacle. This lends a little more colour to the ancient tales, although as noted in a recent work of marine zoology, the sucker marks sometimes found on sperm whales are 'probably inflicted when the whale tries to eat the squid rather than the other way round'.

ST COMBS, ABERDEENSHIRE

See ROSEHEARTY.

ST MONANS, FIFE

Unlucky swine

The pig was generally considered an ill-omened animal by sailors, and in St Monans during the eighteenth and nineteenth centuries the taboo seems to have been a positive obsession, at least among the inhabitants of the lower town. In the upper town swine were kept, and this distressed the lower-town fishermen, who could not go to sea after seeing a pig until the tide had ebbed and flowed to remove the bad luck. According to John Jack's *Historical Account* of 1844, they at last decided to attack the pigs' owners, and set off armed with boathooks. They were foiled, however, by a general release of the swine, which sent the fishermen running back downhill to hide from the terrible animals. Such bad feeling was created that an edict went out from the lord of the manor forbidding pigs to be kept in the town, and Jack reports that no pig was to be seen in St Monans for nearly a century afterwards.

By the mid nineteenth century, pigs were once more raised and eaten in St Monans but were still not to be named. They were called 'the beast' or 'the brute', and if the word 'pig' or 'swine' were accidentally used, the bad luck had to be counteracted by an exclamation of 'Cauld iron!' (cold iron being a time-honoured protection against enchantment of all kinds), and preferably by touching iron as well.

This leads Jack to an account of a clergyman who came to St Monans unaware of the local prejudice against pigs, and elected to preach a sermon on the subject of the Prodigal Son. Naturally, in the course of the story, he used the fatal word. 'And he sent him into his fields to feed swine,' said the minister. Instantly 'Cauld iron!' was whispered by the whole congregation, who at the same time stretched out their hands to find a nail to touch. The minister was rather astonished, but thought that perhaps they were saying 'Amen', and resumed his sermon at the point he had abandoned it. 'Well, to feed swine,' he continued. 'Cauld iron!' his parishioners shouted. The unlucky parson pressed on, and got to where the Prodigal Son feeds on 'the husks that the swine did eat'. The congregation had had enough, and rose as one: 'the church, in one instant, was cleared of the whole seafaring population – and many of their descendants, up to the present day, never see more than the outside of it.'

To support this anecdote, Jack adds an incident that occurred in his own time, when a practical joker put some dead piglets among a boat's baskets and tackle. The prank was not discovered until the boat was launched, and when the captain saw the 'ominous grunkle' (snout), he at once returned to harbour and did not venture to sea again for a fortnight.

These accounts may be exaggerated, but they undoubtedly reflect real beliefs among the fishermen. A correspondent to the *Weekly Scotsman* in 1898 wrote that every year she or he had spent part of the summer in a Fife fishing village, and although the old men there laughed at the superstitions in conversation, 'one can see they guard well against them'. They would not speak of pigs, and if anyone mentioned pork on board a boat, it was thought to bring a storm. In a 1904 article for *Folk-Lore*, David Rorie likewise noted that 'on the Fifeshire coast the pig is *par excellence* the unlucky being':

> 'Soo's tail to ye!' is the common taunt of the (non-fishing) small boy on the pier to the outgoing fisher in his boat . . . At the present day a pig's tail actually flung into the boat rouses the occupant to genuine wrath.

For reasons why the pig should be so maligned, we can look to the New Testament tale (Matthew 8:28–32) of the men possessed by devils which Jesus cast out: 'And when they were come out, they went into the herd of swine: and, behold, the whole herd of swine ran violently down a steep place into the sea, and perished in the waters.' Before that, the cursed nature of the pig is traceable to the Old Testament dietary prohibition on swine's flesh, and ultimately, perhaps, to the animal's appearance, relatively hairless and either pink or brown, not unlike a human being, so that eating pork is seen as a form of cannibalism. As a sort of pseudo-human, the pig invites identification as a demon.

See also ROSEHEARTY.

SOLWAY FIRTH

Phantom ships

The Solway Firth is a famously haunted stretch of sea. In the early nineteenth century, Allan Cunningham wrote at some length of his experiences visiting a place he calls Blawhooly Bay, near Carsethorn on the northern side of the firth. He witnessed, and was unable to prevent, the drowning of a young fisherman near the remains of two wrecked ships, and heard from Mark Macmoran, a veteran sailor, the tale of these 'phantom' vessels. They had been there, Macmoran said, for countless years, but had neither sunk in the quicksand nor rotted away.

One summer evening, the two ships had been seen approaching the coast, the first steered only by a shadow flitting about its deck, the second crowded with merrymakers. Some bold young men among the onlookers proposed going aboard to join the party, and a boat put out from the second ship to collect the lads, but when it had ferried them across and they jumped aboard, the decks melted like mist from beneath their feet, and down they went. The next morning only two wrecks remained. 'Such is the tradition of the mariners,' said Macmoran, 'and its truth has been attested by many families whose sons and whose fathers have been drowned in the haunted bay of Blawhooly.'

An old woman, Meg Moray, added that the tale of the ships was not yet done. Her own husband and seven other men had drowned in the bay, and she had seen one night on the beach the form of an old miser, doomed to dig for shipwrecked treasure:

> The Form found something which in shape and hue seemed a left-foot slipper of brass; so down to the tide he marched, and placing it on the water, whirled it thrice round; and the infernal slipper dilated at every turn, till it became a bonnie barge with its sails bent, and on board leaped the form, and scudded swiftly away.

Coming to the wrecks, the form struck them with an oar. At once their masts sprang up, sails spread, their decks were full of sailors, and away the phantom vessels sped, leaving a track of fire behind them on the waves. Every year, said Meg, the ghostly ships might be seen sailing up the Solway, only to vanish with a fiendish shriek of laughter.

Not only phantom boats patrolled the area. It was also a haunt of the Kelpie, a water-spirit that could appear as a horse or a man, or might remain invisible while making noises to lure the unwary into the depths.

In 1830, a shocking episode was reported as having occurred a few years earlier, when some ship's passengers were landed by mistake on a sandbank in the Solway Firth instead of on the Cumberland coast. As the mist cleared and the moon rose, they found that they were in mid-channel, with the tide rising fast, and shrieked for help, but none came, because on both the Scottish and English shores, their cries were taken 'for the wailings of Kelpies'. All of them were drowned, though they could easily have been rescued 'but for the rooted superstition of their neighbours ashore'.

It was widely believed that anyone who saved a drowning person would themselves be drowned, as reported for instance at COLLIESTON, and it is very possible that the listeners knew people were in danger, but preferred not to help. Alternatively, the tale of the phantom ships might have been in their minds. When Allan Cunningham saw the young fisherman drown, old Mark Macmoran said it would be useless to go to his rescue: 'Whoso touches those infernal ships, never survives to tell the tale.'

STAFFA, ARGYLL AND BUTE

Fingal's Cave

One of nature's most astounding pieces of architecture, the island of Staffa was created sixty million years ago, when subterranean fires forced a breach in the earth's crust, sending an immense tide of boiling lava to the surface. As the molten rock spread and cooled, it formed hexagonal columns in a line between the Hebrides and the north of Ireland. Most of these have been eroded over millions of years, but outcrops remain at the

The island of Staffa has been associated for centuries with the legendary hero Fionn mac Cumhaill or Fingal.

GIANT'S CAUSEWAY (Northern Eire & Northern Ireland), the Treshnish islands, and Staffa. It takes a leap of the imagination to believe that such a complex structure can really be natural, its multitudes of clustered pillars, up to fifty-six foot tall, looking like the work of some crazed designer who meant to build a cathedral and got carried away.

The sea has worked caves stretching deep into the island, some of which can be explored at low water when the weather is calm. One is even said to have been inhabited, long ago, when Abbot MacKinnon of Iona made it his hermitage after having broken his monastic vows by having a love affair, but he found that the cries of seabirds and the booming of the ocean made it too noisy for meditation, and moved to Mull. From MacKinnon's Cave it is sometimes possible, at low tide, to walk through a dark and winding tunnel all the way to Cormorants' Cave, a recess over sixty foot deep, although from the outside it appears no more than a slit in the rock.

The most celebrated feature of the island was known to the Gaels as *Uaimh Binn*, 'the Musical Cave', from the echoing orchestra of the waves within, but since the eighteenth century at least, it has been associated with the Irish hero Fingal (Fionn mac Cumhaill). Epic tales of Fionn were brought from Ireland to Scotland by the earliest settlers, and lived on in oral tradition and ballads. The warrior was imagined as a colossus, standing thirty foot tall, and it was almost inevitable that this towering figure should be linked with the monumental cave on Staffa, as he was with the other end of the pillars, far away in Ireland at the Giant's Causeway.

TIREE, ARGYLL AND BUTE

St Columba's curses

Magical powers were often attributed to the early saints, who were said to be able to summon gales and otherwise influence the environment. In official accounts of their lives, their skills are usually employed for good ends, as when Columba directed the wind to send his brethren from IONA to their separate destinations, but in folk tales they are sometimes seen to use their wizardry for rather petty purposes. When Tim Severin visited Tiree in 1976 on the first leg of his epic journey across the Atlantic from BRANDON CREEK (Southern Eire), he was told that a rock in the harbour was called Mollachdag, 'the little cursed rock', because St Columba had lost his temper with it. When the saint tried to moor his boat to the weeds that grew on the stone, they broke and the boat drifted away, so Columba declared that weeds would never more grow there. Severin's informant said that all the other rocks in the harbour had thick beards of seaweed, but Mollachdag had been completely bald until a few years earlier, when a few wisps had begun to grow.

It was said that Columba had slipped and fallen on the beach at Tiree when he put his foot on a plaice lying in the shallow water, and had therefore condemned the fish to have both eyes looking upward so it could see anyone who was about to tread on it in the future.

Columba also changed the appearance of the flounder, according to a tradition John Gregorson Campbell heard on Tiree and Iona at the end of the nineteenth century. Meeting a shoal of flounders, the saint asked if they were on their way somewhere. 'Yes, Colum-Kil crooked legs,' said one.

> 'If I have crooked legs,' said St Columba, 'may you have a crooked mouth,' and so the flounder has a wry mouth to this day.

In the Isle of Man, it was said that the flounder's mouth stayed twisted after it sneered at the herring (as children are told, or used to be, that if they make faces the wind will change and they will stay like that).

TOBERMORY, MULL, ARGYLL AND BUTE

The Tobermory galleon

In the autumn of 1588, after the defeat of the Armada, the ships of the failed Spanish expedition retreated north. One of them, the *San Juan de Sicilia*, ended up in Tobermory harbour, where it remained for about a month before being blown up and sunk, an event so dramatic that it reverberates in legend to this day.

The controversies surrounding the vessel and its destruction are complex, and some matters are still in dispute. The very identity of the 'Tobermory galleon' was only established in the twentieth century, and the question of whether or not it was carrying treasure continues to be argued. Although there is no evidence that it was freighted with gold, hopeful hunters believe otherwise, and repeated attempts have been made to dredge the ocean floor for the vast wealth supposedly hidden there, although so far with no result other than a few pieces of pewter plate and some cannon.

Another question that has provoked much speculation is who it was that blew up the *San Juan*. An often-repeated tale is that the Spanish captain owed money to MacLean of Duart, the local laird, who sent his foster-son Donald Glas to collect the debt. Instead of paying up, the Spaniards clapped Donald in irons and prepared to sail away. What followed is described by Ralph Paine in *The Book of Buried Treasure* (1911):

> When Donald Glas learned that he was kidnapped in the galleon, he resolved to wreak dreadful revenge for the treachery dealt his kinsmen . . . Overnight Donald Glas had discovered that only a bulkhead separated his cabin from the powder magazine of the galleon, and by some means, which tradition omits to explain, he cut a hole through the planking and laid a train ready for the match.

In his heroic enterprise, he killed not only the Spaniards but himself.

This resourceful suicide-bomber is probably a fiction, since his name is mentioned in no contemporary sources, and many people at the time believed rather that the ship was destroyed by one John Smallett or Smollett, a spy in the pay of the English government. The truth, however, was probably simpler and sillier, since Spanish documents have revealed that the *San Juan*'s crew were taking gunpowder off the ship to dry it out when the explosion occurred, making the disaster neither an act of desperate vengeance nor one of sabotage, but an accident.

It was said that a few Spanish sailors survived the destruction, and fathered children whose 'black-eyed and black-haired' descendants were still to be met with on Mull, one of several local legends claiming ancestry from Armada sailors (even extending to Armada sheep, as at DRIGG, North-West England & Isle of Man). Also reported to have been thrown on shore by the explosion was a dog, which afterwards sat opposite the place where the ship had sunk 'howling most piteously' for days on end. The Greyfriars Bobby touch evidently appealed to nineteenth-century historians, several of whom repeat this anecdote.

The best-known tale connected with the Tobermory wreck, however, and certainly the most romantic explanation for the *San Juan* being anchored at Mull, was that the ship had come to the Hebrides carrying a Spanish princess, who had fallen in love with a man she saw in a dream. She knew he was a foreigner, but not which country he came from, and so she commandeered a ship and sailed to many different countries. Everywhere she stopped, she gave a dinner party and invited all the gentlemen in the district, to see if she could find the hero of her dream.

When she came to Tobermory, one of the men who came to dinner was MacLean of Duart. As soon as the princess saw him she knew that this was the one, and MacLean for his part was quite willing to embark on an affair, but he already had a wife, who was murderously jealous when she heard the gossip about her husband and the princess. A nineteenth-century verse rendition of the tale, 'The Lady of Duart's Vengeance', says that the wife called in the witch of Ben More:

'Weird woman, that dwellest on lofty Ben More,
Give ear to my sorrow, and aid, I implore.
A lady has come from the green sunny bowers
Of a far southern clime, to the mountains of ours;
A light in her eyes, but deceit in her heart,
And she lingers and lingers, and will not depart.'

The witch agreed to sink the Spanish ship, and at dead of night the 'agents of evil' gathered round the poop and prow:

And long ere the morning, a loud sudden shriek
Was heard o'er the bay 'Sprung a leak! – sprung a leak!'
Oh! then there was gathering in tumult and fear,
And a blanching of cheeks, as the peril drew near;
A screaming of women – a shouting of men,
And a rushing and trampling, again and again!

Everyone died, including the princess, and there the poem stops, but local legends added that MacLean took revenge on his wife by exposing her on a rock, a tale current at least from the mid nineteenth century and later adapted in oral tradition (*see* LADY'S ROCK).

A version of the story told in Archibald Campbell's *Records of Argyll* (1885) adds yet another episode, saying that MacLean found the princess's body and buried it in Morvern. A while afterwards, the princess's ghost appeared and implored him to dig her up and rebury her in Lismore, which he did, but later she visited him again and asked for her bones to be sent home to Spain. Why should she first have wanted to be moved to Lismore? This puzzling detail may give a clue to one source for the legend, since also included in Campbell's book is a tale of a Norwegian princess buried in Lismore, whose spirit begged for her body to be returned to her homeland in order to be near that of her lover. Although Campbell makes no cross-reference between the two stories, and does not seem to have noticed that two foreign noblewomen asked for their remains to be dug up from Lismore, it seems clear that the tale of the Norwegian princess must have come first and influenced the one told of the Spanish lady.

Be that as it may, MacLean did as his dead beloved wished. The king of Spain was grateful at first, but then when he looked more carefully he found that one toe had been left behind. Enraged by the deficiency, he sent ships to destroy Mull, and now MacLean himself called in the witches. Nine of them set to work on their magical looms in Castle Duart, making wind to drive the Spanish ships ashore. The Spanish captain, seeing the witches flying round his mast like crows, said he could beat them as long as Gormsuil, a blue-eyed witch from Lochaber, did not join them. Hardly had he said this when Gormsuil appeared in the shape of a cat, and the Spanish frigates were wrecked.

'I understand,' Campbell concludes, 'that a few years ago one of her Majesty's gunboats lifted some of the guns that were on board these frigates,' bringing us back by a roundabout route to history, and the attempts to salvage the Tobermory ship.

TRESHNISH ISLES, ARGYLL AND BUTE

Witches as rats

Rats were one of the animals sometimes counted as unlucky by sailors. A boatman bringing a load of peats from the Ross of Mull to Tiree was met at the Treshnish Isles by some rats, sailing on dried cowpats. Rashly, he threw a piece of peat at them, upsetting their rafts, and at once a storm sprang up that nearly wrecked his boat. These rats, of course, were witches, as he should have realised from their unusual mode of transport, and he should not have interfered with them.

Sailing near the Treshnish Isles, a boatman encountered a crew of witches in the shape of rats.

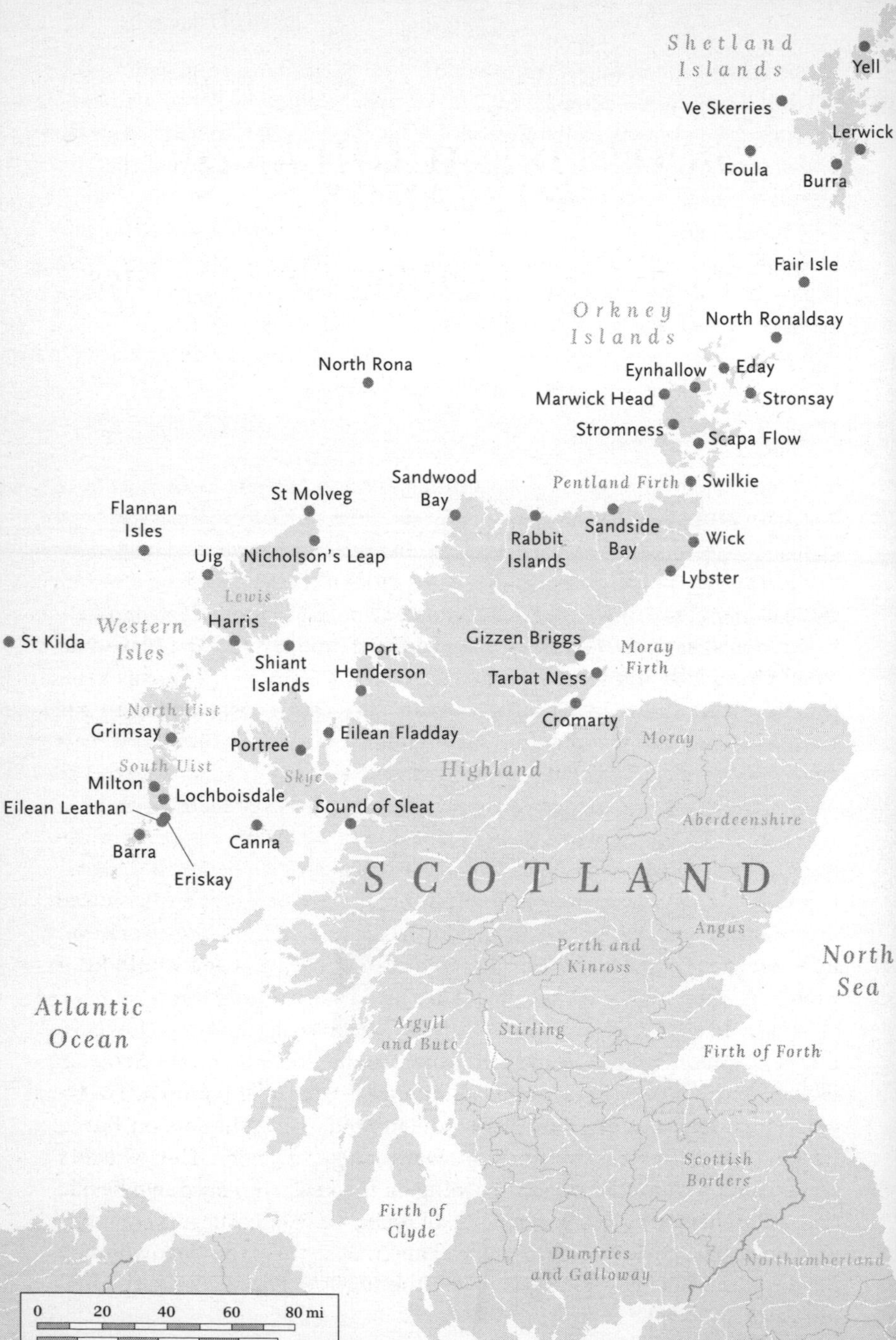

Shetland Islands
Yell
Ve Skerries
Lerwick
Foula
Burra
Fair Isle
Orkney Islands
North Ronaldsay
North Rona
Eynhallow
Eday
Marwick Head
Stronsay
Stromness
Scapa Flow
Pentland Firth
Swilkie
Sandwood Bay
St Molveg
Flannan Isles
Rabbit Islands
Sandside Bay
Wick
Uig
Nicholson's Leap
Lybster
Lewis
Western Isles
Harris
St Kilda
Gizzen Briggs
Moray Firth
Shiant Islands
Port Henderson
Tarbat Ness
Cromarty
North Uist
Grimsay
Eilean Fladday
Portree
Moray
South Uist
Skye
Highland
Milton
Lochboisdale
Eilean Leathan
Sound of Sleat
Aberdeenshire
Barra
Canna
Eriskay
SCOTLAND
Angus
Perth and Kinross
North Sea
Atlantic Ocean
Argyll and Bute
Stirling
Firth of Forth
Scottish Borders
Firth of Clyde
Dumfries and Galloway
Northumberland
0
20
40
60
80 mi
0
40
80
120km
ENGLAND
Solway Firth
Cumbria

SCOTTISH HIGHLANDS & ISLANDS

Highland, Orkney, Shetland, Western Isles

BARRA, WESTERN ISLES

Seal-song

In about 1910, the musicologist Marjory Kennedy-Fraser, collecting traditional melodies in the Hebrides, noted down a tune sung by an old woman from South Uist which the performer called a 'seal song'. It was later printed under the title 'The Sealwoman's Sea-Joy', and attached to the traditional legend of a magical seal-maiden whose enchanted skin was stolen. Without it she could not return to the sea, and so she agreed to marry the man who had taken it, but after several years living with him, when she was toiling over a hot stove and longing for the coolness of the waves, her little son came in carrying a strange thing he had found, softer than mist. It was her own skin, and immediately she slipped it on and dived into the ocean, telling her child that if he and his father were ever in want, they should cast their net off a certain rock and she would fill it with fish. Then off she went, 'lilting her joy in the cool sea-water'.

This is like the tale of John O'Glin (*see* ARRANMORE ISLAND, Northern Eire & Northern Ireland), and many others told in Ireland and the Scottish islands, but Kennedy-Fraser adds a unique experience of her own. Twelve years or so after she first heard the tune in South Uist, she was on Barra, walking along the western shore on a warm, calm afternoon. Here she met two friends, also musicians, sunbathing on the sand, and lay down beside them. Parallel to the beach ran a line of rocks, and along these were many grey seals, basking in the sun like the women. 'Sing them the "Sealwoman's Sea-Joy",' her friends suggested, and without getting up she sang the first phrase.

> Instantly there was a response from the seal rocks. Like a fusillade, single note after single note came from each seal in succession from the southerly end of the reefs to the north.

Then, after a moment's silence, a beautiful solo voice sang a phrase they had never heard before.

> The voice of the seal was so beautiful (of a rich mezzo-soprano quality) and the *cantabile* so perfect, that I should almost have believed I had been dreaming but for the corroboration of my two musician fellow-hearers.

Mrs Kennedy-Fraser wondered if the seals had mistaken her for one of themselves, as she lay full-length on the sand.

A yet more remarkably musical creature was described in Rowena Farre's *Seal Morning* (1957). Her pet seal Lora not only sang, but loved to play the mouth organ and the xylophone, holding the beater between her teeth, and learned tunes including the national anthem and 'Danny Boy'.

Responses to both accounts were sceptical. Kennedy-Fraser's tale provoked a suggestion from *Punch* that the British National Opera Company might capture a seal prima donna, and doubt has been cast on Farre's memoir. It is, however, perfectly possible to teach tunes to animals, as has been demonstrated with songbirds, and it is quite true both that seals like listening to music (*see* CANNA), and that they themselves can sing. Geoff Sample, a birdsong expert who has also recorded seal-song, writes that grey seals are the most vocal around the Northumberland coast:

> My experience from recording them is that the females are the singers in the wild. The males I've watched just grunt and blubber – strange, since usually males are more vocal in non-human animals . . . Weddell's Seals in the Antarctic produce amazing eerie vocalisations underwater, like avant-garde electronic compositions.

In *The People of the Sea* (1954), David Thomson notes that seal-song can best be reproduced on a fiddle or the bagpipes, traditionally the favourite instruments in the Scottish islands.

BURRA, SHETLAND

The Kraken

In the early nineteenth century, a report was made of a Kraken seen off Burra in Shetland. According to the men who saw it, it appeared 'like the hull of a large ship, but on approaching it nearer, they saw that it was infinitely larger, and resembled the back of a monster'. Samuel Hibbert cites their affidavit in his *Description of the Shetland Islands* (1822), and adds that such beasts were thought to possess 'tentacula as high as the masts of a ship', an idea he probably derives from Erich Pontoppidan's eighteenth-century description quoted at ROTHESAY (Scottish Lowlands). Hibbert heard several more accounts of Krakens and sea-serpents around the islands, remarking that 'their occurrence is much connected with the demonology of the Shetland seas'.

Confusion between various sorts of real animals and fabulous beasts led to considerable crossover between legends. References to the Kraken as looking like an island surrounded by weeds may be semi-rationally explained as referring to the tentacles of a Giant Squid, but sound suspiciously like traditional descriptions of the 'whale island', so huge and covered with trees that sailors think it is solid, land on it and light a fire,

A Giant Squid stranded on the seashore could easily be misidentified as that mythical monster the Kraken.

provoking the creature to dive (*see* ELACHNAVE, Scottish Lowlands). The gigantic whale famed in medieval Bestiaries as Fastitocalon was thought to have fragrant breath that lured little fishes into its mouth, and a similar thing was reported by the Norwegian writer Pontoppidan of the Kraken: 'the great Creator has also given this creature a strong and peculiar scent, which it can emit at certain times, and by means of which it beguiles and draws other Fish to come in heaps about it.' This scent, however, was not emitted by the creature's breath, but from the other end. The Norwegian fishermen observed that during some months the Kraken only ate, and during other months he only 'voids his excrements':

> During this evacuation the surface of the water is coloured with the excrement, and appears quite thick and turbid. This muddiness is said to be so very agreeable to the smell or taste of other Fishes, or to both, that they gather together from all parts to it, and keep for that purpose directly over the Kraken: he then opens his arms, or horns, seizes and swallows his welcome guests, and converts them, after the due time, by digestion, into a bait for other Fish of the same kind.

The phenomenon was probably based on observation of clouds of plankton in the water.

In natural history up to the eighteenth century, Krakens were thought of as a species (generally identified by later naturalists as the Giant Squid), but in literature the animal was elevated to *the* Kraken. One variant of the Scandinavian tradition stated that there were only two such beasts, which had been created at the world's beginning and would live until its end, when they would come to the sea's surface and die. On this, and on the medieval legend of Fastitocalon the whale, roused from his long slumbers by fire, Tennyson founded his poem 'The Kraken' (1830):

> Below the thunders of the upper deep;
> Far, far beneath in the abysmal sea,
> His ancient, dreamless, uninvaded sleep
> The Kraken sleepeth . . .
> There hath he lain for ages and will lie
> Battening upon huge seaworms in his sleep,
> Until the latter fire shall heat the deep;
> Then once by man and angels to be seen,
> In roaring he shall rise and on the surface die.

So pervasive an influence has the poem had on subsequent accounts of the Kraken that imaginative details from it are sometimes cited as legend, and Tennyson's apocalyptic vision has coloured popular culture. The novelist John Wyndham, for example, when he wrote of the near-destruction of mankind by extraterrestrials colonising the deeps, called his novel, ominously, *The Kraken Wakes* (1953).

CANNA, HIGHLAND

'The Seal-Woman's Croon'

All seals are really enchanted royalty, the sons and daughters of the king of Lochlann (the old Gaelic name for Scandinavia). Their wicked step-mother, jealous of their beauty, wisdom and bravery, spent seven years with a magician, learning the black arts, so that she could put a spell on the princes and princesses to make them half-fish, half-beast. Three times a year, when the full moon is at its brightest, they must return to their human shape, so that they will feel the fullest sorrow at having lost their kingdom and their natural state, and anyone who sees one of them at such a time will lose his heart to them.

Once, long ago, a man of Canna was wandering by the seashore on an autumn night, when by the light of the full moon he saw a seal-maiden in her human form, and at once fell in love. Putting her to sleep with a charm, he took her in his arms and carried her home, but when she awoke, what should he find but a seal. Full of pity, he carried her back to the sea and let her swim away, and all that night she spent near the shore, singing to him.

Story and preamble are given by Kenneth Macleod in *Songs of the Hebrides* (1909), as a note to a tune he calls 'The Seal-Woman's Croon'. Its words are partly meaningless and partly in Gaelic – for though the seals are of the race of Lochlann, Macleod says, Gaelic is the language they like best. His explanation of the seals' metamorphosis is a hybrid, mixing traditional tales of the seals as shape-shifters with the Irish legend of the swan-children of Lir (*see* INISHGLORA, Northern Eire & Northern Ireland), and the Canna tale is a somewhat cleaned-up version of the usual narrative in which a man steals a seal-woman's skin, and, far from returning her to the water, marries her and has children with her before she finally escapes.

The real point of the story is the seal-song, and it is an entirely authentic tradition (and fact) that seals are musical animals. Some people, as described at BARRA, have heard tunes sung *by* seals, and in tales of music sung *to* seals, they are said to be responsive listeners. David Thomson, who travelled Ireland and Scotland in the mid twentieth century in search of seal legends, was told everywhere he went that the seals loved music. A Kerry man told him how his great-uncle had seen a seal about fifty yards from shore, and began to sing to her, 'Come ashore, come ashore, O seal!' He had a sweet voice, and the seal enjoyed it so much that she went to sleep, rocked by the waves, and drifted in on the tide until she was lying high and dry, fast asleep, on the sand. Evidently many people thought that they could lure seals with a tune. In a late eighteenth-century collection of Highland vocal airs is a wordless melody entitled 'fisherman's Song for attracting the Seals'.

The Uist poet John MacCodrum (1710–96) put Gaelic words to an air described as 'one of the old North Uist seal songs'. It might originally have been another tune sung or played by fishermen hunting seals, but MacCodrum's lyrics are a lament sung by the seal itself. The associated tale is that some hunters on the rocks of Hasker had killed many seals, and spent all day melting the fat off the animals. At night they retired to their small hut and began their supper of seal flesh, when they heard a mournful voice outside singing (in translation):

> It is a pity that in this land
> They eat human beings in the form of food.
> Do you not see the leader of the Seal Host
> Boiling fiercely on a fire?

MacCodrum's sympathy for the seals was explained in local tradition as family feeling. The MacCodrums were known as 'the MacCodrums of the seals', and said to be descended from a union between a seal and a human being. Other Highland clans were once associated with particular emblematic animals (the MacLeods with the horse, the MacDonalds with the dog), but accounts linking the MacCodrums with seals remained current until at least the mid twentieth century, when MacCodrums living in Canada reported from knowledge of their family history that no one of that name would ever have killed a seal or eaten seal's flesh. The nineteenth-century editor of John MacCodrum's poetry notes that 'a woman of the same surname, and probably lineage, as the bard, used to be seized with violent pains at the time of the annual seal hunt', sharing the suffering of her supposed relatives.

CROMARTY, HIGHLAND

A drowned man's wraith

The scientist and poet Hugh Miller (1802–56) wrote extensively on the geology, history, and folklore of his home territory around Cromarty, including several personal experiences of psychic phenomena. His sensitivity dated back to an early age, when he and his family sustained a great tragedy marked by a terrifying apparition.

Miller's father was a sailor, and in 1807 was on a journey around the north coast of Scotland. Having cleared Cape Wrath and the Pentland Firth, he was overtaken by a storm and sought shelter in Peterhead harbour, from where he wrote a letter to his wife. He then set sail again and Captain Matheson, a neighbour from Cromarty, reported seeing his sloop making for the open sea as darkness fell, buffeted by violent gales. Matheson believed that his friend's seamanship was equal to the dangers, but night brought worse tempests and the sloop was never seen again.

At home in Cromarty, meanwhile, the weather was calm, and having just received Miller's letter sent from Peterhead, his wife was in no anxiety. She was sewing by the fire when the front door swung open, and she sent her little son to shut it. What followed is recounted by Hugh himself, with the caveat that this is 'simply the recollection, though a very vivid one, of a boy who had completed his fifth year only a month before'.

> Day had not wholly disappeared, but it was fast posting on to night, and a grey haze spread a neutral tint of dimness over every more distant object, but left the nearer ones comparatively distinct, when I saw at the open door, within less than a yard of my breast, as plainly as ever I saw anything, a dissevered hand and arm stretched towards me. Hand and arm were apparently those of a female: they bore a livid and sodden appearance; and directly fronting me, where the body ought to have been, there was only blank, transparent space, through which I could see the dim forms of the objects beyond.

Hugh ran screaming to his mother to tell her what he had seen, leaving the door open. The maid was sent to shut it, and returned saying that she too had seen the woman's hand. 'And finally, my mother going to the door, saw nothing, though she appeared much impressed by the extremeness of my terror and the minuteness of my description.'

This was, they later realised, the night his father drowned.

EDAY, ORKNEY

Finns and Finn-men

In the seventeenth century, the Orkney islanders sometimes saw strange boats rowed by what they called 'Finn-men'. James Wallace reports in his *Account of the Islands of Orkney* (1700) that one was seen off the south end of Eday in 1682, but when a local boat tried to intercept it, it sped away across the sea. Wallace describes the peculiar vessels, sealed around their pilots:

> they have this advantage, that be the Seas never so boisterous their Boat being made of Fish Skins, are so contrived that he can never sink, but is like a Sea-gull swimming on the top of the Water. His shirt he has is so fastned to the Boat, that no Water can come into his Boat to do him damage, except when he pleases to unty it, which he never does but to ease nature, or when he comes ashore.

This sounds rather like an Inuit kayak, a one-man canoe made of sealskin constructed to fit tightly around the occupant so that man and boat form a virtually continuous surface. If the boatman was wearing something made of the same material, it might have appeared that his garment was attached to the vessel, even if that was not literally so.

Wallace seems to understand the term 'Finn-men' as meaning 'men from Finland', commenting that 'it seems a little unaccountable, how these Finn-men should come on this coast, but they must probably be driven by Storms from home, and cannot tell when they are any way at sea, how to make their way home again.' Tales of these mysterious folk, partly based on real encounters, became mingled with tales from Sweden and Norway that men of the far northern regions, Lapland and Finland, were wizards and could control the winds (*see* IRVINE, Scottish Lowlands). For a while, in Shetland and Orkney, the term 'Finn' became more or less synonymous with 'merman' or 'sea-spirit', perhaps partly because in the old Scandinavian-based language of the islands, as in English, the word 'Finn' sounds the same as the fin of a fish, so that a 'Finn-man' could have been imagined as a man with fins. In the late nineteenth century, the scholar Karl Blind was informed that 'Sea monsters are for most part called "Finns" in Shetland':

> They have the power to take any shape of any marine animal, as also that of human beings. They were wont to pursue boats at sea, and it was dangerous in the extreme to say anything against them. I have heard that

> silver money was thrown overboard to them to prevent their doing any damage to the boat. In the sea-form they came ashore every ninth night to dance on the sands.

There was still a non-specific link with northern lands. Several sources mention 'Norway Finns' without comment on the discrepancy, while the Shetland folklorist Jessie Saxby heard the Finns or 'Mukle Men' spoken of as 'Denschmen', presumably meaning Danes.

In the international seafaring community, meanwhile, rumours of the 'Finn-men' and their supernatural powers came full circle, and were once more attached to the natives of Finland, widely believed in the nineteenth century to be a nation of sorcerers. In *Two Years Before the Mast* (1840), Richard Dana's account of his time in the American service, he writes that the cook on his ship was concerned to know what country another member of crew came from. Dana said the sailor was German, at which the cook was much relieved: 'I was mighty 'fraid he was a Fin. I tell you what, I been plaguy civil to that man all the voyage.' Dana learned that the cook 'was fully possessed with the notion that Fins are wizards, and especially have power over winds and storms', and similar beliefs persisted into the twentieth century. In C. S. Forester's novel *The Commodore* (1945), there is an exchange between Horatio Hornblower and Captain Bush, who asks how Hornblower gets on with his Swedish clerk. 'He's a Finn, not a Swede,' is the reply.

> 'A Finn? You don't say, Sir! It'd be better not to let the men know that.'

'In a sailor's mind,' Forester continues, 'every Finn was a warlock who could conjure up storms by lifting a finger.' Even in the 1960s, according to the experienced sailor Horace Beck, 'the fear of the Finn remains deep-seated and many sailors do not like going to sea with them.' They were supposed to be ill-tempered and unpredictable, and it was bad luck to have them aboard.

EILEAN FLADDAY, HIGHLAND

A cold ghost

In *Witchcraft & Second Sight in the Highlands & Islands of Scotland* (1902), John Gregorson Campbell gives a story which he sets on Fladda-chuain 'on the east coast of Skye'. The island now called Fladda-chuain is north-west of Skye, but there are several Scottish islands called Fladda and Flodaigh (from Norse words meaning 'raft', 'float', or 'flat'), and Campbell probably means Eilean Fladday, which is in the right place, between Skye and Raasay.

On this solitary islet, Campbell writes, there was a burial ground, and nearby lived a man from Mull and his wife. A corpse came floating ashore one day, and the couple stripped its clothes off before burying it. After this, the dead man's spirit came every midnight to the Mull couple's hut and sat by the fire, and as it held out its ghostly hands and feet to the blaze, it said, 'I will softly warm myself, I will softly warm myself.' Then it added a verse:

> Wife, who took my trousers off,
> And my nice black shoes from me,
> And the shirt my sister gave me, –
> To it, to it, cold feet of mine,
> Many a sea you've traversed.

The couple left the island after a while – perhaps scared off by the ghost – and their cottage remained deserted until a party of fishermen arrived one night and decided to stay in the hut. One of them, Donald, went on ahead to light a fire and cook dinner, and as he bent over the hearth, something struck him on the head and knocked him down. He tried to get up, and was immediately struck down again. Donald had heard stories about the haunting, and was convinced that this must be the chilly ghost trying to get to the fire, so without daring to look at what was assaulting him he crawled across the floor to a post, and managed to climb up into the rafters. When his companions arrived, the ghost was found to be a ram, 'addicted, like its kind, to butting'.

EILEAN LEATHAN, WESTERN ISLES

'The Weaver of the Castle'

The largest of the Stack Islands, Eilean Leathan, is shaped roughly like a figure eight, its waist a narrow isthmus leading steeply down to the sea. Visiting the island in 1951, the folklorist John Lorne Campbell found on top of the cliff the ruins of a small stronghold, and also saw two small cairns at the top of the path leading up the south side of the cliff, features he took to confirm in part the traditional story of 'The Weaver of the Castle'.

As recorded in about 1893, the tale begins with a man's banishment from Barra for unspecified crimes. He moved to the nearby Stack Islands, where with the help of his white horse he started to build a castle on the summit with stones from the foot of the cliff. The last trip proved too much for the beast, which died, dropping the final load from the panniers hung either side of its back, and 'to this day you can see where he tipped the pack of stones with the white pony' – the twin cairns of stones that Campbell saw. Having finished his castle, the man decided to take a wife, and went to South Uist, where the crofters and their wives and daughters used to take their cattle to the hills in July. The man looked around the women, made a quick choice, 'and without much debating he flung this young lady on his shoulders and made a bee line for the boat, which carried them both safely to the castle.'

Despite this unceremonious courtship, the girl settled down happily with her husband. She was skilled with the loom, and from his wife's trade the man got the nickname of 'The Weaver'. The couple had several sons, and made their living from fishing and piracy. When ships anchored near the island, the Weaver used to row out to cut the hawsers, so that the vessels would drift on to the rocks where they could be plundered. He trained his wife to be as skilled as himself at wrecking, and when the little boys were old enough, they helped too.

Orders were given to apprehend the Weaver, and one day when he and his elder sons were out fishing, they were pursued to Eriskay by a government ship, whose commander slew the Weaver and the boys, leaving their blood to dry on his sword as proof of their death.

The sequel is known as 'The Little Weaver', and its hero is the Weaver's youngest son, a baby when his father and brothers were killed. When he was fourteen, his mother told him about the tragedy, and he vowed that he would go to sea and never rest until he had found the man who murdered his father. He got to Greenock, where he joined a merchant ship, and eventually he became a captain. For several years he learned nothing of his father's killer, but after a long voyage he arrived in London, and went to a club where old sailors were telling tales and drinking whisky. One of these men boasted of how he had destroyed a pirate and his sons in the Western Isles, and said that the bloodstains were still on his sword. The captain made a point of applauding his bravery, and gained an invitation to the man's house. When he arrived, he was offered refreshment, but said that he would neither eat nor drink until he had seen the sword with which the dangerous raider of the Western Isles had been destroyed. The sword was shown to him, and he killed his father's slayer with one blow of his fist.

The Mary Celeste

In November 1872, the *Mary Celeste* was found floating in the Atlantic, west of the Azores. She was intact and seaworthy but completely deserted, and no evidence has ever emerged as to what became of her captain, his wife and daughter, or the seven-man crew.

What made the *Mary Celeste* famous was, in the first place, a suspicious official in Gibraltar, where the ship was brought in by the barque *Dei Gratia* for salvage. The island's attorney general was convinced that there had been dark doings aboard, either a drunken and homicidal riot on the part of the crew, or a fraudulent conspiracy between Captain Briggs of the *Mary Celeste* and the captain of the *Dei Gratia*. Nothing was proved, but the case caught the imagination of the young Arthur Conan Doyle, whose fictional account was printed in 1884 in the *Cornhill Magazine*.

His tale begins with information attributed to an article that, 'unless my memory deceives me', appeared in the *Gibraltar Gazette* in January 1874. Although there may have been such an article, almost certainly it did not include the spurious and highly significant detail, mentioned quite casually by Conan Doyle, that *no lifeboats were missing* from the vessel. In fact the *Mary Celeste* had set off with one lifeboat, which was gone when the ship was found abandoned. Conan Doyle's suggestion to the contrary, however, has been widely believed ever since, making the ship's desertion truly inexplicable.

That has not stopped many people trying to explain it. Theories advanced have included mass hallucinations, a giant octopus that picked everyone off with its tentacles, and a fit of religious mania that led Captain Briggs to slaughter everyone on board before committing suicide. A purported eyewitness account from one Abel Fosdyk asserted that the captain had leapt into the sea to prove how well a fully clothed man could swim. The rest of the company crowded to watch, standing on a platform that had been newly added to the ship, and fell in when the planks collapsed. Everyone was then eaten by sharks – except Fosdyk himself, who clung to a piece of driftwood and fetched up in Africa. Mythical information accumulated, for instance that breakfast had stood half-eaten on the table, and that those on board had fatefully numbered thirteen.

According to one particularly far-fetched theory, the Mary Celeste *was found abandoned because a giant octopus had removed the whole crew.*

Folk tales have developed around other unexplained desertions, such as that of the FLANNAN ISLES lighthouse (Scottish Highlands & Islands), but the *Mary Celeste* remains the most famous derelict of all time, her very name equated with insoluble mystery. The whole conundrum, however, stands and falls on the misinformation first given by Conan Doyle. Once we are aware that the lifeboat had been taken, the sequence of events becomes intelligible: some false alarm led crew and passengers to abandon ship far from land, the crowded lifeboat sank with no survivors, and the empty vessel floated on her way. There are many nineteenth-century records of drifting ships found empty, several in perfectly good condition. The truly remarkable thing about the *Mary Celeste*, therefore, is – that there was nothing very unusual about her.

After this, the story of the Little Weaver veers into fantasy. Taking a job as a huntsman, he shot at a deer which turned into a princess, and eventually married her. This epilogue, which appears as a separate tale in some sources, contains classic fairy-tale motifs, and has evidently been tacked on to the more historical part of the narrative. The earlier episodes of the Weaver's building of his castle, his wrecking activities, and his son's quest for revenge are what caught the attention of John Lorne Campbell, who thought that these were 'very probably founded on fact'. Although it is always risky to deduce the truth of a tale from the physical remains on the ground, historical evidence does support the Weaver's piratical technique, since in the seventeenth century ships were commonly anchored by ropes rather than chains, and there are references in the records to ships deliberately wrecked by having the cables cut. The destruction wreaked by the Weaver, his death, and the Little Weaver's vengeance are certainly not impossible, and in any case it is a fine, rousing, amoral tale.

ERISKAY, WESTERN ISLES

Whisky Galore

On 3 February 1941, the *Politician* set sail for the States from Liverpool. The next night, in vile weather, the ship went aground near Eriskay, having wedged herself so securely between the rocks and sandbanks that there was no possibility of getting her out again. The Barra lifeboat eventually took all the men off the wreck, and it was observed that several of them were very jolly, and some had bottles in their pockets. When they got to land, they were questioned about their cargo. 'What's aboard her?' they were asked. 'Whisky – twenty thousand cases of it!' was the reply.

So began the saga of the whisky wreck, an episode that resonated around the world, inspiring Compton Mackenzie's novel *Whisky Galore* (1947), a film based on the book (released in the States as *Tight Little Island*) and several songs.

The actual retrieval of the whisky from the wreck was an epic task. After the official salvors left on 12 March, having decided that the cargo hold was too deeply submerged to penetrate, the locals set about proving them wrong. News had travelled of the wreck and its precious contents, and from all over the islands boats set out. The journey was far from safe in the wild spring gales. One man reported later that it had taken him four hours to row twelve miles, and he had nearly been swept out to sea. When he reached the wreck, he saw that nothing could be done at high tide, since the waves were right up to the deck, preventing any access to the hold, and even when the water had ebbed, conditions were horrible. The ship had spilt her oil, which covered everything in a slippery film, and the hold containing the whisky was still deep under filthy water. To reach what lay below, the men had to get down to the lower deck, and then use long home-made 'spears' to hook the cases.

None of these difficulties deterred the seekers. Night after night, up to fifty men at a time scaled a rope ladder dangling fifty feet above the waves, and then set about their dirty work, seeing their way by Tilley lamps. They brought old clothes, but ruined them beyond repair, and some resorted to their wives' worn-out dresses and skirts to protect themselves from the all-pervasive oil. Every so often the police launch would arrive, but a lookout always gave the alert, although one man was caught in the act of changing his clothes, and had to hide naked for several hours in a ventilator.

Liquor was not all they found. The *Politician* had been carrying a remarkably varied cargo, which supplied the islands for years with all sorts of commodities, from cotton cloth to bicycles. The most highly valued booty, however, was the whisky. A determined man could get twenty cases a trip, and many ended up with a total of around three hundred cases each for their trouble.

A whole genre of folk tales developed on the themes of finding, hiding, and drinking the whisky. The ship became affectionately known as the *Polly*, while the whisky itself (different brands being involved) was also generically 'the Polly' or 'the Politician'. 'I would give him a drop of the Politician,' a man prescribed for a sick bullock, and an immediate cure was the result. Many islanders had stories of successfully smuggling a boatload of cases hidden in sacks of peat, outwitting the revenue men by hiding their bottles in coffins, or pouring the spirit into a chamber pot when the authorities came calling. Sometimes the very excess of the liquor foiled their efforts. One man, trying to stow his cache on a hillside riddled with rabbit holes, found every hole already occupied by a bottle.

This state of affairs continued for about two months, and every drink the islanders took was prefaced by a toast: 'To the Captain of the *Polly* – God bless him.' Around early May, however, the mood began to turn bitter. Those who drank their whisky or gave it away disapproved of the profiteers who had set up as traders; the Eriskay residents looked on the *Polly* as exclusively their own, and resented the other islanders' incursions; and raids by the authorities had begun, fostering suspicion of possible informants.

The man largely responsible for ending the fun was Charles McColl, a customs officer from Barra. A Presbyterian among the mainly Catholic islanders, and a man of staunch principle, he was determined to put a stop to the 'rescuing' of the whisky. With only a police constable to help him, he challenged several boats and impounded their loads, and in June he organised a wholesale search across Eriskay, finding so many bottles that they filled the cells, the police garage, and the policeman's house at Lochboisdale.

Forty men were taken to court for defrauding the customs, thirteen were imprisoned, and thirteen fined. The convictions caused fury and humiliation, particularly as the islanders did not see themselves as having committed any crime: the whisky had been abandoned by the salvors, so who else was going to use it? McColl's car was set on fire, and his life was threatened, but violence was useless. The days of the *Polly* were numbered. In September, an engineer succeeded in floating the wreck, meaning to tow her to Lochboisdale, but just north of the tiny isle of Calvay she sank again, and was shattered on a rock. The wreck was cut completely in half, the front section was dragged to the shipbreakers' yard, and the remainder – including the hold containing the rest of the whisky – was dynamited.

Of the twenty thousand cases carried by the *Politician*, perhaps two thirds were destroyed off Calvay, but the rest, seven thousand cases or so, had been retrieved by the islanders. It was a grand party while it lasted.

EYNHALLOW, ORKNEY

Enchanted islands

> Eynhallow frank, Eynhallow free,
> Eynhallow stands in the middle of the sea;
> With a roaring roost on every side,
> Eynhallow stands in the middle of the tide.

The 'roaring roost' in this rhyme, printed in John Gunn's *Orkney Book* (1909), is a line of white water where the tides crash against uneven rocks beneath, making a landing on Eynhallow dangerous in the extreme. Even so, the island was once inhabited, and remains of a twelfth-century monastic foundation attest to the determination of medieval hermits to settle in the most remote locations. Its name means 'Holy Island', and it is also known on Rousay and the western mainland of Orkney simply as 'The Isle'.

According to tradition, it was once the territory of the sea-people, and was under a spell that held it submerged most of the time. At rare intervals it would rise suddenly from the waves, but then vanish before anyone could reach it. Like other intermittently visible islands, such as the magic realm seen from RATHLIN O'BIRNE ISLAND (Northern Eire & Northern Ireland), it could under certain circumstances be disenchanted. 'When Eynhallow was still a vanishing island,' writes Gunn,

> it became known in Rousay that if any man, seeing the isle, should hold steel in his hand and, taking boat, go out through the tides, never looking at aught but the island, nor ever letting go the steel till he leaped on to its virgin shore, that man should break the spell and win the isle from the sea-folk for his own people.

Many tried and failed, but at last a man succeeded in the attempt, and since then the island has stood firm amid the raging billows, although it kept some of its supernatural qualities. Neither iron nor steel would remain on it overnight, and any metal stake used to tether cattle would spontaneously 'leap from the ground at the moment when the sea swallows the sun'. Its soil, like that of TORY ISLAND (Northern Eire & Northern Ireland), was proof against vermin, making it impossible for rats to live there, and it could be used elsewhere as a charm against infestation.

Gunn attributes his account to Duncan J. Robertson, who was so fascinated by Eynhallow that in the early twentieth century he bought it, and later turned it into a bird sanctuary. Robertson had also heard of another magical island called Heather-Bleather, somewhere off Orkney, home to the mysterious Finn-men of whom tales were told in EDAY and elsewhere. It was said that a Rousay girl had once been abducted from her home, and years later, when her father and brothers were lost at sea in a fog, they came ashore at a place they did not know and found her there, happily married.

> When the time came for the men to leave for home, the woman refused to accompany them, but she gave her father a knife, and told him that so long as he kept it he could come to the isle of the waters whenever he pleased. Just as the boat put to sea the knife slipped from the old man's hand into the water; in a moment the fog swallowed the island, and no man has set foot on it since.

FAIR ISLE, SHETLAND

Memories of the Armada

The knitting patterns of Fair Isle, it is sometimes said, were first adapted from a beautifully woven shawl given to his sweetheart by a sailor returning from the Baltic, and were therefore Scandinavian or Russian in origin. Another tale is that they imitated the colourful clothes worn by Spanish sailors, shipwrecked on the island when *El Gran Grifon*, a vessel of the Armada, ran aground at Stromshellier in 1588. The latter story is one of many pieces of folklore relating local customs, clothes, families or animals to wrecks of the Armada. Skye terriers, for example, are sometimes reported to be bred from animals that found their way ashore from Spanish vessels, and in several places there are or have been tales that dark-haired, dark-eyed inhabitants are descended from Armada sailors (*see* for example TOBERMORY, Scottish Lowlands).

Knitted sweaters have their own more fact-based traditions. They have taken their names of jerseys and guernseys from the Channel Islands, where knitting was an important activity for the islanders, and formed their staple means of making a living after the decline in the fishing trade

A ship of the Armada was wrecked on Fair Isle in 1588, and the island's famous knitting patterns are sometimes said to have been inspired by the Spanish sailors' clothes.

there, but Scottish jumpers were just as distinctive. Indeed, almost every fishing district in Britain once had its own type of sweater. The traditional explanation is that this would help identify any strange bodies washed up on the shore, a factor that may well have been relevant, although local pride in established patterns or sets of colours must also have played a large part in their continued use.

FLANNAN ISLES, WESTERN ISLES

The cursed lighthouse

One of the simplest but spookiest tales of the sea is that of the lighthouse on the lonely Flannan Isles, eighteen miles west of Lewis in the Outer Hebrides. In December 1900, it was reported that the light was out, although the three lighthouse keepers in residence were supposed to keep it burning at all times. When a ship called at the islands, it was found that the building was empty. The living quarters were in good order, the lamp was filled, ready to light, and there was nothing to show what had become of the three men, nor was anything ever seen of them again.

Like the case of THE MARY CELESTE (p. 334), this inexplicable disappearance provoked speculation and fabrication. The Northumberland poet Wilfrid Gibson produced a sexed-up verse account in 1912, given in the character of one of three men who arrive to see what has become of the lighthouse keepers. As he and his companions approach the island, they feel a sense of creeping dread, and see three enormous 'queer black ugly birds':

> Like seamen sitting bolt-upright
> Upon a half-tide reef:
> But as we neared they plunged from sight
> Without a sound or spirt of white.

In the lighthouse, they find the table spread for dinner, and a chair at its head overturned as if the occupants have made a hasty exit. A search of the island reveals no trace of the vanished keepers, and the men recall how others have perished on the island:

> We thought how ill-chance came to all
> Who kept the Flannan Light:
> And how the rock had been the death
> Of many a likely lad:

How six had come to a sudden end,
And three had gone stark mad;
And one whom we'd all known as friend
Had leapt from the lantern one still night,
And fallen dead by the lighthouse wall.

As a matter of history, the Flannan Isle lighthouse had only been completed in 1899, and had certainly not been the downfall of ten men in the following year. Gibson did not allow fact to stand in the way of his drama, which concludes on a haunting note:

We seemed to stand for an endless while,
Though still no word was said,
Three men alive on Flannan Isle
Who thought on three men dead.

The isolation of a lighthouse makes it a particularly effective setting for a story of a curse, and the Flannan incident has inspired an opera by Peter Maxwell Davies, a song by Genesis, and an episode of *Doctor Who*.

FOULA, SHETLAND

The pirate-gull

The skua gull is an impressive bird, about two feet long, with a wingspan of nearly five feet, and it once had a fearsome reputation to match. In the Orkneys and Shetlands, as reported in Thomas Pennant's *British Zoology* (1776), it was said that it preyed on animals as large as lambs, and could be overcome only by using its own strength against it:

> It has all the fierceness of the eagle in defending its young; when the inhabitants of those islands visit the nest, it attacks them with great force, so that they hold a knife erect over their heads, on which the Skua will transfix itself in its fall on the invaders.

Pennant heard from the minister of Birsay, Mr Low, that he had tried to approach the skuas' nest, but 'they attacked him and his company with most violent blows; and intimidated a bold dog of Mr Low's in such a manner, as to drive him for protection to his master.' The local shepherds too were often harassed by the gulls.

On Foula, however, the birds were privileged, since they protected the

The huge and ferocious skua gull can fight off eagles, men, and dogs.

sheep and lambs from the still more dangerous eagle, 'which it beats and pursues with great fury'. For this reason it was a crime to kill a skua gull, as mentioned by Samuel Hibbert in *A Description of the Shetland Islands* (1822):

> I ascended several of the highest points of the rocks of Foula. One of them is occupied by the Bonxie, or Scua Gull, the terror of the feathered race, but so noble minded, as to prefer waging war with birds larger than himself. The Eagle forbears to make an attack upon lambs while Scua is present; on which account he was long considered a privileged bird, the act of destroying him being visited with a severe penalty.

The Foula natives denied that the gull injured their flocks or their poultry. They maintained that it ate only the dung of other gulls, which, writes Pennant, 'it persecutes till they mute [shit] for fear'.

Other animals were reported to survive in peculiar ways. The inhabitants of St Kilda, as reported by Martin Martin in 1703, said that the fulmar seabird 'picks its Food out of live Whales', a neat trick, while herring, as mentioned at DUART CASTLE (Scottish Lowlands), were thought to live purely on the foam they made by thrashing the sea with their tails. In the case of the skua, however, the legend has a factual basis, since this piratical gull does indeed chase other birds until they disgorge what they have eaten. In the nineteenth century, after the extinction of the White-tailed Eagle in Orkney and Shetland, the skua lost its privileged status, and its population declined sharply, but in the later twentieth century it was protected, and the last few decades have seen an encouraging increase in its numbers.

GIZZEN BRIGGS, HIGHLAND

Sea-sprites and mermaids

Jane Hutchins moved to the shores of the Dornoch Firth in the 1940s, and reports her experience in *Discovering Mermaids* (1968):

> I was amazed to be told, quite seriously, by several of the older inhabitants with whom I had become friendly, that if I cared to walk some miles along the shore at the right time, I might be able to catch a glimpse of a mermaid on a certain spit of land known as the Gizzen Briggs.

The Gizzen Briggs, a stretch of quicksand, had quite a reputation for sea-spirits of one kind and another. It was said to have been created when imps tried to build a bridge across the firth, which collapsed when somebody walked across it and called on the name of God, and shipwrecks there in the nineteenth and twentieth centuries were widely attributed to the influence of the wicked sprites. The place was also said to be haunted by a spectre ship, the *Rotterdam*, whose crew could be heard on calm days singing psalms from beneath the waves.

Hutchins saw no mermaid (nor even a seal, which she thought might have been mistaken for a mermaid, although this suggestion offended her informants), and in fact she never took the long walk to the Gizzen Briggs, thinking that it would be a waste of time. 'Now, I wonder . . .' she writes, having spent much time investigating tales of the sea-people. She was told of more than one family who claimed a mermaid as an ancestor, or that a mermaid had once promised them that none of their family would ever be drowned.

A final charming note from Hutchins deals with the statue of Hans Andersen's 'Little Mermaid' in Copenhagen, which shows her with two legs (ending in fins) rather than with one tail. This, Hutchins was told, was 'So that she could ride a bicycle – like the rest of the Danes!' It also suggests a seal, since pictures of swimming seals often show their tails separating at the end, looking very like a pair of human legs underwater.

GRIMSAY, NORTH UIST, WESTERN ISLES

'The Tailor and the Fishing Wives'

A collection of *Scottish Traditional Tales* published in 1994 includes a story told by Peter Morrison of Grimsay, North Uist, about a tailor, a recurrent figure in Scottish lore. Until the nineteenth century, tailors were travelling workmen who stayed wherever they happened to find employment, and the trade was often adopted by weakly men, ill-suited to a life of warfare, hunting, or fishing, who were therefore looked down on by more muscular types. In folk tales, however, they are often seen as witty and resourceful characters who can outwit a fairy or a witch. This was partly, perhaps, because tailors often doubled as storytellers, bringing new tales from place to place as they travelled around the country. A man who told a memorable story might easily become identified as its hero – if he had not already given himself a starring role.

This particular tailor, said Morrison, 'was like the rest of his kind: he had sharp eyes and open ears for everything that might be going on in the houses where he was working', and also he was particularly fond of fresh herring. Coming to a new town, he had his favourite dish for breakfast, and the treat continued day after day, wherever he went in the parish. The tailor was amazed at his good luck, since he had seen no one going fishing, nor anyone visiting to sell fish, and after three days he decided that he would stay awake that night to see where the herring came from. His hostess tried to get him to sleep, saying, 'It's time you went to bed, tailor,' but he insisted on staying up, and in the end she told him that she had some women friends, and at night they went out for herring.

The tailor asked if he could go too, and reluctantly she agreed. With the wives he went down to the shore, where some sieves were hidden. They all put out to sea in their sieves, and the women's sieves were fastened to the tailor's sieve with heather ropes (known to be used by witches), while his was moored to a rock. The women said their spells, and told him that every heather tip on every rope would catch a herring.

When the ropes were fully paid out, the tailor shouted, 'With the Lord's blessing fish well.' At these fateful words, the sieves began to sink, and the women began to scream. The tailor, nearest the shore, pulled himself to land and ran for the hills. He didn't stop until he reached his own house, and he was convinced that all the women must have drowned. Much to his surprise, however, he heard that they had survived by pulling themselves to shore on their heather ropes. 'I wouldn't much have cared to eat that sort of herring,' Morrison finished, 'but all the same the tailor enjoyed it.'

The collector Donald MacDonald comments that this tale, describing a witches' coven as 'something like a Women's Guild trip to the seaside', is a fairly flippant version of a tale treated more seriously by other narrators. In a very similar nineteenth-century legend, a tailor causes the death of most of the Lewis witches, including his own wife.

HARRIS, WESTERN ISLES

A duck raises a storm

One way that witches caused a shipwreck was to fill a big vessel with water to represent the sea, and then float a small round dish on top, as the ship. With spells, they would set their miniature ocean swirling until the small dish upset, when the ship it symbolised would sink. J. M. McPherson's *Primitive Beliefs in the North-East of Scotland* (1929) describes the technique, in the tale of a lady who wanted her husband drowned on a sea voyage, and asked her maid, skilled in the black arts, to see to it. The maid ordered a large tub to be filled with water, placed an empty wooden dish on the surface, and began her incantations:

> At first there was a slight movement of the water; then waves appeared on the surface: the wooden vessel began to rock: water dashed over its sides. The lady looked on with pale face and bloodshot eyes. The motion of water and vessel increased till one big wave broke over it and engulfed the little barque.

Soon afterwards, the lady learned that her husband's ship had sunk, leaving no survivors.

In *Witchcraft & Second Sight in the Highlands & Islands of Scotland* 1902), John Gregorson Campbell shows what could go wrong with the process. Three Harris witches, he writes, went out one night leaving their big milk-pan full of water with a little bowl floating on top, and giving their servant strict instructions not to let anything come near it.

> The girl's attention was, however, called away for a short time, and a duck came in and took to squattering about in the water on the floor. The witches on their return in the morning, asked if anything had come near the milk-pan.

The girl said no, since she had not been present when the duck did its dirty work. One of the witches then exclaimed, 'What a heavy sea we had last night coming round Càbag head!'

LERWICK, SHETLAND

Up-Helly-Aa

A spectacular procession takes place in Lerwick every winter, in which a Norse longship is dragged through the town by hundreds of men dressed as Vikings and carrying blazing torches, with which they ultimately set light to the boat. These days, the ceremony happens on the last Tuesday in January, but it used to be held on 29 January, the date marking the end of the Yule festivities and called Up-Helly-Aa, meaning 'the end of the holiday', a name that has been transferred to the boat-burning festival.

Before the end of the nineteenth century, the ritual was like an informal version of the fire festival at BURGHEAD (Scottish Lowlands). Barrels mounted on wheeled platforms were filled with straw and tar that was then set alight, and pulled blazing up and down the main street by gangs of young men and boys. The police tried to crack down on the custom, leading to some groups trying to outwit the law by producing a small barrel to distract attention, then dragging out their big barrel in a hurry and rushing it along the street before anyone could stop them.

Fears over safety, and complaints from housewives fed up with having sticky black tar tracked into their houses, led to modifications. First a torchlight procession took the place of the barrel-burning, and then in 1889, under the influence of an enthusiast for Norse customs, the ship made its debut in the proceedings. It is now the star performer, a thirty-foot-long galley whose preparation takes months of work, and is destined for destruction in the flames.

In the treeless Shetlands, worn-out boats were customarily used as the foundation for seasonal bonfires, but the pageant is a conscious revival of an older Scandinavian tradition, Viking funerals that involved setting the corpse afloat and sometimes alight in a wooden vessel (*see* SUTTON HOO, East Anglia). Although the Lerwick ritual in its present form can be dated to fairly recent times, it has its roots in ancient fire ceremonies of death and renewal.

Sea-Serpents

The gargantuan sea-snake, imagined here in an illustration to Olaus Magnus's History of the Goths, Swedes, & Vandals *(1555) as far larger than the ship it is about to overwhelm.*

Are there really unnamed and unknown creatures in the sea? Certainly there are. The depths of the ocean are stranger territory to man than outer space, and large new species recently identified include the Colossal Squid (*see* ROTHESAY, Scottish Lowlands), while the coelacanth, thought to have been extinct for around sixty-five million years, was found alive and well in the 1930s. None of this means that the 'sea-serpent', however you might define it, is a scientific fact, but it remains at least possible that a large and hitherto undiscovered animal exists in our seas.

Modern surveys usually take their starting point in Scandinavia. The sixteenth-century Swedish archbishop Olaus Magnus described the giant beast said to live near the Norwegian coast (*see* SOUND OF SLEAT, Scottish Highlands & Islands), but as the archbishop was in exile in Rome at the time, and based his account on exaggerated information from other expatriates, we need not take his statements too literally. The next important accounts come from the eighteenth century. Erich Pontoppidan, bishop of Bergen in Norway, heard from many sources of a vast 'sea-snake' supposed to haunt the fjords, and Hans Egede (another cleric, this time a Protestant) wrote in 1741 of a serpentine beast seen off Greenland:

This Monster was of so huge a Size, that, coming out of the Water, its Head reached as high as the Mast-Head; its Body was bulky as the Ship, and three or four times as long.

In the early nineteenth century, reports of enormous snake-like creatures were received regularly over several years from around Nahant, on the north Atlantic coast of America. Some scientists accepted that a very large elongated animal had indeed been seen, but other commentators remained sceptical. The New England poet John Brainard composed a 'Sonnet to the Sea-Serpent', beginning with a lyrical encouragement to 'Welter upon the waters, mighty one', and ending satirically:

> But go not to Nahant, lest men should swear,
> You are a great deal bigger than you are.

A particularly well-attested sighting was made in 1848 by the crew of HMS *Daedalus*, who watched a hundred-foot serpent for twenty minutes between St Helena and the Cape of Good Hope, and many similar animals were seen, or said or thought to have been seen, around the British Isles in the nineteenth and twentieth centuries (*see* for example NEWHAVEN, South-East England; THORPENESS, East Anglia; COLL, Scottish Lowlands).

The French zoologist Bernard Heuvelmans devoted a book of nearly six hundred pages to the subject, *In the Wake of the Sea-Serpents* (1968, in English translation), which is essential reading for anyone interested in the history, natural or otherwise, of the great ocean beast. His final example (there are, of course, more recent ones now) dates from 1966, when a couple saw a humped creature, 'something like the Loch Ness monster', swimming near Skegness. A supporting statement, received from another man who had seen a huge dark shape moving quickly off shore, concluded incontestibly: 'I still believe there are things which have yet to be discovered.'

LOCHBOISDALE, SOUTH UIST, WESTERN ISLES

A Water-horse claims a victim

The Water-horse was a particularly unpleasant mythical beast of the Highlands, which could appear as a horse or as an attractive young man (or, on rare occasions, as a woman). Anyone unwise enough to go off with a Water-horse in any shape would commonly be dragged into the water and eviscerated, only a few bits of their insides floating to land. Similar things were said of the Kelpie, and although some folklorists distinguish between the two beings, classifying fresh-water horse-monsters as Kelpies and sea- or loch-dwelling beasts as Water-horses, many others use the terms interchangeably.

Horace Beck, an American folklorist and sailor who collected Scots and Irish sea tales in the 1960s, heard an unusual legend of the Water-horse in Lochboisdale. There was a girl whose father had married again, and her stepmother gave her all the hard work to do while her own daughter spent her time flirting. The stepdaughter was fishing one day, having no luck at all, and as she dared not go home without fish, she was in despair. Then the Water-horse appeared in the form of a handsome young man, and he told her that he knew all about her misfortunes, and had fallen in love with her. With his help she caught more fish than she had ever netted in her life before, but when she found out her admirer was a Water-horse, she refused to have any more to do with him, and he had to return to his watery realm.

Not long afterwards, a dance was held in the village, and the Water-horse told his submarine courtiers that he was going up to the lands of men. Before the night was out, he said, they would have a mortal with them. He put on fine clothes and took his chariot to the surface of the sea, and from there he went to the dance. When the idle, flirty sister saw such a richly dressed stranger, she had no eyes for anyone else. As Beck's informant told the story, 'he danced with her and he flattered her and he courted her and the end of it was that he took her down to the shore':

> And she was so bewitched she didn't know what was happening until it was too late and he had taken her down below the surface of the sea. And that was the last they saw of her. Nobody saw her again.

We assume that the Water-horse meant to do the good sister a favour by removing her unpleasant stepsister, but he may have intended to abduct whichever young woman he could persuade to go with him.

Beck associates the story with the ballad of 'James Harris', a song also known as 'The Daemon Lover', in which a sailor finds his sweetheart has married another man, but persuades her to run off to sea with him. Once they are a good way from land he lets her see the fact that he has hooves:

> They had not saild a league, a league,
> A league but barely three,
> Until she espied his cloven foot,
> And she wept right bitterlie.

Telling her that they are both bound for the snowy mountains of hell, he strikes the mast of the ship, which splits and sinks. Harris has probably become a devil rather than a Water-horse, but the difference is not all that great, and in many tales it is a glimpse of the Water-horse's hoof that alerts some girl to the real nature of her company.

LYBSTER, HIGHLAND

Sailors don't carry corpses

Sailors were traditionally unwilling to carry a dead body on a ship, as was well known in Shakespeare's day. In *Pericles* (1609), Act III scene 1, when Thaisa is thought to have died in childbirth while travelling by sea, a sailor tells her husband: 'Sir, your queen must overboard: the sea works high, the wind is loud, and will not lie till the ship be cleared of the dead.' 'That's your superstition,' Pericles responds, and gets the reply: 'Pardon us, sir; with us at sea it hath been still observed, and we are strong in custom.'

The taboo was still going strong in the nineteenth century, as reported by Walter Gregor in an article of 1885. A man from ROSEHEARTY (Scottish Lowlands) told Gregor that he had once found a floating corpse near Lybster, Caithness, and retrieved it. The boat he used, however, was not his own, but had been borrowed from a Lybster man, who was furious and frightened when he heard what had happened. He scrubbed and rubbed his boat for three days running, and if it had not been almost new, would probably never have used it again.

The same Rosehearty man was once fishing with someone from a neighbouring village when a corpse was caught on one of their lines. A suggestion that they should lift it on board was vetoed by the other man, and they left the body in the water. Next day they were fishing in the same place, the body came up again, and again the neighbour could not be persuaded to carry it to land. It was quite common, Gregor notes, for a boat to return to the same spot several times, when plenty of fish were to be found there, and so once again the boat went to its former ground, and for a third time the body rose to the surface, 'jest as gehn it wisst t' come t' them to be beeriet [just as though it wished to come to them to be buried]', as his informant's daughter put it.

> But it could not be. The fisherman was as relentless as the waves, and 'somebody's darling' sank again.

Other east-coast fishermen, however, assured Gregor that they would carefully bring ashore any dead body they found. The frequent finding of bodies at sea gives an idea of how fraught with danger were the lives of these fishermen.

MARWICK HEAD, ORKNEY

Kitchener's death

The Kitchener Monument at Marwick Head commemorates the loss of HMS *Hampshire* on 5 June 1916. The ship had sailed from Scapa Flow to take Field Marshal Lord Kitchener, then Secretary of State for War, to a conference in Russia. Kitchener had insisted on setting out in a storm, and off Marwick Head the ship struck a German mine and sank at once. Kitchener was drowned along with more than six hundred others, the high death toll partly due to the navy's wartime obsession with secrecy, which led them to block access to the area and prevented the rescue of survivors.

Kitchener was a heroic figure, the champion of Khartoum and the scourge of the Boers, and his death was significant and mysterious enough to provoke enquiry. In the 1920s, a series of hysterical articles in the *Referee* purported to expose everything from a conspiracy between the Admiralty and the Germans to a South African revenge plot, with hints of the supernatural thrown in. The author Frank Power declared among other things that the disaster had been predicted by a fortune-teller, and that Kitchener had had premonitions of death.

The more tangible allegations were refuted, but a flavour of the unexplained hung about the affair, and persistent rumours circulated that Kitchener was not really dead. His body had never been recovered, the navy remained reticent, and whispers that he had gone undercover to Russia vied with reports that he languished a prisoner in German hands.

MILTON, SOUTH UIST, WESTERN ISLES

Flora MacDonald

One of the best-known historical romances of all time is the tale of how Flora MacDonald helped Bonnie Prince Charlie escape his enemies. Whatever your feelings about the Jacobites and the '45 rebellion, or the personal qualities of the Bonnie Prince himself, the young fugitive's flight across the Hebrides makes a thrilling story. In April 1746, Charles and his followers suffered a crushing defeat at Culloden, a thousand men killed, another thousand taken prisoner. With government forces on his heels, the prince took to the islands, slipping from one hideout to another on the islands of Benbecula, Eriskay, and North and South Uist, while the danger grew ever greater, with every port and ferry guarded, and every bay patrolled. Towards the end of June he was in Corrodhale, South Uist, where a plan was made to get him to Skye in the character of a young gentlewoman's maidservant.

Flora MacDonald was twenty-four, two years younger than the prince, when she first met him, and after ten days in his company she would never see him again. She was a member of a well-connected Milton family, and was with her brother in South Uist at the time, but her mother, widowed and remarried, lived on Skye, and so there was a ready-made excuse for Flora to travel between the islands. Her stepfather, Hugh MacDonald, a captain of the militia, provided passes for Flora, her manservant, and an 'Irish maid' named Betty Burke, in order to get them past troops on the lookout for the prince. MacDonald knew very well that his stepdaughter had no such maid, but no more than anyone else who protected the prince, at the risk of their own lives, was he tempted by the vast reward of £30,000 that had been offered for his capture.

In women's clothes hastily cobbled together to fit his big frame, Charles embarked on the next stage of his odyssey. Reaching Skye, he hid in a cave while Flora sought further help. Next day they set out for Kingsburgh, from where the prince would be ferried on to Raasay, and here Flora's part was ended. The prince ardently thanked her, gave her his portrait in a golden locket, and said he hoped to reward her – but never in the forty years or so he lived afterwards, wrote to her, or about her, or in any way acknowledged her help.

From Raasay, the prince returned once more to Skye, then back to the mainland, and finally in September made his escape to France. Flora, meanwhile, was arrested by government troops, and after being kept on a ship at Leith for nearly three months, was taken to London in November.

Flora MacDonald, the heroine of the Highlands.

She was briefly a prisoner in the Tower, but was later placed under house arrest with friends. Her heroism was celebrated even among the anti-Jacobites, and she became the toast of London. After nearly a year, an Act of Indemnity was passed and she was allowed to return to Skye, where she married her childhood sweetheart Allan MacDonald (from whose father's house at Kingsburgh the prince had taken boat for Raasay). The couple emigrated to America, but returned a few years later, and Flora lies in Kilmuir churchyard, where a monument stands to her memory.

To those who live on Uist and the neighbouring islands, the journey has become a cycle of local legends, in which the outcast prince's sufferings and stratagems have left their marks on the landscape. Popular tradition tells how the prince landed on Eriskay, found it deserted, and like Robinson Crusoe single-handedly built a large stone house for himself, planting around it flowers that still flourish there, but will grow nowhere else. In South Uist he is said to have been befriended by 'a woman called Flora' who disguised him in one of her own dresses, but he was found out when somebody threw an apple into his lap, and he caught it not as a woman would (by spreading her skirts) but by bringing his legs together like a man – a time-honoured way of testing somebody's sex, used in Mark Twain's *Huckleberry Finn* (1884).

Flora's memory is celebrated in songs and ballads, of which the 'Skye Boat Song' is the best known ('Speed, bonny boat,' etc). 'Flora Macdonald's Farewell' by James Hogg (1770–1835) is written in much the same spirit, describing the 'bonny young Flora' alone on the shore:

> She look'd at a boat wi' the breezes that swung
> Away on the wave, like a bird of the main,
> An' aye as it lessen'd she sigh'd and she sung,
> Fareweel to the lad I shall ne'er see again!
> Fareweel to my hero, the gallant and young,
> Fareweel to the lad I shall ne'er see again!

NICHOLSON'S LEAP, LEWIS, WESTERN ISLES

'Revenge!'

A ghastly story recorded by the twelfth-century cleric and historian Gerald of Wales tells of a prisoner who had been both blinded and castrated by the lord of Châteauroux in France. Thinking that in his condition he could do little harm, the guards allowed the man to wander about freely, and he became familiar with the passages and steps leading to the castle battlements. One day he seized the lord's only son, carried the child up to the top of one of the towers, and threatened to throw him to his death unless the lord had himself castrated. The lord pretended to agree, struck himself below the waist and cried out. The man asked where the pain was. In the loins, came the reply, at which he dragged the child to the edge, saying he knew this was untrue. A second time, he was told that the pain was in the heart. That too was wrong. Then the lord had no choice but to carry out the terrible order, and when the question was asked a third time, he shrieked in agony that it was his teeth that hurt most.

> 'This time I believe you,' said the blind man, 'and I know what I am talking about. Now I am avenged of the wrongs done to me, in part at least, and I go happily to meet my death. You will never beget another son, and you shall certainly have no joy in this one.'

So saying, he flung himself and the boy from the parapet to their deaths.

Such a memorable tale was bound to be repeated, and versions have been localised in many places besides Châteauroux. The nineteenth-century American poet Sidney Lanier set the scene of his 'Revenge of Hamish' on the Inner Hebridean island of Rhum. MacLean of Lochbuy, in Lanier's poem, wants to kill a red deer, but his henchman Hamish, winded after running up a steep hill, cannot shout when he sees the deer run. He has to confess to MacLean that he has let three does and a buck pass him, and is whipped for his neglect of duty while the chief's wife and child stand watching. Furious at his unjust punishment, Hamish snatches the child from its mother and clambers up a crag towards the sea. He 'stands bold on the brink, and dangles the child o'er the deep', and MacLean's men follow in vain.

> On a sudden he turns; with a sea-hawk scream, and a gibe, and a song,
> Cries: 'So; I will spare ye the child if, in sight of ye all,
> Ten blows on Maclean's bare back shall fall,
> And ye reckon no stroke if the blood follow not at the bite of the thong!'

The chief is whipped in his turn. Then Hamish turns and springs, holding the child, and screaming 'Revenge!' as he falls.

Reincorporating the more graphically gruesome portion of the legend (though without the interesting detail about the teeth), Horace Beck's *Folklore and the Sea* (1973) gives a version set at Nicholson's Leap (*Leum Mhac Nicol*), which he says, wrongly, is off the Point of Ardnamurchan in Argyll. In fact it is much further north, on the island of Lewis, at a place near Tolsta where a narrow pinnacle rock is separated from the body of the island by a deep chasm. Here a clansman named Nicholson, unjustly accused of murder, was sentenced by his chief to be publicly castrated. He was bound and awaiting his punishment, and the clan had gathered at the headland to witness his fate. Exerting all his strength, Nicholson burst the ropes asunder, seized the chief's only son, and leapt on to the rock, still holding the child. A deadlock ensued, since none cared to risk the jump after Nicholson, who was positioned to kick off anyone else who arrived on his eyrie. He shouted his terms across the water: he would not return until the chief was himself castrated on the spot. At length the chief agreed, the operation was performed, and Nicholson declared, 'You have sentenced me to die without issue and you shall die the same way.' He leapt from the rock with the child in his arms, and both were killed.

NORTH RONA, WESTERN ISLES

St Ronan and the sea-monsters

The remote and rugged island of North Rona slopes at its southern end, the rocks showing deep grooves that look as if they could have been carved by claws. The place is known as *Leac na Sgròb*, 'the Scratched

Slab', and towards the west is *Scròbagan nam Biast*, 'the Scratches of the Beasts'.

In the 1860s, Angus Gunn, a Lewis man, told the folklorist Alexander Carmichael the tale of the clawing beasts. Long, long ago, St Ronan came to convert the Lewis islanders to Christianity, and built a prayer-house in Ness, but nothing went right for him. 'The men quarrelled about everything, and the women quarrelled about nothing, and Ronan was distressed and could not say his prayers for their clamour.' An angel appeared to the saint, telling him to go down to the shore, and when he did so, there was a huge sea-beast, 'his great eyes shining like two stars of night'.

Ronan climbed on his back without fear, and the creature flew north-east over the ocean to North Rona. There the saint found a terrible scene awaiting him: crowds of poisonous adders, roaring lions and savage griffins, but when they saw the man of God they fled backwards before him, and slithered off the rocks into the sea. 'And that is how the rocks of the island of Roney are grooved and scratched and lined with the claws and the nails of the unholy creatures.' Ronan built himself a new oratory on the island, where he could worship in peace.

Several Ronans appear in church calendars, and this one cannot be firmly identified, but a rudimentary stone cross of the eighth century found on North Rona indicates a Christian presence, and the remains of a stone cell show that a hermit once made his home on the island. The island might have been named after the saint, but *ròn* is Gaelic for 'seal', and as seals breed there, 'Rona' may have meant 'Seal Island', the tale of St Ronan's visit thus inspired by the name rather than the other way round.

The benevolent monster is no easier to pinpoint. Angus Gunn called it the *cionaran-crò*, and Carmichael suggests that it may have been the Gaelic equivalent of the Kraken (*see* BURRA). Another version of the name, *Cirean Cròin*, is given in John Gregorson Campbell's *Superstitions of the Highlands & Islands of Scotland* (1900) to the sea-serpent, the largest animal in the world, 'as may be inferred from a popular Caithness rhyme':

> Seven herring are a salmon's fill;
> Seven salmon are a seal's fill,
> Seven seals are a whale's fill,
> And seven whales the fill of a Cirein Cròin.

To this was sometimes added, 'seven Cirein Cròin are the fill of the big devil himself.'

NORTH RONALDSAY, ORKNEY

Brita and the seal

At the harbour in Kirkwall, Orkney, in the 1940s, the writer David Thomson talked to a man who had worked there all his life, and remembered that in his youth all the Kirkwall people would watch the boats arrive, and exchange joking insults with the other islanders. 'Sheep' was what they called the men from Shapinsay, 'limpets' those from Stronsay, while the Papa Westray islanders were 'dundies' (cod), and Eday men were 'scarvs' (cormorants). In reply, the newcomers would shout their name for the Kirkwall men, 'Starlings!'

None of these epithets is explained, but the nickname for people from North Ronaldsay, 'seals', must have originated with Gaelic-speakers, *rón* being Gaelic for 'seal'. There was a tale, too, that might have contributed to the name, although Thomson's informant Osie Fea asked him not to identify the family concerned nor even the island where they lived. Their distant ancestor was an Orkney girl named Brita, a laird's daughter, who married when the moon was waning, on the ebb of the tide, an inauspicious time for a wedding, since it was said to make the match barren. Sure enough, the years passed and no child came. Early one morning Brita went down to the sea at high tide, and shed seven tears into the sea. It was said that these were the only tears she ever shed, for she was a proud woman, but she knew this was what she had to do if she wanted to speak with one of the selchies (seal-people). As dawn broke, a big seal came to her, and told her to come again to the rock 'at the seventh stream', a particularly high tide that came only twice a year, for that was the only time he could appear to her as a man.

She did as he said, and met the seal many times afterwards. Several children were born to her, and all of them had webbed fingers and toes, which the midwife used to clip, and that was the reason, so the gossip went, that her descendants had hard horny patches on their hands and feet, although Osie Fea insisted, perhaps too vehemently, that this was skin trouble, and 'nothing to do wi' the selchies'.

This is a twist on the common tale of the 'seal-wife', as told for instance at BARRA. Seal-husbands were rarer, but not unknown. An Orkney ballad told of 'the Great Silkie of Sule Skerry' who got a single woman pregnant, and predicted that he and his child would be shot by her husband when she married:

An thu sall marry a proud gunner,
An a proud gunner I'm sure he'll be;
An the very first schot that ere he schoots,
He'll schoot baith my young son and me.

PORT HENDERSON, HIGHLAND

A mermaid's blessing

Some time in the mid nineteenth century, a young boat-builder named Roderick Mackenzie was gathering fishing bait on the shore when he saw a mermaid asleep among the rocks. Creeping up on her, he grasped her by the hair and she awoke. The mermaid begged him to let her go, and promised him whatever he wanted in return for her freedom. What he asked for was a blessing on his boats, that no one should ever drown while using any boat he built. This she promised, and he let her go.

J. H. Dixon, who reports this story in *Gairloch* (1886), knew Mackenzie as an old man. He writes that in Port Henderson the story was said to be quite true, and that Mackenzie was much respected.

> The promise has been kept throughout Rorie's long business career; his boats still defy the stormy wind and waves. I am the happy possessor of an admirable example of Rorie's craft.

While one does not wish to seem too cynical, the tale was certainly an excellent advertisement.

PORTREE, SKYE, HIGHLAND

Witches as whales

In the late nineteenth century, two brothers were once catching herring at Portree. They noticed that out of all the boats fishing that season, only one was having any luck, and all the others were returning nearly empty to shore. One night the brothers' boat sprang a leak, and they had to take it back into the harbour and haul it up on the beach for repairs. The crew went back to the village, but one of the brothers stayed by the boat, and what he saw next he told to John Gregorson Campbell, who printed the story in *Witchcraft & Second Sight in the Highlands & Islands of Scotland* (1902). A young girl, he said, came out of a nearby house and tapped on the window next door. Another girl came out to join her, and the two of them walked on together.

Superstitions

Sailors and fishermen are so much at the mercy of the elements and the fates that there are naturally many superstitions dictating things that must or must not be done, not only by the mariners themselves but by their families. For example, in the Scottish Highlands it was said that no woman should comb her hair at night if she had a brother at sea, in case the hairs met him in the dark and entangled his boat.

All along the east coast of England, it was maintained that a sailor's wife who wound wool by candlelight would be inviting doom, either because it was like spinning a shroud, or because it would lead the boats astray, making them steer in circles, or, more simply, because she would be 'winding the husband overboard'. In a nineteenth-century Yorkshire tale, a young wife wound her skein after sunset although her neighbours begged her not to do it. Three times the ball slipped from her hand, and when she told her husband what had happened, he realised that his third voyage would be his last. After two more trips he left the sea, but a day came when volunteers were urgently needed to rescue a shipwrecked crew. The

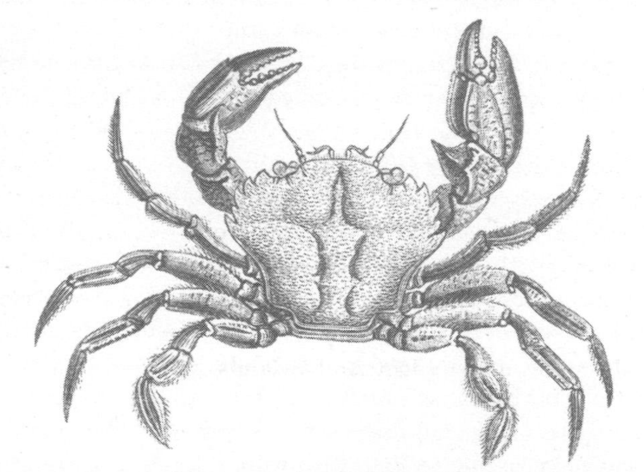

The names of some animals were taboo among superstitious sailors. On the east coast of Scotland, for example, crabs were called 'snifltie fits', meaning 'dawdling feet', a reference to their scuttling sideways motion.

sailor went with the lifeboat, and although he saved many lives, he lost his own.

Pins were said to 'part love', and so it was a bad omen to give anyone a pin before they went on a journey. A ship's captain, about to go aboard, asked his wife for a pin because his clothes were torn, and she remarked, laughing, 'Do ye no ken that preens [pins] pairt love?' She gave him the pin, though, and he was drowned on the voyage.

An obvious symbolism informs certain beliefs, such as that it was unlucky to lose a bucket overboard (representing the loss of the boat), or to wear the clothes of a dead sailor during the voyage when he died, although this only applied to the merchant service. On a battleship, many of whose crew might be killed, the clothes of the deceased were auctioned off almost at once.

Not only actions but even some words were forbidden to sailors, and strange euphemisms used instead. On the east coast of Scotland, crabs could not be explicitly mentioned, and were called 'snifftie fits', roughly meaning slowcoaches, 'fit' meaning 'foot' and 'snifftie', 'dawdling'. In the seventeenth century, the island of Hirt was never to be named by local fishermen, a taboo that may have helped the adoption of its erroneous other name ST KILDA, although the island could also be called the High Country. Other prohibited words are mentioned at RABBIT ISLANDS, and some terms are so unlucky that they remain a mystery. Humberside fishermen interviewed in the mid twentieth century said that it was a death omen to find 'a particular kind of fish' in their crab pots, and claimed to have forgotten what the fish was called, but it seems more likely that even its name spelt disaster.

See also ROSEHEARTY (Scottish Lowlands).

> On reaching the green, the two girls began to disport themselves, then of a sudden became hares, and chased each other round and round. After this they made their way to the shore, and at the edge of the water, leapt into the sea and became whales. They went out from land spouting the water as high as a ship's mast.

Next morning the 'lucky boat' was full once again, but all the others came in empty as usual, and the fishermen said that during the night's fishing they had seen two whales 'throwing up the sea in a dreadful manner'. The link between what the girls were doing and the good catches made by the one boat is left unstated, but the implication is that the magical whales somehow herded the fish into whichever nets they chose, presumably those belonging to their own families.

Turning into whales is very unusual, but witches were often supposed to appear as hares. So closely were the animals associated with sorcery that the very word 'hare' was forbidden, particularly among sailors. A thirteenth-century source calls the hare 'The animal that no one dare name', and the antiquary Richard Carew wrote in 1602: 'we will launch out into the deepe, and see what luck of fish God there shall send us, which (so you talke not of Hares or such uncouth things, for that proves ominous to the fisherman . . .) may succeed very profitable.'

RABBIT ISLANDS, HIGHLAND

Forbidden words

Sailors and fishermen had taboos on a vast range of words: salt, salmon, knives, eggs, cats, pigs, hares, foxes. There were euphemisms by which some things could be mentioned, such as 'the white stuff' for salt (*see* EYEMOUTH, Scottish Lowlands), while a pig, the supreme example of a forbidden animal as described at ST MONANS (Scottish Lowlands), was 'grumphie', 'curlie-tail', 'guffey', 'the Grecian' or 'the article'. A hare should not be named (*see* PORTREE), and neither should a rabbit, which was known as 'the gentleman', so that the Buckie fishermen always called the Rabbit Islands 'the Gentleman's Islands'.

Several more alternative names for rabbits are mentioned in Iona Opie and Moira Tatem's *Dictionary of Superstitions* (1989). A Dorset boy in 1953 reported that the local sailors referred to a 'Wilfred' or 'one of them furry things', unless he wanted to wish someone bad luck, in which case he would say 'Rabbits to you.' An even more recent example comes from a

Brighton fisherman in the 1980s, who said that if you needed to talk about rabbits, 'You had to call them Bexhill Runners.'

See also SUPERSTITIONS (p. 360).

ST KILDA, WESTERN ISLES

No such person

St Kilda is a mystery. Who was she or he, and when did this holy hermit come to the Hebrides?

The answer is that there never was a St Kilda. She is a mapmaker's mistake. The original Gaelic name for the island group was Hirt, but in the early sixteenth century it was also known as 'Skildar' or 'Skilder' ('shields', in Old Norse). Where today 'Saint' is abbreviated to 'St', the common form then was 'S.', often without a space before the proper name, so that St Mulvay on Lewis, for instance, would have appeared as S.Mulvay. In 1573 and 1578, two maps were produced where 'Skilder' appeared as 'S.Kilder'. Such mistakes were by no means uncommon – a 1592 map gives Skarbo (now Scarba) as 'S.Karbo' – but perhaps because 'St Kilda' sounds so like the genuine St Hilda, she took hold, and by the late seventeenth century 'St Kilda' was firmly established as the name of the island. Martin Martin, whose *Description of the Western Isles of Scotland* was published in 1703, asserts that the name 'is taken from one *Kilder*, who lived here', and his explanation was repeated by many later writers.

See also SUPERSTITIONS (p. 360).

ST MOLVEG, LEWIS, WESTERN ISLES

Shony the sea-god

Seaweed used to be valuable as fertiliser, and special ceremonies were followed around the western coast and islands of Scotland to encourage a good crop. The Thursday before Easter, generally known as Maundy Thursday, was in the Hebrides called 'the Day of the Big Porridge', since on this day, if insufficient seaweed had been thrown on shore during the winter, pots of porridge made with butter were poured into the sea, to the accompaniment of incantations. The object of throwing the produce of the land into the sea, says the nineteenth-century folklorist John Gregorson Campbell, was to make the sea generous in throwing its produce on land. Porridge was recommended for pregnant women, and the sea too needed nourishment to make it fertile.

A more elaborate ritual was followed in Lewis, where in the seventeenth century the people gathered each All Hallows Eve (a traditional time for supernatural rites, and also around the start of the stormy season) at the church of St Molveg or Mulvay, and lit a candle on the altar. They then went from the church to the shore, where they joined in the Lord's Prayer, and one of them waded into the sea with a cup of ale, saying words to this effect: 'Shony, I give you this cup of ale, hoping that you will be so kind as to send us plenty of sea-ware [seaweed] for enriching our ground the ensuing year.' He poured the ale into the sea, and the people then returned to the church and put out the candle they had left burning. Drinking, singing, and dancing followed.

'Shony' was evidently a sea-god, and the folklorist Ronald Hutton calls this 'one of the most blatant examples recorded in the British Isles of a pagan rite surviving into the Christian epoch'. The ceremony was notionally brought to an end in the 1670s by two determined Presbyterian ministers, but in practice it seems simply to have been moved to a different date. In the nineteenth century, Alexander Carmichael records a very similar ritual taking place on Maundy Thursday, with a verse he translates as:

> O God of the sea,
> Put weed in the drawing wave,
> To enrich the ground,
> To shower on us food.

A nineteenth-century correspondent to the *Fortnightly Review* recalled her Hebridean nurse telling her about a mysterious sea-god named Shony, and the folklorist Anne Ross, writing in 1976, records that she met people in the Outer Hebrides who remembered the custom, as described by Carmichael, taking place in their childhood.

SANDSIDE BAY, HIGHLAND

The schoolmaster and the mermaid

In about 1797, William Munro, schoolmaster of Thurso, had a close encounter with a mermaid. Some years later, a Dr Torrence asked him to describe his experience, and his reply was printed in *The Times* on 8 September 1809.

Munro's letter shows that he is sensitive on the subject. He fears that Torrence is mocking him, and notes that his own testimony is unlikely to convince anyone who has already rejected other accounts. Having said that, however, he goes on to give a detailed report.

One fine summer's afternoon, the teacher was walking along Sandside Bay towards Sandside Head, when he saw 'a figure resembling an unclothed human female' sitting on a rock and combing its hair. He knew it could not be an ordinary bather, because the rock it was perched on was a very dangerous one, and therefore he took a good look at it, seeing its plump face, rosy cheeks and blue eyes. Its torso and arms were like those of a fully grown woman, and its fingers did not seem to be webbed, although he could not be certain. For three or four minutes the creature stayed in view, combing its long light brown hair, 'of which it appeared proud', before it looked at the hill where Munro was standing. Perhaps it caught sight of him, because immediately after that it dropped into the water and disappeared.

Munro adds that although he had heard about mermaids from truthful people, he had never believed them until he saw for himself. His letter is long-winded in the extreme, and he ends with a flourish:

> If the above narrative can in any degree be subservient towards establishing the existence of a phenomenon, hitherto almost incredible to naturalists, or to remove the scepticism of others, who are ready to dispute every thing which they cannot fully comprehend, you are welcome to it.

The schoolmaster's pedantic style somehow makes him a persuasive witness. Walter Scott was one of those who read the account, and was more than half convinced by it that mermaids were real. It demonstrated, he wrote, 'either that imagination played strange tricks with the witnesses, or that the existence of mermaids is no longer a matter of question'.

SANDWOOD BAY, HIGHLAND

The Land of the Mermaid

Sandwood Bay, near Cape Wrath in the far north of Scotland, is a scene well suited to weird encounters. Its wild desolation gives it an otherworldly atmosphere on even the sunniest of days, and in a storm it appears a haunt of demons. 'I am informed,' writes R. Macdonald Robertson in *Wade the River, Drift the Loch* (1948), 'that even to-day some of the local inhabitants will not venture near it after sun-down.'

A 'foul thing' with a hairy face and devilish grin was seen near Sandwood Bay by three fishermen in the early twentieth century, and was said to have caused the death of two of them.

Robertson met a landowner called Alexander Gunn, who told him of an experience he had there on 5 January 1900. He had climbed down a gully to the seashore to rescue one of his sheep, and when he reached the bottom his collie dog howled, crouching at his feet, all its hair on end and its tail between its legs. Gunn looked up, and to his astonishment, on a ledge of rock only six or seven feet from where he was standing, he saw a mermaid.

> So impressed was I that I can to this day distinctly recall her appearance which left a vivid picture on my mind which I can never forget, old man as I am. She was no grey seal; she was a real mermaid – a bonnie lassie, clear in complexion as ever I saw. Her hair was reddish-yellow in colour, and curly; and she had a wreath of seaweed round her neck. She had greenish-blue eyes, and arched eyebrows, and she stared at me with a kind of frightened expression on her face.

The mermaid stayed stock-still, not even moving 'her wee short arms', and did not speak. Gunn could see her fish-like tail dangling on the other side of the rock. He realised that she was marooned there until the tide rose, and that she was scared and angry. For several minutes they gazed at each other, and then 'I took to my heels in terror.'

Gunn wished that his dog had been alive and gifted with speech, so that

Draped entirely in ocean, St Cuthbert prays, while the inquisitive monk of Coldingham (Scottish Lowlands) drops off to sleep. Inset to the right is a second image of Cuthbert with his robes back on, his feet warmed by two sketchy little otters.

One of the famous snakestones of Whitby (North-East England), a fossil ammonite with a serpent's head carved on. The tale went that St Hilda had turned all the local snakes into stone, and that their heads broke off in their fall from the cliffs to the sea.

In the first picture, St Ursula sets off with her virgin companions. Next we see the barbarous Huns slaying the maidens, and finally St Ursula herself, having refused the love of the heathen king, is martyred. A version of her legend is associated with Llangwyryfon (Wales).

At the annual ceremony at Lerwick (Scottish Highlands), a magnificent Viking ship goes up in flames.

An aerial shot of the excavations at Sutton Hoo (East Anglia) shows the outline of the great rowing boat buried there with its load of gold, jewels and weapons, and perhaps with a royal occupant.

J. M. W. Turner (1775–1851), greatest of British sea artists, painted several pictures of the 1805 battle of Trafalgar. This version from 1822–4 shows the chaos of wounded men and falling masts.

In John Millais's Boyhood of Raleigh *(1870), two lads drink in the sea-tales of a veteran, inspiring at least one of them to a career of exploration and adventure.*

In real life, Captain Kidd was a fairly incompetent pirate, but imaginary portraits such as this one show a glamorously sinister figure of buccaneer myth.

George Frederic Watts's Sea Ghost *(1887) is marvellously evocative of a phantom ship, perhaps the* Flying Dutchman, *advancing through the pearly mist.*

A nineteenth-century engraving of a contraband cargo being brought ashore. The man holding the lantern cuts a more respectable figure than the rest, and may even have a clerical collar. It was often said that clergymen assisted the smugglers in their midnight work.

it could corroborate his story. 'What I saw was *real*,' he insisted. 'I actually encountered a mermaid.'

Robertson was told some other stories of sea-creatures in the district, including one from a guest at the Garbert Hotel, Kinlochbervie, who saw what she at first thought was 'a bunch of yellow seaweed' near her in the water, while she was boating off Achriesgill Bay on 24 June 1939. It then turned to reveal 'a beautiful face'. Speechless with astonishment, she pointed out the mermaid to her companions, and as she did so the creature sank, with a flip of her tail above the surface. When she mentioned this to a local man, he told her that this mermaid had been seen several times, either floating in Loch Inchard or lying on the rocks.

A less lovely denizen of the sea described to Robertson was seen near Cape Wrath by John Falcony and two other fishermen, who were rowing close to the shore when a 'grotesquely hairy, semi-human figure' rose out of the waves. It had beady black eyes and 'a diabolical grin on its face', and a powerful tail was visible underwater.

> Whether merman or not, what the three men saw was a foul thing. Two of the fishermen died shortly afterwards, of peculiar maladies, aggravated by the memory of their frightful encounter; and John Falcony alone remains alive to tell the tale.

With so many sightings, it is not surprising that Sandwood Bay was sometimes called 'The Land of the Mermaid'. Even in the 1960s, writes Jane Hutchins in *Discovering Mermaids* (1968), 'some friends of mine, visiting relatives in that area, were assured that mermaids were still occasionally seen just off the shores there.'

SCAPA FLOW, ORKNEY

'Drake's Drum'

On 21 June 1919, the German High Seas Fleet sailed to Scapa Flow, preparing to surrender to the British. On board the British flagship HMS *Royal Oak*, a sound like drumming was heard by Captain John Luce and another captain who was on board with him. All the crew were at action stations, making it unlikely that any of them was beating a drum, and a search was made throughout the ship to see who or what was responsible, but nothing was found. Luce was tempted to suppose that he had heard 'Drake's Drum'.

The underlying story is a classic example of how folklore changes and develops over time. When Sir Francis Drake died in 1596 in Panama, his drum, sword and Bible were sent home to his widow at Buckland Abbey in Devon. Here the drum was hung on a wall, and it hangs there now. Even in his lifetime, Drake had a reputation as a wizard (*see* PLYMOUTH, South-West England & Channel Islands), but nothing special seems to have been said about his drum until the mid nineteenth century, when Robert Hunt, investigating West Country folklore, was told by the old housekeeper of the abbey that 'if the warrior hears the drum . . . he rises and has a revel'. In 1895, Sir Henry Newbolt's poem 'Drake's Drum' was published, and became enormously popular. In it, Drake is said to have promised on his deathbed that if ever England is in danger and the drum is beaten, he will descend from heaven to his country's defence.

Newbolt may have drawn on Devonshire oral lore, although it is now hard to disentangle which came first, the tradition or the poem. It was certainly soon accepted as a genuine folk tale that the drum had been heard before the battle of Trafalgar, and that Trafalgar's hero Nelson was a reincarnation of Drake. In the First World War, the poet Alfred Noyes wrote in *The Times* that the drum had rolled not only for Trafalgar, but for the recent battle of Jutland, and events at Scapa Flow a couple of years later were seen as confirmation that it could be heard at fateful moments.

Sir Francis Drake, England's defender against the Spanish Armada, was said to have been reincarnated as the later champion Nelson. According to legend, Drake's Drum sounded at the battle of Trafalgar and other moments of national crisis.

The legend had been transformed, from an instruction to beat the drum in time of national peril, to the idea that the drum would sound, *without* being beaten, at times of war and crisis. Further reports circulated of hearing it in September 1940, around the beginning of the Blitz. At that time, as in 1919 at Scapa Flow, strict security measures were in place forbidding drum signals, and a search failed to discover the instrument, but a precise description was given of 'a very incessant beat, pause, two sharp beats in quick succession, one sharp beat'.

SHIANT ISLANDS, WESTERN ISLES

Blue Men

Travelling around the Hebrides in the 1960s in search of sea legends, Horace Beck asked about the 'Blue Men' or 'Blue Men of the Minch'. His informant had heard of these beings, and that they could raise great storms, but did not know much else about them.

In the late nineteenth century, tradition was much more lively concerning these Blue Men. John Gregorson Campbell writes in *Superstitions of the Highlands & Islands of Scotland* (1900) that he had talked to a man who was convinced he had himself seen one of them, describing a 'blue-coloured man, with a long grey face' who followed a boat he was in, floating half below the water but visible from the waist up, and 'so near that the observer might have put his hand upon him'. So well known were these creatures that the channel between Lewis and the Shiant Islands, now called the Minch, was known as *Sruth nam Fear Gorm* ('the Stream of the Blue Men').

A ship passing through the channel once chanced upon a Blue Man asleep on the waves. The crew took him on board, and before he awoke they bound him carefully in strong ropes. Soon afterwards, however, two more such men came after the ship talking to each other. As soon as he heard their voices, the Blue Man who had been so securely tied, as they thought, leapt to his feet and broke the ropes 'like spider threads'. He jumped overboard and swam away with his two friends.

An explanation sometimes proposed to account for the Blue Men is that they were Picts, who could have continued to live on remote islands long after they had vanished from the Scottish mainland. The name 'Pict' comes from the Latin *picti*, 'painted people', and the inference is that these aboriginal survivors would have coloured their skin. The Picts were evidently a seafaring people, since their distinctive carved stones are found on islands as far as St Kilda, and if they are to be identified with the Blue Men, they would not have been swimming in the water but rowing small craft, perhaps something like the kayaks of the 'Finn-men' (*see* EDAY). Campbell does not try to rationalise the phenomenon, but gives a mythic origin for the Blue Men which he heard on Skye:

> The fallen angels were driven out of Paradise in three divisions, one became the Fairies on the land, one the Blue Men in the sea, and one . . . the Northern Streamers, or Merry Dancers, in the sky.

The 'Merry Dancers' is a name for the aurora borealis or Northern Lights, imagined here as luminous air-dwellers.

SOUND OF SLEAT, SKYE, HIGHLAND

'The great Sea Snake'

A detailed account of a sea-monster was printed in the magazine *Land and Water* in 1872, from a witness identified in later sources as the Reverend David Twopeny, vicar of Stockbury in Kent. He was on a boat trip up Loch Hourn with another minister, John Macrae of Glenelg, and their two families, and while the party was in the Sound of Sleat, just south of Skye, they repeatedly observed 'the creature which has so often been seen in the Norwegian fiords; but nobody knows what it is, as one has never been caught'.

The head, wrote Twopeny, appeared first, then up to eight loops or 'convolutions' that rose and sank in regular sequence. With four pairs of binoculars on board, he and his companions were able to examine it closely, and calculated, from the size and number of the loops, that it must be about ninety-six foot long. It was inky black, 'as dark as a black slug', and quite different from anything else in the water. At one point the creature made straight for the boat and got within about a hundred yards before turning away. One of Macrae's daughters was so frightened she insisted on being put ashore at 2.30 a.m., and walked thirteen miles alone over wild mountain tracks rather than stay around the monster. Macrae had heard of a large beast 'rushing about Loch Derich with great rapidity' for several days the previous summer, and thought it must have been the same thing.

Twopeny finished his letter with a wish that the naturalist Francis Buckland (editor of *Land and Water*) would visit Skye and catch the serpent, to which Buckland replied, 'I need not say how delighted I

The sixteenth-century historian Olaus Magnus wrote of the enormous sea-snake that haunted the coast of Norway and devoured sailors. A contemporary illustration to his book shows the creature at its terrible work.

should be to have the job.' He found Twopeny's account convincing, and was satisfied that neither seals, porpoises, turtles nor barnacle-encrusted logs of wood could have accounted for what the clergymen and their families had seen:

> It is, therefore, evident to my mind that the coast of Norway and the north of Scotland are occasionally visited by a living creature, which, for want of a better name is called 'The great Sea Snake.'

The reason why Buckland and Twopeny mention Norway is that the Norwegian fjords were long supposed to be the particular haunt of the gigantic beast, as they were of the Kraken (*see* ROTHESAY, Scottish Lowlands). All Norwegian mariners, Olaus Magnus reported in the sixteenth century, agreed that a vast serpent, two hundred feet long, and twenty feet thick, lived off the coast of the country near Bergen. It was black, with a long mane hanging from its neck, sharp scales and flaming eyes.

> This Snake disquiets the Shippers, and he puts up his head on high, like a pillar, and catcheth away men, and he devours them; and this hapneth not but it signifies some wonderful change of the Kingdom near at hand; namely that the Princes shall die, or be banished; or some Tumultuous Wars shall presently follow.

About two hundred years later, the Reverend Erich Pontoppidan, Bishop of Bergen, was inclined to doubt such tales, but found them confirmed by hundreds of sailors and fishermen. Many said that the sea-serpent had sunk boats, sometimes quite large ones, by throwing itself across them, while one observer declared that he had seen the serpent raise itself so high out of the water that its head reached above the mainmast of a tall ship. Some of the northern traders, wrote Pontoppidan, thought it very strange when they were asked whether such a creature really existed: 'they think it as ridiculous as if the question was put to them, whether there be such Fish as Eel or Cod.'

STROMNESS, ORKNEY

Bessie Millie the witch

The Reverend Forbes Phillips of GORLESTON-ON-SEA (East Anglia), who used the pen-name Athol Forbes, was the grandson of George Phillips, an exciseman stationed at Orkney in the early nineteenth century. In *The Romance of Smuggling* (1909), Forbes writes at length about his grandfather's encounters with Bessie Millie, a well-known witch of Stromness.

Millie made a living by selling winds to sailors and providing luck charms for whalers, and was an influential figure in the islands. Phillips had been selected to put down the Orkney smuggling trade, and from his first arrival in his revenue cutter HMS *Widgeon*, 'he found himself thwarted and baffled by the witch Bessie Millie'. One day soon after he had come to the islands, he found her standing in his cabin. She was an impressive figure in her hand-woven gown and plaid shawl, with dark eyes, and a strong masculine face 'like discoloured parchment'.

> He sternly demanded from the sentry why the woman had been allowed to invade his sanctum unannounced. The man, a brave and splendid seaman, said she had threatened to blind him, and he believed she could do it. He ordered the man under arrest, and the woman to leave the ship. 'Not before I have had my say,' she said. 'George Phillips, you come from a proud, wilful race, and most of yer brood have met death by sword, bullet, and violence; beware how ye defy Bessie Millie.'

Reluctant to turn her out by force, Phillips told her to say what she had to say, and then go. Quickly and fluently she told him a number of events in his own family history, some of which he thought she might have learned from local gossip, 'but what did astonish him was her knowledge of certain facts which he did not know then, and which eventually turned out to be true.' This preamble, he assumed, was to intimidate him and his crew. She then told him to take his ship back to 'the cursed Government that had sent him'. Phillips took the affair lightly, but he could see that his men were frightened.

Soon it was reported that a cargo of contraband was to be landed on the Orkney mainland. The *Widgeon* stood out to sea, while two pinnaces (small boats) were sent to watch the creek where the landing was supposed to take place, but within hours they were back. While the pinnaces were making their way up the inlet, Bessie Millie had appeared on the headland and ordered the men back to their ship, threatening them with curses, and although the officers had tried to proceed, their crews had rebelled. On returning to the cutter, several of the men took to their hammocks, saying that they had been bewitched.

Phillips went at once to see Millie, who was expecting him. Unless she came to the ship and let the men see her remove her spell, he said, he would burn her house down. She held out until he had actually started the fire, but then did as he asked.

> The men who said they were sick quickly recovered, and something like an armed truce prevailed between the Government ship and Bessie Millie. But she always seemed able to divine the plans of my grandfather, and on more than one occasion she was the means of frustrating them.

Forbes does not give his grandfather's dates, nor does he mention whether this was the same Bessie Millie as the woman Sir Walter Scott visited in 1814. Like the witch Phillips met, Scott's Bessie was famed for her power over the weather, and he notes in his diary that she had 'a corpselike complexion . . . a nose and chin that almost met together, and a ghastly expression of cunning'. He could not resist paying her for a wind to see what would happen, and she assured him that there was nothing evil in what she did, but that he would have a fair breeze from her prayers. He was disappointed to find that the winds next day were dead against his journey.

Scott's Bessie Millie was, she told him, 'upwards of ninety', and his visit cannot have been many years after Phillips went to Orkney. If this was the same woman, it seems rather shocking that the exciseman should have gone to the lengths of setting such an old lady's house on fire. It is possible, however, that Phillips's experiences were later, and that this was a different woman – an idea supported by the fact that Scott describes her eyes as light blue, whereas Phillips says they were dark. Phillips's Bessie might have been a daughter, perhaps, or another wind-seller who had taken the name of her predecessor to give her reputation a boost.

STRONSAY, ORKNEY

The Animal of Stronsa

The renowned 'Animal of Stronsa' was cast ashore on the island now spelt Stronsay in 1808, and appeared at the time to confirm all rumours and reports of vast sea-serpents. The poet Thomas Campbell wrote to a friend in 1809 that 'A snake . . . has been found thrown on the Orkney Isles, a sea-snake with a mane like a horse, four feet thick, and fifty-five feet long – this is seriously true.'

That was certainly how it was described by several observers, in statements sworn before the Justices of Kirkwall. In addition, it had six fins or paws, popularly described as its 'wings', evidently with the idea that this was some kind of dragon. The reason why the Animal could be examined so closely by so many was that it was dead, and in fact it was considerably decomposed, which, according to some naturalists, accounted for some of its less common features. It was, they said, nothing more fantastic than a large and decayed basking shark. Others, some reputable scientists among them, were convinced that this was no less than a sea-serpent. The naturalist Robert Hamilton asserted in 1839 that reports of the Stronsa beast left 'no manner of doubt as to the existence of some such animal', and Dr Barclay, on behalf of the Wernerian Natural History Society, declared in about 1809 that those who maintained it was a shark were mistaken, if not downright liars, given the unanimity of the Orcadian reports. He demanded what could have made the witnesses' solemn declarations 'so widely different from nature and from truth'. Those who cast doubt on the sea-serpent story, he said, had discovered, 'if not a new species of fish, at least a new and remarkable variety of the human species, inhabiting the Orkneys'.

Zoologists now, however, agree that the Animal was indeed a shark, its 'mane' the remnants of its back fin. One of its bristles and a part of its vertebra or cartilage are still preserved in the National Museum of Scotland, but were unfortunately kept in formalin, which has destroyed much of the material that might otherwise have been submitted to DNA analysis.

It is today generally, though not universally, accepted that giant sea-serpents, Krakens and mermaids – where the sightings are not considered to be illusions or lies – are misidentified members of species known to science, although the recent discovery of the Colossal Squid (*see* ROTHESAY, Scottish Lowlands) may encourage serpent-fanciers to hope for more such stupendous creatures lurking in the ocean depths. The zoologist Bernard Heuvelmans, who in the 1960s devoted a long and scholarly book to the subject of sea-serpents, writes that he has personally examined many 'stranded monsters' that proved to be basking sharks or similar creatures:

> But what if one day a real sea-serpent, a big undiscovered animal, breathes its last on some beach and nobody but 'ignorant fishermen' stares at it, none but sea-birds dissect its flesh and only the waves take charge of its mighty bones?

SWILKIE, HIGHLAND

'Why the Sea is Salt'

The word 'maelstrom' is often used more or less as a synonym for 'whirlpool'. Originally, it was the name for a strong tidal current that rips past the Lofoten Islands in Norway, just north of the Arctic Circle, which was thought so terrifying that it was sometimes said to be the entrance to hell, or alternatively to be a whale-like beast of immense size, with a huge round mouth that could suck down whole ships. It is a genuine menace to ships, especially when the current and the wind are in opposing directions. At such times it can sink large vessels, although not whole fleets as described by Edgar Allan Poe in 'Descent into the Maelstrom'. Poe's imagery has had a profound influence on the portrayal of whirlpools in adventure stories and films. He describes it as a circle of more than half a mile in diameter, formed quite suddenly:

> The edge of the whirl was represented by a broad belt of gleaming spray; but no particle of this slipped into the mouth of the terrific funnel, whose interior, as far as the eye could fathom, was a smooth, shining, and jet-black wall of water, inclined to the horizon at an angle of some forty-five degrees, speeding dizzily round and round with a swaying and

Fiddler's Green and Davy Jones's Locker

Fiddler's Green is the sailors' paradise, a place of never-ending grog, tobacco, plum pudding, and steak. It is the nautical equivalent of the medieval 'Land of Cockaigne', a fantasy kingdom where food and drink were bountiful and work unknown, or the American 'Big Rock Candy Mountain', with the added attraction of willing women who would attend to the mariner's every need. In one of Captain Marryat's sea novels, *Snarleyyow* (1837), a parson reproves a sailor for having three wives:

> Parson, says I, in each port I've but *one*,
> And never had more, wherever I've been;
> Below I'm obliged to be chaste as a nun,
> But I'm promised a dozen at Fidler's Green.

It may be significant that Marryat spells it 'Fidler'. This is a nineteenth-century sailor's word for a crab that burrows underwater, making 'Fidler's Green' a submarine kingdom, and a place to which only the drowned could travel. The double-d form 'Fiddler's Green', however, recorded from the late eighteenth century onwards, conveys an image of singing and dancing on the grass, far from the toil and danger of the ocean, and could refer to a real refuge for weary seamen. One folk tale says that a sailor who is tired of the waves should walk inland with an oar over his shoulder, until he comes to a village where people ask what strange thing he is carrying. Then he knows he has arrived at Fiddler's Green, and will never have to work again.

A less desirable destination was Davy Jones's Locker, a name for the bottom of the sea that has been explained in various ways. 'Davy' could be from 'Devil', or from 'Duffy' or 'Duppy', a West Indian spirit, or even from 'duffer', a clumsy ass (Duffer Jones would presumably be a Welsh sailor who falls overboard), although it may be just a familiar name added to a corruption of 'Jonah', the Old Testament prophet swallowed by a whale (*see* JONAH AND THE WHALE, p. 164). An inventive theory mentions Deva Lokka, a Hindu god of destruction,

An illustration to Tobias Smollett's Peregrine Pickle *(1751) shows Trunnion, a superstitious old commodore, drawing his sword to defend himself from the fiendish 'Davy Jones', in fact one of his former shipmates in disguise.*

but 'locker', a common word for a sea chest, seems clear enough without looking for exotic origins.

The name 'Davy Jones' for a marine demon was first mentioned in print by Tobias Smollett in *Peregrine Pickle* (1751), but this 'fiend who presides over all the evil spirits of the deep' must have been well known already among seafarers. In the novel, a sailor disguises himself in an oxhide and a mask of leather stretched over a shark's jaws, with huge glass eyes, and then grips a firework between his teeth. A terrified commodore declares that he recognises Davy Jones 'by his saucer-eyes, his three rows of teeth, his horns and tail, and the blue smoke that came out of his nostrils'.

Later nautical expressions elaborate the idea of the sea-devil. 'Davy Jones's natural children' (his bastards) is recorded from the nineteenth century to mean pirates or smugglers; 'Davy's dust', from 1830, meant gunpowder; and in the twentieth century, 'Davy Jones's shocker' was applied to an underwater explosion.

> sweltering motion, and sending forth to the winds an appalling voice, half-shriek, half-roar, such as not even the mighty cataract of Niagara ever lifts up in its agony to Heaven.

In fact this Maelstrom is not a whirlpool proper like CORRYVRECKAN (Scottish Lowlands), since there is no vortical action, but the image contained in the Norwegian name – *male*, 'to grind or churn', and *strøm*, 'current' – is clearly of a revolving whirl of water. An old Norse myth associated with the Maelstrom tells of Grotti, a magic mill that would produce whatever it was asked for. The king of Denmark had two big, strong women slaves called Fenia and Menia, who turned the millstones to make peace and prosperity, but they were captured by a Viking, who carried them off on his ship, along with the mill, and ordered the women to grind salt (a valuable commodity, when it was the only means of preserving meat). After several hours they asked him if he was not tired of salt, but he told them to go on grinding, and so they did, until the ship sank under the weight. The mill Grotti kept on grinding away underwater, making a whirlpool, and turning the sea water to brine.

The tale of 'The Magic Mill', or 'Why the Sea is Salt', brought to the British Isles by Viking settlers, has become common in Irish and Scottish tradition, and is particularly attached to Swilkie, a place where four or five opposing tides meet off Stroma in the Pentland Firth. Like the Maelstrom of Norway, it is not exactly a whirlpool, but its swirling action readily suggests the working of a giant mill underwater, and the story of its operation by two witches, Grotti Minnie and Grotti Finnie, derived from the Scandinavian tale of Menia and Fenia and the mill Grotti, was current until at least the late nineteenth century.

TARBAT NESS, HIGHLAND

Winds in a bottle

In *Scenes and Legends of the North of Scotland* (1835), Hugh Miller describes the promontory of Tarbat Ness, a narrow headland encircled by ocean, sterile and solitary, where one is overwhelmed by 'wide, measureless, interminable space'. During the winter storms, he writes, 'On the north a chain of alternate currents and whirlpools howl, and bellow, and toss, and rage, as if wrestling with the hurricane; on the east the huge waves of the German ocean come rolling against the rocky barrier.'

Here, in the early eighteenth century, a small community of fishermen and their families had lived in four huts, until one of them found his mother engaged in witchcraft. He and his father intended to denounce her, but before they could do so their boat was swallowed up by a quick-sand. The remaining residents soon departed, terrified by 'supernatural sights and noises', leaving only the witch herself in occupation. She was thought to be in league with the Devil, and was often consulted by sailors who paid her for favourable winds, but terrible things could happen to those who offended her. A shipmaster who had made her his enemy moored his craft one evening in a bay at Tarbat, but next morning found that the vessel was in a muddy ditch far inland, having been carried by the infernal powers over rocks and beach, a meadow, two cornfields and a moor. He managed to conciliate the witch, and his ship was returned to the bay.

Another captain offered the witch two bottles of spirits in return for a good wind, but forgot to deliver them. On his way home, what had begun as a favourable breeze soon turned into a hurricane. Catching sight of the bottles, he remembered his promise. At that moment two ravens appeared overhead, croaking hoarsely, and a great wave broke over the ship, carrying off nothing but the two bottles, which were swooped upon and borne away by the ravens.

Stories like these haunted the imaginations of the local people, and when in the autumn of 1738 a group of Cromarty fishermen found themselves stranded at Tarbat Ness by a prolonged storm, it was more than two weeks before they plucked up courage to approach the witch. At last, however, their provisions were nearly exhausted, and one of their number, Macglashan, set out with two companions to walk the mile to the witch's hovel. They found a desolate scene, three huts ruined and deserted, and the fourth almost as weather-beaten, but with clouds of smoke coming from the door. A heavy, oppressive smell hung in the air.

Macglashan entered the cottage, and found the hag feeding the fire with dried seaweed, muttering in Gaelic the while.

> The thick smoke circled round her as she bent over the fire, and when the flame shot up through the eddies, Macglashan could see her long sharp features, but when it sunk her eyes were alone visible. Her grizzled hair escaped from a red coif, and fell over her shoulders, round which there was wrapped a square of red tartan, held on by a large silver brooch.

Macglashan asked her for a fair wind, and offered her all the money he and the other fishermen had between them. She told them to give her a water-stoup (a big bottle) and sent them outside to wait. After a while she came out, holding the flask stopped up with a bunch of straw, which she told them they must not remove until they had reached home.

Next morning conditions were perfect for their sailing, a calm sea with a light north-west breeze, and almost before they knew it they found themselves approaching Cromarty Bay. One of the fishermen wondered what magical substance could be in the water-stoup, and another said they had better throw out the straw and the contents now, before the minister could see what they were doing. Macglashan pulled out the stopper and flung it into the sea. All peered into the stoup, which proved to be quite empty, and as they did so the man at the helm cried, 'For Heaven's sake, lads, mind your haulyards! lower, lower, a squall from the land!'

The gale struck with the fury of a tornado, and almost forced the boat out of the water. Straight back to Tarbat Ness they were blown, and found the witch sitting on the shore. She was not best pleased to see them again, but in the end agreed to provide a new straw stopper for their water-bottle, and the next day they arrived in Cromarty once more without mishap.

Miller comments that Macglashan was 'probably acquainted with the tenth book of Homer's *Odyssey*'. He is referring to the episode when Aeolus, guardian of the winds, summoned a gentle westerly breeze to waft Odysseus back to Ithaca, and gave him all the other winds sewn in a leather bag. When home was in sight, Odysseus relaxed and went to sleep, and his crew, supposing the bag to contain treasure, opened it. At once the winds leapt out and blew the ship all the way back to Aeolus' island, where the wind-master declared that he could give no more help to a man whom the gods were determined to thwart.

Although the stories are closely matched, it is not necessary to suppose that Macglashan was himself a classicist, since he could easily have heard one of many similar Scottish tales of a witch who sells winds closed in a bag or tied in knots on a string, which are then released by unwary sailors who find themselves at the mercy of a tempest.

UIG, LEWIS, WESTERN ISLES

The Lewis Chessmen

At Uig, on the western seaboard of Lewis, there is a magnificent beach, windswept and bleak on a stormy day, breathtakingly beautiful in fine weather. Here, in early 1831, a local man unearthed a stone vault buried beneath the dunes, and inside it he found a collection of small white figures. It is said that he fled in terror, thinking that he had stumbled on an assembly of the elves. He or somebody else, however, returned to collect the little people, now known as the Lewis Chessmen.

Little more is known for certain about the discovery, neither the name of the man, nor the exact site of the vault, nor how many chessmen were originally there. Via dealers and antiquaries in Stornoway and Edinburgh, most of them ended up in the British Museum in London, some in the National Museum of Scotland in Edinburgh, and it may be that a few remain in private hands. Certainly some are missing – the museums' holdings contain pieces from at least four distinct sets – but without precise details of what was found in the first place, it is impossible to guess when and where the others went astray.

Quite soon in the modern history of the chessmen, a tale was set going to account for their long concealment, a legend that was repeated many times, gaining details as it went, and which appears at length in Donald MacDonald's *Tales and Traditions of the Lews* (1967). According to this account, one day in the early seventeenth century, a herdsman known as the 'Red Gillie' saw a dinghy pulling away from a ship at anchor off Uig beach. When it reached the shore a boy jumped out holding a big bundle. He told the Gillie that he was running away from the ship, and had stolen some 'playthings' belonging to the sailors, which he hoped to sell. The covetous Gillie murdered the lad, buried him, and hid the bundle in the sand. He then approached his master, Calum Mór Mackenzie, with a suggestion that they should rob the ship still moored in the bay. Mackenzie was appalled, and immediately banished his wicked servant, who was thus prevented from digging up what he had taken from the sailor-boy. Some years later, the Gillie was arrested for other crimes and hanged, and before his death he confessed to the murder. Nobody, however, went to the trouble of searching the dunes of Uig, and so the twice-stolen treasure stayed where it was for more than two hundred years, until a cowherd saw one of his beasts rubbing herself against a dune on the sandy shore, and looked at what she had dug up with her horns.

However colourful the legends, they cannot match the chessmen

This engraving of two of the Lewis Chessmen was made in 1834, very soon after the pieces were discovered in the sand-dunes of Uig.

themselves for artistry and interest. They were carved from walrus ivory, one or two possibly from whales' teeth, between 1150 and 1200, in Scandinavia or the British Isles, and each piece is a work of master craftsmanship, ornate and brimful of personality. They are our best record of twelfth-century armour, and incidentally confirm accounts from the Norse sagas that Viking warriors used to bite their shields in battle-frenzy, in which action four of the rook or castle figures are portrayed. Although it is a poignant thought that other such sets must have existed and vanished, we must be grateful for the circumstances that led some long-dead traveller to hide his treasure, and for the chance that later brought it to light.

VE SKERRIES, SHETLAND

Sons and daughters of the sea

In *A Description of the Shetland Islands* (1822), Dr Samuel Hibbert devotes many pages to a description of creatures who revelled by moonlight on the Ve Skerries, rocks to the west of the Shetlands not far from the small island of Papa Stour. These 'sons and daughters of the sea', as Hibbert calls them, looked like human beings and were mortal, but were of surpassing beauty and possessed of some supernatural powers. They lived in the depths of the ocean, far below the region of the fish, in an airy realm, since they could not breathe water. 'I could obtain little satisfaction from the Shetlanders relative to the nature of the country beneath the sea,' Hibbert writes, 'but a native of the Isle of Man once visited it by means of a diving bell, that drew after it a rope double the distance of the moon from the earth.' The Manxman's journey is related at PORT ERIN (North-West England & Isle of Man).

The sea-people, Hibbert goes on, could only rise to the world above by putting on the skin of some amphibious animal, 'which they are enabled to occupy by a sort of demoniacal possession'. Sometimes they appeared as mermen and merwomen, human above the waist with fishes' tails below, but their preferred disguise was in a sealskin. Once on land they would take off this borrowed hide and return to their natural shape, but the covering was entirely personal to the wearer. If one of them lost their own particular skin, they were unable to return to the sea and would have to remain forever among men on land.

Of their origins, Hibbert writes in a somewhat similar vein to reports of the Blue Men of the Minch (*see* SHIANT ISLANDS):

> These inhabitants of a submarine world were, in the later periods of Christianity, regarded as fallen angels, who were compelled to take refuge in the seas: They had, therefore, the name of Sea-Trows given to them, as belonging to the dominion of the Prince of Darkness.

'Trow' was the word used in Shetland and Orkney for a supernatural being, derived from the Scandinavian 'troll', but quite different from the huge rock-dwelling monsters usually called trolls. *Troll* in Norwegian and related languages can denote that creature, but also signifies magic in general (a *trollman*, for example, is a wizard), and the Shetland and Orkney Trows were sprites not unlike British fairies. A 'Sea-Trow' is therefore a sea-fairy.

If a Sea-Trow was wounded in the shape of a seal, as soon as their blood mingled with the briny ocean it caused fearful waves. A story was told of a crew from Papa Stour who went to the Ve Skerries to kill seals, and stunned a number of animals. Having skinned them, they left the carcasses on the rocks and were about to re-embark when the sea rose in tremendous billows. The hunters ran to their boats, but one man was left behind, and saw a company of seals swimming to the Skerries. They took off their skins and, revealed in their own beautiful shapes, revived their friends, who lamented the loss of their hides, without which they would be exiles forever. Among them was a Sea-Trow named Ollavitinus, whose mother Gioga offered the man a bargain: she would carry him to Papa

Stour on her back, if he would recover her son's sealskin. Seeing the stormy weather and fearing to slide off the seal's slippery back, the man asked if he might cut holes in her hide, so that he could get a hold 'between the skin and the flesh'. This was permitted, and Gioga took the man to Papa Stour, where he found Ollavitinus's skin and handed it over, so that one Sea-Trow at least was able to return to his own underwater world.

The ultimate point, from a narrative point of view, of the holes in Gioga's hide is left unstated by Hibbert, but in a slightly later version reported by Lieutenant F. W. L. Thomas in 1852, the truth of the man's story is proved some weeks afterwards to his incredulous neighbours when the body of a large seal is found, 'and behind the neck were the two holes in the skin by which the fisherman had held on during his strange ride across the sea!'

WICK, HIGHLAND

The pirate John Gow

John Gow was born in Wick, in about 1698. A little later his family moved to Stromness in Orkney, and tales of his notorious career were found for over two centuries in both Wick and the Orkneys. Gow learned seacraft in his youth, and in the 1720s his thoughts turned to piracy. Serving on board a Guernsey ship (it is not entirely clear whether it started out as the *Caroline* or the *George Galley*), he stirred up a mutiny, resulting in the murder of the captain and other officers, and then took charge of the vessel, renaming it the *Revenge*.

According to the contemporary account in Captain Charles Johnson's *General History of the Pyrates* (1724), Gow announced to his crew, 'If hereafter, I see any of you whispering together, you shall be served in the same Manner, as those that are gone before.' Having thus insured himself against a second mutiny, he set off in search of loot.

His first prey were fishing vessels, which had only herring and salmon to offer. This was all very well, and Gow took the fish and sank the boats, but what he and his men wanted was wine, and even more importantly water. At Porto Santo, not far from Madeira, he presented the governor with three barrels of salmon and six of herring (taken from his previous victims) and asked permission to take on water and buy provisions. The governor visited the ship, which was at this point flying English colours, and all was courtesy until he tried to leave. No goods having yet appeared, he found himself held at gunpoint. Next morning the *Revenge* was duly provisioned with livestock and water, and soon afterwards she took her first valuable prize, a French ship carrying wine, oil, and fruit, from which Gow also liberated several mounted guns and other arms.

Having taken a couple more vessels, Gow felt that it was time to move on, and suggested a return to Orkney, where he knew of several rich houses that he thought would offer easy pickings. In January 1725 the ship, rechristened the *George*, anchored near Stromness, where Gow, posing as a successful merchant, was welcomed with open arms by his old neighbours, and cut a romantic figure among the young women, forming an attachment to a Miss Gordon, the daughter of a well-off resident. Quite soon, however, whispers of felony began to circulate. One of the *George*'s more reluctant crew members made his escape from the ship and gave the game away, and it became known that a house on Graemsay had been looted by Gow's men.

Gow knew that word was out against him, but he resolved on another raid. The target was a house on Eday belonging to James Fea, one of his old schoolmates. Here the pirate's luck deserted him, since his pilot miscalculated the tide, and the ship had to be anchored far too near the little island of the Calf, north-west of Eday, in danger of running ashore. Gow was forced to ask his intended victim for help in getting the *George* back out to sea, and Fea played for time, sending friendly notes to Gow but meanwhile arranging for all the boats on the island to be put out of action.

Five of the pirates came ashore, and were captured in an ambush after being decoyed into the pub. A day or so later, Gow was persuaded to meet Fea on the Calf and was taken prisoner, and within a short space of time the rest of the crew found themselves in the hands of the law. Gow, it was said, was held overnight in a room at Fea's house, which was later haunted by his ghost rattling its fetters.

The pirates were taken to London, and their trial was one of the most celebrated events of 1725. Most were executed. Gow at first refused to plead and was tortured by having his thumbs tied with whipcord, and threatened with 'pressing', a horrible procedure described by Johnson in some detail:

> The Prisoner is laid in a low dark room in the Press Yard at *Newgate*, all naked but his Privy Parts, his Back upon the bare Ground, his Arms and Legs stretch'd with Cords, and fasten'd to the several Quarters of the Room. This done, he has a great weight of Iron and Stone laid upon him. His diet, till he dies, is only three Morsels of Barley Bread without Drink the next Day; and if he lives longer, he has nothing daily, but as much foul Water as he can drink three several Times, and that without any Bread, till he expires.

Rather than undergo these torments, Gow pleaded not guilty, but was condemned to death and hanged. 'Thus ended the short Reign of this Pyratical Crew,' Johnson concludes, but their fame lived on, partly through the creative efforts of Sir Walter Scott, who visited Orkney in 1814 and collected traditions which he incorporated into his novel *The Pirate* (1821). His main informant was the weather-witch Bessie Millie (*see* STROMNESS), who claimed that she personally remembered Gow. One point that particularly took Scott's fancy was the tale that Miss Gordon had signified her betrothal to Gow with a handclasp, considered such a binding engagement that she had to go to England and hold his corpse's hand in order not to be plagued by his ghost if she should ever marry another.

Even two hundred years afterwards, the islanders still told tales of Gow's life and death. In the early twentieth century, the postmaster on Eday said that his own grandfather had seen Gow's body hanging in chains, and used to point out to visitors the place where the pirate had buried his treasure.

See also PIRATES AND PRIVATEERS (p. 232)

YELL, SHETLAND

A mermaid caught

Discussing mermaids in *The Natural History of the Amphibious Carnivora* (1839), Dr Robert Hamilton writes:

> Not many years ago the Wernerian Natural History Society (and to its praise we tell it) publicly offered a prize of considerable pecuniary value to the individual who would first present them with one of these far-famed animals; and by many this offer was regarded as a proof of weakness and credulity. Not long afterwards, however, the following statements appeared in one of the periodicals of the day, for the general truth of which, from personal knowledge of some of the parties, we can vouch.

In 1823, a fishing boat near Yell 'captured a mermaid by its getting entangled in the lines'. The creature was about three feet long, its upper body like that of a woman, with breasts, the lower part like a dogfish's tail. Face and neck were short and monkey-like, with a low forehead, but with distinct lips. On top of its head it had a crest of shoulder-length bristles, which it could raise or lower. It had separated fingers, not webbed, and it kept its small arms folded across its breast. Its skin was smooth

A meek and harmless mermaid was caught by some fishermen off Yell in 1823. This picture, from later in the nineteenth century, shows a creature neither threatening nor seductive, and may have been based on sightings of real marine mammals.

and silvery-grey shading to white below, hairless and without scales, and it had neither gills nor fins. 'It offered no resistance, nor attempted to bite, but uttered a low plaintive sound.'

The six-man crew at first took it into their boat, and kept it there for about three hours, but 'superstition getting the better of curiosity', they then released it from the lines, and a hook that had pierced its body, and returned it to the sea, where it instantly dived straight down.

The fishermen told their story to the Professor of Natural History at Edinburgh University, and to Mr Edmonston, a naturalist, who noted:

> That a very peculiar animal has been taken, no one can doubt. It was seen and handled by six men, on one occasion, and for some time, not one of whom dreams of a doubt of its being a Mermaid . . . the Mermaid is not an object of terror to the fisherman; it is rather a welcome guest, and danger is apprehended only from its experiencing bad treatment . . . It is quite impossible that, under the circumstances, six Shetland fishermen could commit such a mistake.

One must agree with Edmonston and Hamilton that this sounds like a truthful report of a real animal, possibly one of the sirenians, large aquatic mammals including the manatee and the dugong. No supernatural gloss is offered or required, except to note that sightings of such creatures must at least have contributed to legends of the mermaid.

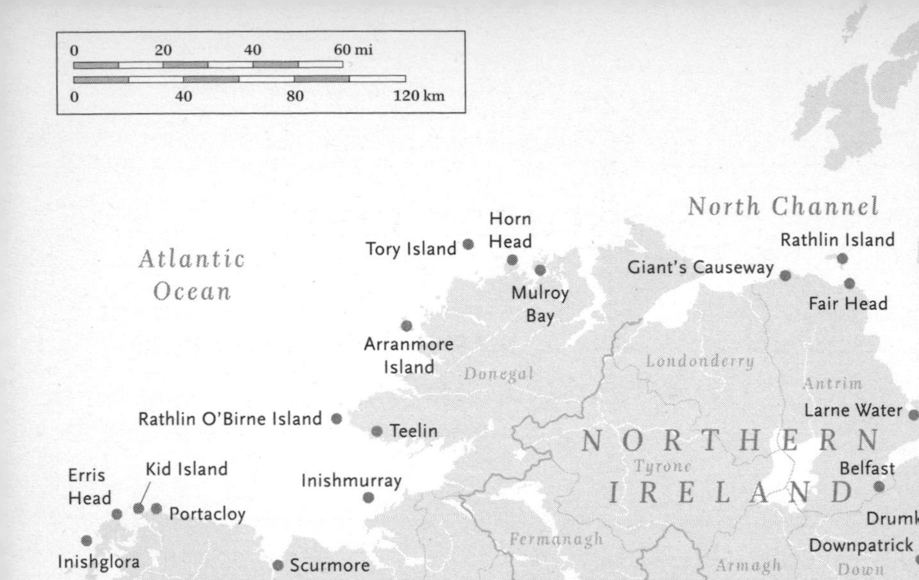

0
20
40
60 mi
0
40
80
120 km
North Channel
Atlantic Ocean
Horn Head
Tory Island
Mulroy Bay
Rathlin Island
Giant's Causeway
Fair Head
Arranmore Island
Donegal
Londonderry
Antrim
Rathlin O'Birne Island
Teelin
Larne Water
NORTHERN IRELAND
Tyrone
Erris Head
Kid Island
Portacloy
Inishmurray
Belfast
Drumkirk
Inishglora
Fermanagh
Downpatrick
Armagh
Down
Scurmore
Castle Gore
Sligo
Monaghan
Dundrum Bay
Clare Island
Mayo
Leitrim
Cavan
Carlingford
Louth
Roscommon
Inishbofin
River Boyne
Longford
Teltown
Drogheda
Meath
Westmeath
Saint MacDara's Island
Galway
Skerd Rocks
Galway
Dublin
Irish Sea
Inishmore
Offaly
Inishmaan
Kinvarra
Kildare
IRELAND
Gregory's Sound
Clare
Laois
Wicklow
Carlow
Limerick
Tipperary
Kilkenny
Wexford
Kerry
Waterford
St. George's Channel
Cork
Celtic Sea

NORTHERN EIRE & NORTHERN IRELAND

Counties Antrim, Donegal, Down, Galway, Louth, Mayo, Meath, Sligo

ARRANMORE ISLAND, COUNTY DONEGAL

The wounded seal

A Donegal fisherman named Rodgers was once driven ashore in the Scottish Highlands. Although he had never been there before, an old man addressed him by name and reminded him that he had once injured a seal in a cave on Arranmore. 'I was that seal, and this is the mark of the wound you inflicted on me,' he said, pointing out a scar on his own forehead, but since Rodgers had acted in ignorance the Highlander forgave him.

This story was included in a series of letters about Donegal written in the 1870s by a schools inspector, who reported of Arranmore that 'Until lately the islanders could not be induced to attack a seal, they being strongly under the impression that these animals were human beings metamorphosed by the power of their own witchcraft.' The notion of seals as enchanted beings was very widespread, and may not have completely died out yet. It was commonly reported that they could shift their shape between human and animal, and tales of men who married seal-wives were told in many places (*see* BARRA, Scottish Highlands & Islands).

In Donegal, it was often said that the seal-people came originally from Scotland, presumably because seals would be seen swimming down from the north. This is hinted in the story of Rodgers, and stated explicitly in an account from Rosguill recorded in 1941, which explained that long ago fish were so plentiful off the Irish coast that the fishermen could make their catch from the beaches, and did not even have to put out to sea, but at the same time there were no fish at all around Scotland or England. 'This enraged the Scottish people so much that their wise women turned themselves into seals and came over to the Irish coast to drive away the fish.'

In the south of Ireland, however, there was a tradition that the seals came from Donegal, and another view of their origins is implied in a tale from Mayo, quoted in Mr and Mrs S. C. Hall's *Ireland* (1841–3), of a seafood seller called John O'Glin who once fell asleep on the seashore and woke to see a crowd of dancing men and women. They were seals who had taken off their skins, a pile of which lay on the sand. Most were grey or black but one was pure white, and this John took. Soon the seal-people put their fur back on and went back into the water, but a beautiful girl was left behind, looking for her white skin. John persuaded her to marry him, and they lived happily together, she helping him with his fishing, although there were some comments about her habit of always eating her own fish raw. One day, however, having been away for a few days at a fair, he came home to find her and the white sealskin gone, and he knew she had gone back to the sea. The couple's three children stayed with their father, and grew up small and sallow, 'only very sharp indeed at the learning . . . particularly at the Hebrew' – suggesting that the teller believed the seals to be Jewish.

BELFAST

The wreck of the Titanic

Built in the Queen's Island shipyard, Belfast, *Titanic* was the largest vessel in the world when she entered the service, and it was claimed that she was 'practically unsinkable'. She began her maiden voyage on 10 April 1912, and four days later she sank, having struck an iceberg off Newfoundland.

The main factor to blame was that the ship was going too fast, and the disaster was compounded by the fact that there were too few lifeboats for everyone aboard. About 1,500 lives were lost in one of the worst catastrophes ever to occur in peacetime, and almost at once folklore began to gather around the event. The place where the wreck occurred was avoided long afterwards as a haunted spot, and it is said that *Titanic* has been seen there on 14 April, like other phantom ships that reappear on the anniversary of their destruction (*see* for example SHOREHAM-BY-SEA, South-East England).

One of the most outrageous myths concerned 'The Shipwrecker', an Egyptian mummy or mummy case said to have been on board. The story goes that it was bought from an Arab dealer in the late nineteenth century by an Englishman who almost immediately had an accident with a gun, shattering his arm. He sold the artefact to an Egyptologist. A few days later the purchaser lost his fortune in an ill-fated investment, and he subsequently died of a mysterious disease. The mummy case meanwhile was taken to England, and after changing hands a few times, always with unfortunate results for those who acquired it, found its way to the British Museum. Strange phenomena were reported from the First Egyptian Room where it was displayed, and several of the staff were taken ill, some fatally. In 1912 the case was sold to an American and shipped aboard *Titanic*. When the ship struck the iceberg, the owner - a man of immense wealth - was able to bribe the officers to put the mummy case on a lifeboat. Reaching the States, it was sold yet again, to a Canadian, and sent to Montreal, where it continued to cause trouble until the new owner sent it back to England on the *Empress of Ireland*, which also sank. This time the mummy went down with the ship, and now lies fathoms deep in the St Lawrence river.

Little or none of this story is true, although the artefact itself is real. It is neither an actual mummy nor even a complete case, but the lid of a coffin dating from the 21st Dynasty, around 950 BCE, which remained on display in the British Museum until 1990, when it was lent to a temporary exhibition, and will return to the Ancient Egyptian rooms in the near future. The item was presented to the museum in 1889 by A. F. Wheeler, and its bad reputation seems to have sprung from the imagination of the newspaperman W. T. Stead, an enthusiast for psychic phenomena, who saw the coffin lid and decided that the face painted on it showed 'a living soul in torment'. The story of the mummy's misdeeds was broadcast in the press, and its adventures on *Titanic* were added later, as detailed in J. G. Lockhart's *Curses, Lucks and Talismans* (1938), probably in response to the fact that Stead himself died in the wreck.

Stead's daughter Estelle believed that her father had communicated with her via a medium after his death, and she published the result in 1922 as *The Blue Island: Experiences of a New Arrival beyond the Veil*. Although the deceased Stead is reticent about his experiences just before and just after *Titanic* went down, one detail rings irresistibly true to character. On his arrival among the other dead souls, the inveterate reporter exclaims, 'Oh, how badly I needed a telephone at that moment! I felt I could give the papers some headlines for that evening.'

More remarkable, because irrefutable, is the publication of a 'premonitory' book by Morgan Robertson, *Futility*, in 1898, fourteen years before the wreck. Robertson, himself a sailor, said later that he had written it in one sitting, by 'automatic writing', a process of unconscious creation that is sometimes claimed to channel dictation from dead spirits or, as in this case, from the future. The short novel deals with a ship named *Titan*, 'the largest craft afloat and the greatest of the works of men'.

> From the bridge, engine-room, and a dozen places on her deck the ninety-two doors of nineteen watertight compartments could be closed in half a minute by turning a lever . . . With nine compartments flooded the ship would still float, and as no known accident of the sea could possibly fill this many, the steamship *Titan* was considered practically unsinkable.

The 'watertight compartments' were a feature of *Titanic*, broached when she hit the iceberg. Like the real ship, the *Titan* carries 'as few boats as would satisfy the laws', and 'would steam at full speed in fog, storm, and sunshine'. In case of a collision with an iceberg, 'the only thing afloat that she could not conquer', her bows might be crushed, but three at most of the air-filled compartments would be flooded, leaving thirteen to keep the ship buoyant. *Titan* docs strike an iceberg, but rather than being crushed she runs half out of the water up the sloping ice and then overturns. Only one boatload survives (plus a sailor and little girl who take refuge on the iceberg itself), and the rest of the three thousand or so passengers and crew perish.

The correspondences between Robertson's novel and later history led to the book's republication after 1912 under a more pointed title, *The Wreck of the Titan; or, Futility*. Many tales rolled in after the wreck from passengers who claimed they had cancelled their tickets on the *Titanic* in response to premonitions, but Robertson was the only person who could prove that his foretelling really did pre-date the event.

RIVER BOYNE, COUNTY LOUTH

The Daghdha expels an octopus

The Daghdha was the principal god of the ancient Irish, a deity of the sun and sky, and was described in medieval literature as one of the leaders of the Tuatha Dé Danann, the fairy people said to have been among the first settlers of Ireland.

A tale of the tenth century, translated in Edward Gwynn's *Metrical Dindsenchas* (1903–35), relates how the Daghdha used his club to banish a sea-monster, a giant octopus which had the most terrible powers of suction:

> It would suck in a man in armour till he lay at the bottom of its treasure-bag. The Dagda came with his 'mace of wrath' in his hand, and plunged it down upon the octopus, and chanted these words: 'Turn thy hollow head! Turn thy ravening body! Turn thy resorbent forehead! Avaunt! Begone!'

The Daghdha, a Celtic deity, once overcame a giant octopus under the sea. When the creature retreated it sucked out the tide, leaving the land around the Boyne estuary above the waves.

The octopus retreated and with it retreated the waves, exposing the plain of Muirtheimne, meaning 'darkness of the sea', or 'under the sea's roof'. This was the old name for the land between Dundalk and Drogheda, around the mouth of the river Boyne.

The word that Gwynn translates as 'octopus' means literally 'sea-snail', but he comments that this must be 'a monster cuttle-fish'. There was a very widespread idea that some octopuses or cuttlefish were so enormous that they could devour human beings, a legend used to great dramatic effect by writers including Victor Hugo (*see* BRECQHOU, South-West England & Channel Islands). The naturalist Francis Buckland wrote in the mid nineteenth century that 'The existence of some enormously large species of the cuttle-fish tribe . . . can hardly be doubted.' He cites a report of a monster specimen in the seas around Africa that seized three men from a ship and drowned them. One of its arms, cut off by the other sailors in an attempt to save their comrades, was as thick as the ship's mizen mast, 'and the suckers of the size of pot-lids'.

Some distrusted accounts like this. Writing in 1874, the naturalist

J. G. Wood commented with disapproval on the great amount of sensational writing about the octopus, 'which the newspapers *will* call the devil fish, whereas it is, as a general rule, a very harmless creature'. The octopus, he writes, 'is nothing but a big sea-slug', which sounds rather like the 'sea-snail' of the Irish legend.

See also ROCKABILL (Southern Eire).

CARLINGFORD, COUNTY LOUTH

A whale with gold teeth

Gerald of Wales or Giraldus Cambrensis was one of medieval Britain's most enthusiastic travel-writers. Born in Pembrokeshire in about 1145 into an aristocratic family, part Norman and part Welsh, he joined the church and took part in missions around Wales and Ireland, recording his impressions and the stories he heard. He was fascinated by anything supernatural or mysterious, and even his descriptions of his own land are full of wonders. In Ireland, he listened eagerly to the most unlikely yarns, and it was largely due to his influence that the country became regarded abroad as a place of marvels.

One prodigy reported by Gerald as having occurred a few years before his visit was the stranding of an enormous fish – probably in fact a whale – in the bay of Carlingford in 1167. This he interprets as a portent of the English invasion in 1169. Even when whales were seen off our coasts more frequently than they are now, their vast size made them seem significant of momentous events, as when a terrible battle of 'Sea Monsters' near EDINBURGH (Scottish Lowlands) in 1707 was understood as an image of the struggle between England and Scotland.

Gerald goes on to report of the beast that 'Among other wonderful things' (which sadly he does not describe) 'it had three teeth of gold of about fifty ounces' weight in all.' We may be reasonably sure that it had no such thing. Gerald's appetite for miracles must have been obvious to all he met, making it tempting for them to see how much he would swallow.

CASTLE GORE, BALLINA, COUNTY MAYO

Lord Tirawley and the ship of death

Ruined Castle Gore, near the northern end of Lough Conn, may be the setting for a macabre Mayo tale reported in *Folk-Lore* in 1918, dealing with the fate of 'a sensual, reckless young man', Lord Tirawley. This title was in fact borne only by the eighteenth-century MP James Cuff, and the legend more probably relates to his illegitimate son, who scandalised the neighbourhood in the early nineteenth century by keeping a French mistress at Castle Gore.

Whether or not this was the man, his revels were interrupted one night when a black coach drove up to his door, and a tall black-clad stranger warned the lord that he had but a year to live. Tirawley was briefly inspired to repentance and reform, but before the year was out he had relapsed into his old ways. On the anniversary of the ominous visit, when a large party had been invited to the castle, the dark stranger returned. He beckoned to the doomed lord, who followed him like a sleepwalker into an unfurnished room, leaving the door open so that all could see what followed:

Some legends tell of 'death ships', which flew across the sea to fetch the souls of the wicked.

> The stranger drew a ship on the wall; it became solid and moved out; he got on board, Lord Tirawley followed, and the ship sailed round and passed through the wall, which closed upon it, and neither of its occupants was seen again on earth!

Some evildoers were supposed to be fetched by 'death coaches', and as in Celtic myth the dead were believed to travel across water to an Otherworld island (sometimes said to be located off the Kerry coast at THE BULL, Southern Eire), the vehicles were occasionally ships or carriages that travelled across the waves. One such was reported at COLLIN (Scottish Lowlands), and when Cromwell died, according to a story in Sean O'Sullivan's *Folktales of Ireland* (1966), a ship's captain near Liverpool saw a fiery chariot drawn by mastiffs racing above the sea, and heard a voice cry, 'Clear the way for Oliver Cromwell!'

CLARE ISLAND, COUNTY MAYO

Grace O'Malley

Making conversation with a taxi driver in Galway in the late 1960s, the folklore collector Horace Beck asked if he had heard of the pirate Grace O'Malley. 'Well, I know she was a grand, great woman,' came the reply. 'Queen Victoria invited her to her Diamond Jubilee and she came in a dress without no top.'

Gráinne Mhaol, or Grace O'Malley as she was called by the English, lived far too early to have had anything to do with Victoria, but was said to have held her own against another English queen, Elizabeth I, and was herself known as 'Queen of the West'. She was born in about 1530, the daughter of a chieftain of the O'Malleys, a family famous from the twelfth century onwards as master mariners. Gráinne married twice, and both husbands brought her property, including two strongholds in Clew Bay, one on the mainland and one on Clare Island. Her fleet of ships used to raid along the west coast of Ireland and as far as Scotland, but she was also in her own way a diplomat, keeping her feet in the shifting alliances between the Irish and the English, although she was imprisoned for periods in Limerick and Dublin.

In July 1593 Gráinne sailed for London, her aim being to get the English queen's support against her main enemy in Connacht, Sir Richard Bingham. In order to gain her ends she submitted to Elizabeth, and Bingham was restrained from attacking her further. The interview between the two powerful women attracted many rumours: it was said that Gráinne arrived barefoot in the dress of a Gaelic chieftain (or, less credibly, that she went naked to the waist – the story the Galway cab driver must have heard), and talked to Elizabeth in Latin. At one point during the conversation she needed to blow her nose, and was given a delicate cambric handkerchief. Having used it she threw it in the fire, and was disgusted when Elizabeth suggested she might have kept it.

On her return from this visit it was said that she stopped at Howth, where the lord of the castle was at dinner and refused to open his gates to her. Furious, she captured his son, and kept him prisoner until Howth promised that his castle would always give hospitality to any who needed it. The story of the young lord's kidnapping was wrongly attached to Gráinne (it happened, if at all, in the fifteenth century), but is entirely suitable to her reputation. In the early nineteenth century, it was reported that a picture of the abduction hung at Howth Castle, although this was a highly imaginative interpretation of the painting concerned, which apparently showed the Flight of the Israelites from Egypt (and has now been sold or destroyed).

After her death, Gráinne passed into legend as a symbol of Ireland, and her name is often used in poetry as a synonym for her country. Oral tradition embroidered her tale with many picturesque details. Legend has it that she gave birth to her son Theobald on board ship, and leapt from her bed shortly afterwards when Turkish pirates attacked, bearing a pair of blunderbusses which she discharged at the Turks, putting them to flight. She was keen for Theobald to be a great fighter, which he later proved to be, but once when he hung back she taunted him, 'Are you trying to hide in my backside, where you came from?' Another tale of Gráinne's mothering skills related that when one of her sons fell into the sea and seized the gunwale, she cut his hand off. 'I was only going to get back aboard,' he protested. 'If you had been a true O'Malley, you'd not have fallen overboard in the first place,' she replied.

Gráinne's skull was apparently displayed for many years in the church on Clare Island. According to Horace Beck in *Folklore and the Sea* (1973), it was once stolen by some English scientists, but as soon as they got into their boat a terrible storm sprang up which blasted the culprits until they put back to the island and returned the skull. The local people then hid Gráinne's skull and replaced it with somebody else's. When the next lot of English robbers fell for the fake and took it to London, no storm arose, and the islanders quietly provided another substitute skull. Neither the original nor any counterfeit is on show now, but on Inishbofin, south-west of Clare Island, a huge pile of rocks is pointed out as the remains of one of Gráinne's castles, and a chasm on the island is said to mark the place where she started a canal, tales classing her with legendary heroes like Fionn mac Cumhaill, who was said to have altered the landscape at the GIANT'S CAUSEWAY, for example.

See also INVERARAY (Scottish Lowlands).

DOWNPATRICK, COUNTY DOWN

St Columba's coffin

When Norse raiders came to Iona in 825, they broke open the shrine of St Colm Cillé (St Columba) and stole the coffin, hoping that it contained treasure. Once at sea they broke it open, and finding nothing but a body, they threw the coffin into the sea. By a miracle it floated to Down, where it was found on the shore by monks – or, according to another account, by a cow, which licked the coffin, and whose yield of milk afterwards increased so greatly that the coffin was opened to see what wonderful thing lay inside. The body of Colm Cillé was revealed, and was reburied next to St Patrick and St Brighid, whose graves moved apart to allow the third great saint of Ireland to lie between them. Colm Cillé himself had prophesied his reburial, according to a life of the saint compiled in the sixteenth century, in a verse:

> Though buried I be in Iona,
> According to my King, not offended,
> In Down I shall dwell anew,
> O King of Hosts, in due season.

A nineteenth-century version of the legend makes no mention of Vikings, saying instead that although Colm Cillé wished his grave to be beside those of Patrick and Brighid, when he died the brethren of Iona were unwilling to move his body. One evening, however, they found his coffin had disappeared, and a boat too was missing. Three monks were sent to look for the holy corpse, and at last they anchored in the loch of Down, where they met a former companion of theirs from Iona who had a wonderful story to relate.

Some weeks ago, he told them, some men working on the shore had noticed a small object out at sea. As it approached the harbour they saw it was a boat with its sails furled. 'No oars were visible, but still it came swiftly onward, leaving a long, straight line of foam behind.' When it came to land nobody dared approach it until the bishop had come from the cathedral and scattered holy incense on the decks to bless the craft while hymns were sung. A coffin had then been brought up from below decks, and the bishop declared that it contained the uncorrupted body of Colm Cillé. The saint having made his wishes abundantly clear, his body was buried in Down church and the boat was taken back to Iona.

These traditions have historical roots in the twelfth century, when the Anglo-Norman knight John de Courcy established English monks in the religious houses of north-eastern Ireland, driving out many of the native clerics. To conciliate local opinion, he claimed to have deposited the bones of Patrick, Brighid and Colm Cillé in the cathedral at Down, then renamed Downpatrick in honour of Ireland's patron saint.

DROGHEDA, COUNTY LOUTH

Macaldus the robber-bishop

In the mid fifth century, a brigand named Macaldus was chief of all the villains around the river Boyne, but this was the time of St Patrick's mission, and more and more of his followers were converting to Christianity. In fury, Macaldus laid an ambush, meaning to kill the saint, but when the holy man appeared the robbers could not bring themselves to hurt him. Macaldus, however, determined at least to fool St Patrick, got one of his men to lie down and pretend to be dead, and then begged Patrick to restore him. 'I dare not intercede for him,' said the saint, and passed on his way. Macaldus then found that his henchman really was dead, and ran after the saint to implore his help, telling him the whole truth.

Patrick relented and brought the man back to life, but imposed a penance upon Macaldus for his deception. He took the robber chief down to the river, where they found a boat fastened with a chain. Patrick wound the chain around Macaldus, padlocked it and threw the key into the Boyne. He then told Macaldus that he must stay in the boat until the key was found. The boat drifted down the river to the old port of Colpa (now Drogheda), out into the open sea and across to the Isle of Man.

Still chained, Macaldus told the whole story to the Bishop of Man. Repenting of his crimes, he began to study Christianity and prepared to receive holy orders. One evening the bishop's cook slit open a fish and found inside the key to Macaldus's chains. Next day Macaldus was ordained, and on the death of the bishop the former robber became Bishop of Man.

This tale is first reported in a seventh-century biography of St Patrick. Stories in which a lost or stolen article is found inside a fish are very common, with the earliest example the classical Greek myth of Polycrates, and other peculiar discoveries made in fishes are described at KING'S LYNN (East Anglia). In legends of the saints, such anecdotes take on an extra resonance, the fish being an ancient symbol of Christ.

See also RATHLIN ISLAND.

DRUMKIRK, COUNTY DOWN

Rains of fishes

A newspaper report appeared on 30 May 1928 of a rain of fishes in Drumkirk. Dozens of tiny red fish (probably sticklebacks) were found lying on the roof of a bungalow belonging to James McMaster, a farmer:

> In the course of enquiries it was ascertained that just before the discovery of the fish there had been an exceptionally violent thunderstorm with heavy rain. There is no river in the neighbourhood, the nearest sheet of water being Strangford Lough, two miles distant, and the theory advanced by an expert was that the fish had been lifted from the sea in a waterspout.

In *A History of Fishes* (1931), James Norman commented that although such occurrences are rare, he knew of more than fifty accounts of 'rains of fishes' from various parts of the world, and many more have been recorded since. The phenomenon is nearly always described as accompanied by thunderstorms and heavy rain, and it appears that the action of a waterspout passing over shallow coastal water, or of a tornado over inland pools and lakes, can lift small fish and transport them some distance before depositing them again. There is a case on record from 1896, of an enormous hailstone that fell during a heavy storm in Essen, Germany, which contained a frozen baby carp, indicating that the fish not only entered the clouds but must have been lifted to the great height necessary for the formation of hail. This carp was only about an inch and a half long, but occasionally bigger fish are involved: in India, one specimen was described as over a foot in length and weighing more than six pounds. Falls in Europe have included herring, sprats, trout, smelts, pike, perch, and sand eels.

DUNDRUM BAY, COUNTY DOWN

The king of Ulster and the sea-monster

According to a tale from the early eighth century, Fearghus mac Léide, king of Ulster, once fell asleep on the shore and was carried out to sea by water-sprites. Waking, he seized three of the elves, and before he would release them he demanded that he be given the power of swimming underwater. This was granted, but the sprites warned him that he must not try to swim in Dundrum Bay in his own territory.

King Fearghus ignored or forgot the prohibition, and one day he went swimming in the bay, where he met a dreadful monster that grew larger and smaller, inflating and contracting like a pair of bellows. At the horror of this sight 'his mouth was turned to the back of his head', and he fled to land, where he passed out.

When he woke up, he knew nothing of the fact that his mouth was now on the wrong side of his head. The wise men of Ulster counselled that his disfigurement should not be mentioned, since a blemished man was not allowed to be king in Ireland, and they did not want to lose Fearghus as their ruler. For seven years Fearghus reigned in ignorance, until a woman who was serving him lost her temper when he struck her, and told him the truth. In fury he assailed her with his sword, slicing her quite in two, and then ran to Dundrum Bay to confront the monster once more. For a day and a night the king and the beast fought, and at last Fearghus emerged with its head, but the struggle had been too much for him and he died. The bay was red for a month from all the blood that had been shed in the water.

See also ROSSBEIGH STRAND (Southern Eire).

ERRIS HEAD, COUNTY MAYO

'Spare your auld grandfather'

A man from Erris once asked his neighbour to go seal-hunting with him. The neighbour's wife was out and he was in charge of the baby, but he wrapped the child up in a shawl and took him along in the boat. When they got to the seal cave, the father put the baby on a ledge and then the two men went in to kill seals. Before they could start, a huge wave flooded in, and the men only just managed to get back to their boat. They had no chance to rescue the child, and the sea was now so rough that it was quite impossible for them to get back into the cave.

The storm continued to rage for two or three days, and only when the wind and waves dropped could they go back to find the baby's body.

> It was quite calm and as they were going into the cave they saw a big, huge, old seal, an old mother seal, nursing what they thought was a baby seal, nursing it at the breast in the manner of human beings.

Cauls

A newborn baby sometimes has its head draped in a portion of the amniotic sac that encloses the embryo before birth. These fragments of tissue have gone by various names, some of which, such as 'shirts' and 'cloaks', indicate that they have been known to cover most of the infant's top half, but mostly the membranes are found on the head alone, and are known as 'cauls', meaning caps.

An infant born with a caul was considered particularly fortunate, safe from death by water as long as the caul itself was preserved. Belief in the caul's buoyant properties is shown, for example, in a letter sent to *Notes and Queries* in 1899, recording a baby born with a caul that:

> ... when his mother tried to bathe him he sat on the surface of the water, and if forced down, came up again like a cork. There seems no doubt that this was fully believed and related in all seriousness.

Now that most births happen in hospital, cauls are very rarely kept, but when babies were born at home, the membrane was often cherished for years. The Folk Museum in Cambridge has a caul dating from 1891 that was preserved in a Norfolk family, and was lent to anyone about to make a sea voyage.

The protection conferred by a caul was said by some to be unique to the original infant owner or transferable only to relatives, but this opinion was evidently not universal, since there was a brisk trade in the fortunate objects as early as Roman times, and cauls continued to be sold until the twentieth century. As faith in talismans and folk remedies declined, the value of cauls dropped, and in the early 1900s the folklore collector Edward Lovett bought two very cheaply from shops near the London docks. When the First World War broke out, however, the situation changed. Seeing a notice in a shop window that a caul was for sale, Lovett went in and asked the price, which was two pounds.

> 'Why! A year or two ago I bought one down here for eighteenpence.' 'Yes! I dare say you did, but that was before the War,' was the shopman's reply.

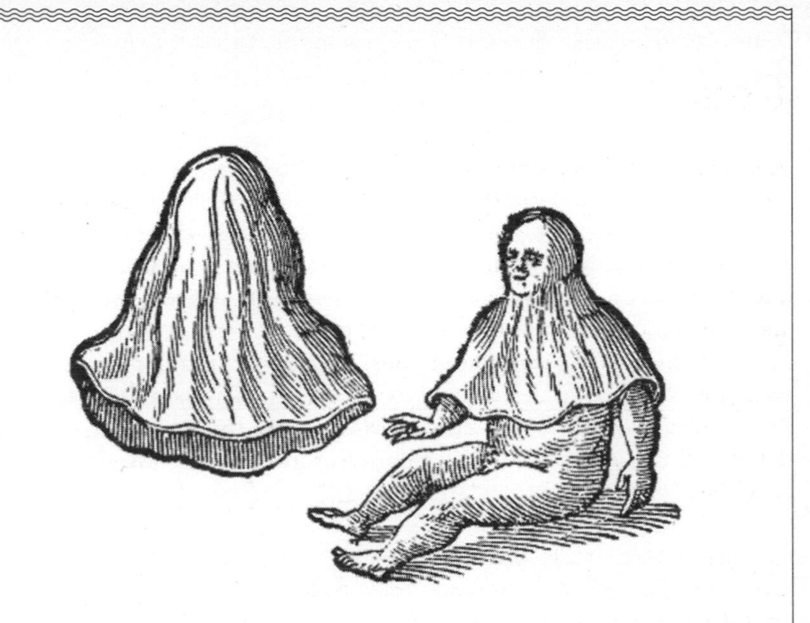

The physician Cornelius Gemma described the caul in De Naturae Divinis Characterismis *(1575), and included an illustration of an* infans galeatus *('helmeted baby'). Although the caul usually covers only the head and face, like a cap or veil, here it is shown stretching to the elbows.*

He explained that everyone who went to sea was so frightened of submarine attacks that 'all the old superstitions have revived', and he regretted that he had not more cauls, since he could have sold plenty. Even in the 1950s cauls occasionally changed hands, but a search on the internet reveals no current market.

There is a possible link between beliefs about cauls and the widespread legend that mermaids and seal-people had a special garment that let them live underwater (*see* for example VE SKERRIES, Scottish Highlands & Islands). Although this is often described as a sealskin, a cloak, or even a detachable tail, in the story of GALLARUS (Southern Eire) and a few other Irish examples it is a magic hat or cap, suggesting the idea that these mermaids had an extra-powerful version of a caul.

When they got closer, the mother seal dived into the sea, dropping what she was holding. They found it was the man's son, who seemed in perfect health, and with great rejoicing they took him home. 'And he grew up to be a great man, a fine young man, and a remarkable swimmer,' finished the narrator, and one of his audience agreed that this was a true story, something that had happened about 150 years earlier.

The tale, printed in David Thomson's *People of the Sea* (1954), typifies one strand of seal legend, what one might call natural rather than supernatural lore, a category that includes the singing by and to seals reported at BARRA and CANNA (both Scottish Highlands & Islands). A borderline case is the story told in 1955 by a Clare man, Pádraig Ó Briain, about a big black seal that used to follow one particular mackerel boat, until one of the fishermen tried to hit it with an oar. That night the man fell and broke five ribs, and was bedbound for twelve years until he died. There is nothing impossible here, although Ó Briain interpreted the accident as retribution: 'They said around here that the seal was a good omen, and the fact that this man struck it caused the good omen to turn into a curse, and sure 'twas true enough.' Similarly, David Thomson was told of a man who killed a young white seal to make a waistcoat, and died exactly a year later. Killing seals was generally thought to be unlucky, although clearly this did not stop a brisk trade in sealskins.

More common, however, particularly in the west of Ireland and the Western Isles of Scotland, is the notion that seals had a dual nature, animal and human. Tales such as that told at ARRANMORE ISLAND describe someone who has injured a seal then meeting it in its human shape, and another frequent report is of a man who steals a sea-maiden's sealskin and marries her. It was often thought that seals could speak even when they were in animal form. In a tale from south Kerry recorded in 1952, a veteran seal-hunter found a seal giving birth in a cove. She addressed him by name, and asked him not to kill her until she had had her baby. 'I won't, or ever again!' he said, and never hurt another seal. Another popular anecdote was of some youths who entered a cave at Erris, intending to kill an old seal, but were terrified when it cried out, 'Och, boys! spare your auld grandfather Darby O'Dowd.'

The O'Dowds were traditionally said to be descended from seals or mermaids, and sometimes teased for their supposedly webbed feet. Other families had the same reputation, including the O'Sheas and O'Sullivans of Kerry, the Conneelys or Connollys of Galway, the O'Haras of Sligo and the Gallaghers of Donegal. In the 1940s, David Thomson heard that the MacNamaras were seal-people, their name a corruption of Mac Con Mara, meaning 'the son of the sea-hound', but that they did not like it to be spoken of. Peadar MacNamara, a travelling man Thomson met, would 'nearly destroy you, if you spoke of that', he was told.

FAIR HEAD, COUNTY ANTRIM

The Grey Man

Fair Head is Northern Ireland's tallest cliff, rising some six hundred foot above sea level. A waymarked path leads along the top of the cliffs from Coolanlough, and courageous walkers can follow Fhir Leith (the 'Grey Man's Path') down a spectacular ravine bridged above by a fallen pillar of rock, but be warned: not only is the going steep, but the chasm is famously haunted.

Mr and Mrs Hall visited the spot in the mid nineteenth century,

The Grey Man's Path is said to be haunted by a gigantic wraith that appears before a storm.

and heard local reports of ghosts that 'come from out the deep before or after the rising of the moon, and climb, or rather stalk up the rocks'. One boatman told them that just before the great winter storms of 1796, his father had been out on the sea and seen 'a wreath of smoke passing over the waters'. As it came nearer the coast, 'it grew into the shape of a giant, folded in its cloak', and climbed the path, spreading its arms forward as if praying – or cursing. That very night, the boatman went on, the colliers of Fairhead had been terrified by what they thought at first was thunder, followed by lightning, clouds of dust and a raging sea.

> In the morning the effects of the Grey Man's curse were sufficiently plain; rocks had been detached that no earthly power could move, and they had crushed in the collieries, so that more than a thousand ton of coals were buried past recovery. Columns were hurled into the sea, which had stood erect in the sight of heaven since the world was a world.

In *Ulster Folklore* (1951), Jeanne Cooper Foster suggested that the Grey Man might have been invented by parents to keep children away from this dangerous place, but she knew many adults who claimed to have seen the phantom, including a Belfast schoolmistress who was 'very perturbed because she believed that she had seen the Grey Man, and someone had told her that he brought misfortune'.

GALWAY, COUNTY GALWAY

The knife in the wave

'I did hear of the souls of the drowned living under the sea long ago,' an Aran islander told the writer David Thomson in the 1940s. There was a fisherman, he began, who went out one night from the Claddagh – the old fishing district of Galway city – with his three sons. Before they left the shore they were approached by a stranger riding a white horse, who advised them to take an axe, a hook and a knife. They saw no harm in doing as he said, went back to find the tools, and then set sail, but before long a storm blew up and they took to their oars, making for home. They saw a huge wave coming towards them, and when it was almost swamping their boat one of the sons threw out the hook. The wave split in two and passed either side of the boat without harming them.

Soon a second wave loomed above them, high as a hill. The second son took the axe and cast it at the wave, and like the first it divided in two and left them safe. Then a third wave threatened, and they were sure it would finish them, but the third son threw the knife at the wave, it split, and their boat was cast up on their home strand. Mightily relieved, they went home to their supper.

Late that night a knock came at the door, and there was the strange rider. He called out the sons and took them on his white horse to a big town. The boys had no idea where they were, but it seemed a jolly place, with crowds of people apparently coming from a dance. The horseman took them to a mansion, where he warned them that they should speak no word, nor answer any question put to them, until they came out again.

The doorman of the mansion addressed the eldest son, telling him that he should go to the top room where he would find somebody waiting for him. Up he went, and found his way to a fine parlour.

> When he looked around what should he see but a young woman stretched on a bed, a woman so beautiful that he had never seen her like with the light of his two eyes before. Stuck in her forehead was the axe!

She asked him to pull out the axe, which he did, but he never spoke a word to her. The second son was then sent up to another room, where he found a woman with a hook stuck in her shoulder. He pulled out the hook as she asked him, but said nothing. Finally the third son was shown a room where he saw a third woman with a knife stuck in her head behind her ear. When he had pulled it out she said to him, 'My blessing on you and my curse on those who have given you orders. There isn't a young woman in the town to-night, except the three of us, who hasn't got a husband. We're three sisters. My curse on the man of the white horse. Only for him we would have got the three of ye to-night.'

The three sons were taken home by the horseman, who warned them to keep their journey secret, and never to go to sea again. If they did, he said, they would be taken by the women. 'There were thirty-one men drowned to-night and ye saw them up and down the street with their women.'

It was a fairly common idea that those lost at sea were stolen by the sea-people, as mentioned for example at TEELIN, and the legend of 'The Knife in the Wave' is one of the most popular tales of the Irish coast, told among other places at TRAFRASK (Southern Eire).

See also DOLPHINS AND PORPOISES (p. 422).

GIANT'S CAUSEWAY, COUNTY ANTRIM

Fionn mac Cumhaill and the Scottish giant

The Giant's Causeway was formed during the Cainozoic period, sixty million years ago. It is made up of basaltic rock split into polygonal columns like those at STAFFA (Scottish Lowlands), the two structures being the visible remains of the same prehistoric event, a volcanic eruption that extended from Skye in the Hebrides to the northern coast of Ireland.

From the Irish end the tremendous columns stretch for about half a mile. Their grandeur, far beyond the human scale, forcibly suggests association with giants. Mr and Mrs Hall, nineteenth-century visitors to the site, were shown natural features called the Giant's Theatre, the Giant's Pulpit, the Giant's Bagpipes and the Giant's Granny, and they observed that every towering peak was said to have been either used by the giants, or made by them. The Causeway itself was reportedly the work of Fionn mac Cumhaill, the legendary warrior and hero who is often described as a giant, so tall, Elizabeth Andrews was told in Donegal in the early twentieth century, that 'his feet came out of the door of his house', although nobody said how big his house was.

A Scottish giant, even bigger than Fionn, once crossed the Causeway looking for a fight. Fionn was afraid to meet him, and on the advice of his crafty wife he hid in the baby's cradle. The rival colossus (Andrews calls

The great pillars of the Giant's Causeway suggest the ruins of an ancient civilisation, beyond the human scale.

him Goll, but he appears under other names in different versions of this much-repeated folk tale) asked for Fionn. When his wife said he was out, and she was minding the baby, the stranger took one horrified look and hightailed it back to Scotland before he met the father of such a monstrous child.

GREGORY'S SOUND, COUNTY GALWAY

The restless dead

Pope Gregory the Great was known in Irish tradition as 'the Pope of the Golden Mouth', from his association with Gregorian chant. Probably through confusion with another Gregory who founded a monastery on Inishmore, it became said that Pope Gregory had Irish ancestry, being the son of a pilgrim who settled in Rome. When he died, in accordance with his instructions, his body was placed in a golden casket and floated off on the river Tiber. After a long sea journey the casket was cast up on Aran, and there the pope was buried.

A strange folk tale from the Aran Islands, evidently derived from traditions of Pope Gregory, tells of a man called Gréagóir who was given to cursing and blasphemy. As a penance he cut off his mouth, but a golden mouth grew in its place. He directed that after he died his body was to be put in a coffin and thrown into the sea between Inishmore and Inishmaan, but it persisted in floating ashore, and he was finally buried on Inishmore. The channel between the islands was thereafter known as Gregory's Sound.

Another body that could not rest beneath the waves was that of Oliver Cromwell, who according to legend wished to be buried in Ireland, but the earth rejected his corpse and cast it into the sea. It is an ancient idea that a tyrant's body would be refused by the earth and would have to lie at the bottom of the ocean, but in this case even the waves were uncomfortable with the body. The English and Irish tides threw it back and forth, neither willing to keep it, and the sea between Ireland and England has remained rough ever since.

HORN HEAD, COUNTY DONEGAL

Fairies demand poteen

Some smugglers from Tory Island were once heading east with a cargo of poteen (home-brewed spirits), and as they came into the shelter of Horn Head they heard a chorus of voices screeching out for a drink. The cries were those of some thirsty fairies. The sailors did not see why they should share their liquor, but when they tried to leave the bay a wind raised by the fairies forced them back. At last they had to let the fairies have a drop:

> It was not long until the gentry changed their tune: the sweetest singing the boatmen had ever heard began to come from inside the cliff overhead. The wind fell and the sea calmed, and from that on it was the finest night ever wafted out of the air.

This story was told by a Donegal fisherman in the mid twentieth century. Another sea-spirit that liked poteen was the Merrow of DOONBEG BAY (Southern Eire).

INISHBOFIN, COUNTY GALWAY

The white cow

The name Inishbofin (*Inish-bo-finn*) means 'the island of the white cow'. It is said that the island was hidden from human sight by an enchanted fog until one day two fishermen happened to drift ashore here and lit a fire. At once the mist lifted, and the fishermen found themselves on a beach between the ocean and a lake. Into this lake a witch was driving a white cow, and as the animal entered the water she struck it with her wand, turning it to stone. The fishermen in turn struck the witch, and immediately they became rocks, 'which remain to prove the marvel to our incredulous age', wrote the antiquarian T. J. Westropp in the early twentieth century. Despite her stony condition, the cow used to come out of the lake and walk round the island when any great event was about to take place.

Westropp guesses that this is the island referred to in the twelfth century by Gerald of Wales as one of the 'Wonders and Miracles of Ireland'. A mound in the sea, Gerald reported, had one day appeared where none had been before. The locals at first thought that it must be a whale, but then, seeing that it did not move, concluded that it was land. A party of young men set off in a boat, but when they approached the mound it sank. The next day it appeared again, and again the men rowed towards it, and again it disappeared. On the third day, they took the advice of a wise old man, and as they approached the island they fired an arrow of red-hot iron on to its shores. This time they were able to land, and the island has remained above the water ever since.

Tales of phantom lands seen from INISHMORE and RATHLIN O'BIRNE ISLAND mention the idea that a burning spark will break their spell, and magical cows too feature in several Irish legends, including that of DURSEY ISLAND (Southern Eire).

Visiting Inishbofin in the late 1960s, Horace Beck was told that a great treasure lay in the seabed to the west of the island. Many years ago, said the islanders, the *Royal Oak* had lost her rudder and driven ashore here, and went to pieces almost at once. Divers were sent to bring up the valuables the ship had been carrying, but came up empty-handed.

> Once ashore, nothing could induce them to return to the sea, and they refused to say why, other than that there was something down below they could not or would not face. People on the island say it is a serpent that lies coiled about the gold guarding it, and although they are very poor they prefer poverty to bearding the serpent.

INISHGLORA, COUNTY MAYO

The Children of Lir

From the Greek myth of Leda, ravished by Zeus in the form of a swan, to the story of *Swan Lake*, legends have gathered around this most regal and graceful of birds. An Irish tale, 'The Tragic Fate of the Children of Lir', tells of children transformed to swans. Derived from medieval legend, it first appeared in full in a fifteenth-century text and has been retold many times since.

Two pairs of twins – Fiachra, Conn, Aodh, and their sister Fionnghuala

The Children of Lir were changed into swans by their wicked stepmother, and lived for nine hundred years in that shape before the spell was broken.

or Finola – were born to a chief of the Tuatha Dé Danann, the ancient fairy people of Ireland. Their mother died, and their father's new wife, a woman of magical skill, became so jealous of her stepchildren that she changed them into swans, telling them that they must stay in this form until a day long in the future when a wedding would take place between a nobleman from the north and a noblewoman from the south.

As punishment for her crime, the stepmother was turned into a demon and condemned to fly forever through the air, but this did not help the swan-children, who spent nine hundred cold and miserable years roaming the waters, first on Lake Derravaragh in Westmeath, then on the Sea of Moyle, between Ireland and Scotland, and finally off the coast of Erris, before they came to Inishglora. There they met a Christian missionary, St Mochaomhóg or Kemoc, who treated them kindly.

At this time the king of Connacht (in the north) was about to marry the daughter of the king of Munster (in the south), who had heard about the wonderful swans of Inishglora and longed to own them. The marriage, however, signalled the end of the spell, so just as the king arrived to demand the birds, they were disenchanted, and all he found was four very ancient, withered people. Having been baptised by St Kemoc they died happily, and were buried together.

INISHMAAN, COUNTY GALWAY

Sea-horses

> A horse I've seen myself on the sea and on the rocks – a brown one, just like another. And I threw a stone at it, and it was gone in a minute. We often heard there was fighting amongst *these*. And one morning before daybreak I went down to the strand with some others, and the whole of the strand, and it low tide, was covered with blood.

This sinister little anecdote, collected on Inishmaan and included in Augusta Gregory's *Visions and Beliefs in the West of Ireland* (1920), centres on the idea that the sea as well as the land was inhabited by fairy beings – called only *these*, since to name the supernatural creatures directly would have been to invite their attention.

A common Irish and Manx saying held that 'whatever there is on land is to be found in the sea too'. Horses from the sea were often mentioned, and Gregory was told that three fishermen caught one in their nets one night, 'but they let it go, it would have broke the boat to bits if they had brought it in, and anyhow they thought it was best to leave it.' Sometimes

Artists sometimes portray sea-horses with fishes' tails. In folk tales, however, the beasts can run on land like any horse, although in the end they always return to the waves.

the beasts came on shore to eat the oats, and like Welsh water-horses (*see* ST BRIDES, Wales), they could be tamed at least temporarily. One man watched at night to catch his neighbours' cows eating his grass, but what he saw was horses and foals coming up from the sea:

> And he caught a foal and kept it, and set it racing, and no horse or no pony could ever come near it, till one day the race was on the strand, and away with it into the sea, and the jockey along with it, and they never were seen again.

This story has echoes of a fifteenth-century tale describing how St Brendan summoned fifty seals from the sea and transformed them into racehorses. These he gave to the king as ransom for a captive, but after a year and three months the horses' riders used their whips, and immediately the steeds galloped into the sea, where both they and the men became seals.

INISHMORE, COUNTY GALWAY

Hy Brazil

In April 1668, a man named Morogh O'Ley was wandering in the countryside, alone and depressed after falling out with his wife. He met a couple of strangers who kidnapped him, forced him into a boat and carried him off to a magical island they told him was called O'Brazil, from which he could see the Aran Islands and the Galway coast. There he stayed for two days and then found himself back on the mainland, where he lay for some time very ill in a friend's house, not knowing how he had got there. After his recovery he began to practise medicine with considerable success, although he had never studied the subject, and he attributed his new-found skill to a book he had obtained on the enchanted island.

There are clues to a mundane explanation. The quarrel with his wife might have led O'Ley to take off for a few days, and the tale to account for his absence could have been an attempt to get out of trouble. The 'Book of Hy Brasil' or 'Book of the O'Lees', a manuscript dating from the fifteenth century listing various diseases together with their cures, is real, and is preserved in the library of the Royal Irish Academy. It is suggested that O'Ley had inherited it from his ancestors, who had been physicians in the Connaught district, and linked it to the story of his magical abduction in order to attract business.

The uncanny island itself has a long history. Irish legend from the ninth century onward tells of a supernatural land located in the western ocean, towards the setting sun. It is known by various names, including Tir na nÓg ('the Land of Youth'), Eamhain Abhlach ('the Sacred Place of Apples') and Magh Meall ('the Enticing Plain'), but is most often called Brasil, Uí Bhreasaíl or Hy Brazil, either from the Irish word *bres* ('noble') or derived from Latin *brasilium*, a red wood used in dyeing, perhaps referring to the sunset glow around the mythic land.

For centuries, Hy Brazil was supposed to be a real and discoverable place. On maps of the fourteenth and fifteenth centuries, the island was marked in various remote places of the ocean, and even after the name Brazil was given to the New World land discovered in 1499 (from the red-dye trees that grew there, rather than by reference to the legend), the phantom island continued to appear on sea charts. Belief in the existence of Hy Brazil was strong in the seventeenth century, when Morogh O'Ley made his fabulous journey. In 1663, an Ulster MP announced in the Dublin Parliament that he was, according to a revelation, destined to take possession of 'O Brazile, or the enchanted island', which had been seen by many people off the coast of Ulster. A few years later, in 1674, a Captain Nisbet reported rescuing an old Scottish gentleman from an island called O Brazile, where, the man said, he and his servants had been imprisoned by a wicked magician until Nisbet had arrived and lit a fire, breaking the enchantment.

Later accounts tend to be of seeing the island, rather than actually visiting it. Robert Southey wrote in 1815 that the inhabitants of Inishmore, the westernmost of the Aran Islands, 'are persuaded that in a clear day they can see *Hy Brasail*, the Enchanted Island, from the coast', and there are many reported sightings from the nineteenth and twentieth centuries. Fairly typical is an account given to Augusta Gregory and printed in her *Visions and Beliefs in the West of Ireland* (1920):

> There was a woman walking over by the north shore - God have mercy on her - she's dead since - and she looked out and saw an island in the sea, and she was a long time looking at it. It's known to be there, and to be enchanted, but only few can see it.

There is a hint here that seeing the island was fatal, as was said of Cill Stuifín at LAHINCH (Southern Eire).

Hy Brazil was most frequently visible from Inishmore and the Galway coast, but other western districts have traditions of a magical island, which goes under a variety of names - Cill Stuifín in Clare, Mainistir Ladra in Sligo and Mayo, and Tir Hudi in Donegal, while the Galway island is sometimes called Beag-Árainn (Little Aran). Its characteristics are similar everywhere: it appears once every seven years, when seen it is a delightful land, green and fertile, or a magnificent city, and it can be disenchanted with a spark or burning ember, although an alternative way to counter its spell was described in another story told to Gregory. A fisherman landed on the enchanted island, where he found a little house:

> and a very nice-looking young woman came out and said, 'What will you say to me?' and he said, 'You are a very nice lady.' And a second came and asked him the same thing, and a third, and he made the same answer. And after that they said, 'You'd best run for your life,' and so he did, and his curragh was floating along and he had but just time to get into it, and the island was gone. But if he had said 'God bless you,' the island would have been saved.

Perhaps it was just as well that he said the wrong thing. It is sometimes reported that if the phantom island's charm is ever broken so that it remains above the waves, the real Aran Islands will at that moment sink forever.

INISHMURRAY, COUNTY SLIGO

The Well of Assistance

A ceremony performed at a spring on Inishmurray known as Tobernacoragh, or the 'Well of Assistance', was described to Thomas Mason in the 1930s. He was told that, however stormy the weather, a calm next day could always be produced by sprinkling three drops of water from the well into the sea to the east, west, south and north, three times in each direction. If an immediate journey was necessary, three drops sprinkled on the waves at embarking would give a safe passage.

In earlier times a more thorough procedure seems to have been followed, according to W. G. Wood-Martin's *Pagan Ireland* (1895):

> When tempestuous weather prevails, communication between the island and the mainland is sometimes rendered impracticable even for weeks. On such occasions, the waters of the spring are drained into the ocean, upon which – the charm rendered doubly certain by the repetition of certain prayers – a holy calm succeeds the strife of the elements.

Unlike the weather-controlling well on GIGHA (Scottish Lowlands), the Inishmurray spring seems to have had no special guardian who had to be paid for a charm, but to have been free to anyone on the island. Wood-Martin calls the ritual 'probably the most pagan in character still exercised in connection with a holy well', although he records that in the late eighteenth century the islanders used it particularly whenever they needed a priest, for example to perform the last rites for someone on the point of death. Even in the worst winter storm, as soon as the boat touched the waves they became perfectly smooth, and remained so until the priest had come and gone and the boat returned to Inishmurray, when the tempest would begin again.

KID ISLAND, COUNTY MAYO

'The Land Beneath the Sea'

In 1976, the storyteller John Henry told a tale of 'The Land Beneath the Sea', translated by Séamas Ó Catháin in *Stories of Sea & Shore* (1983). Henry's grandfather, then only a lad, was out one day with some other men fishing off the Kid Island bank, the best place for cod at the time. They were using long lines called 'rippers', weighted with lead and armed with six barbed hooks. Henry's grandfather cast out the line and to his surprise brought up a half-boiled cabbage leaf, and immediately a woman stuck her head up out of the sea next to the boat.

> 'Damn your souls,' she said, 'away home with you. I had a pot of cabbage boiling for the man of the house who is out at his work and you threw your rotten old "ripper" down the smoke hole and where did it land only on my pot on the fire. And you have spoiled my pot.'

The terrified men got out their oars and rowed away as fast as they could. A sudden wind caught them, and they were only just able to reach the shore. 'That was a true story,' said Henry, who had heard it many times from his father. 'I suppose that it was a house that they happened on in that spot.'

Another story Henry had heard, though he was not so convinced of its truth, was that a crew was 'ripping' for cod in the same way when one of them pulled up a child on his line, wrapped in a piece of cloth. It was alive and kicking, and its mother popped up her head crying, 'Give me back my child, for if you don't, not a man jack of you will ever blow his nose on the green land of Ireland again.' They threw the baby back into the sea and headed for home, again pursued by a storm that nearly wrecked the boat.

Both tales come from the folk belief that under the sea was a world corresponding closely to that on dry land, so that fishermen might pull up sprigs of flowering heather or domestic articles such as saucepans full of boiling potatoes. Sea-babies were a not uncommon catch, but were always returned to the water.

KINVARRA, COUNTY GALWAY

How Noah got drunk

Long, long ago, Noah set to work in secret to make a big ship. One day a man called on Noah's wife and asked her what her husband was up to. 'I don't know,' she said, 'and no one in the world knows what he is doing.' The man told her that soon a black bear would come past her house with froth on his fur. If she wanted to know what Noah's plans were, she was to take some of the froth and put it in his drink. So she did this, and after Noah had drunk the drink with the froth, he told her everything.

Next day, the first blow Noah struck with his hammer was heard through the whole world, and then everyone knew what he was doing, even the Devil. When Noah's ship was ready he took his wife and children on board, and his dog too, but there was not quite enough room for the dog, so he had to stick his nose outside and that is why the dog has a cold nose ever since.

The Devil asked Noah's daughter to take him along with her, and he turned himself into a brooch that she put in her blouse. The Ark began to sink, and Noah wondered why. He went round the ship with a bottle of holy water, sprinkling it on everything he saw, and when a drop fell on the girl, her brooch flew off in a flame. After that the vessel floated on its way.

The stranger who came to Noah's wife and told her to put the froth in the drink was the Devil. 'That is what has made drink evil ever since,' finished the Kinvarra farmer who told this story, which he had heard in the late nineteenth century from his grandfather.

LARNE WATER, COUNTY ANTRIM

The mermaid-saint

It was once reported as a matter of history that in the sixth century, a mermaid named Liban was captured by the monks of Bangor, in Down. The tale dates back to a medieval text, its events beginning with an overflowing spring that drowned a great territory along with all its inhabitants except one. The disaster had been foretold by Curnan, the husband of Liban's sister Ariu, whose prophecy is translated in a nineteenth-century version:

> Come forth, come forth, ye valiant men; build boats, and build ye fast!
> I see the water surging out, a torrent deep and vast;
> I see our chief and all his host o'erwhelmed beneath the wave;
> And Ariu, too, my best beloved, alas! I cannot save.
> But Liban east and west shall swim
> Long ages on the ocean's rim,
> By mystic shores and islets dim,
> And down in the deep sea cave!

After her entire family was killed by a flood, the maiden Liban survived in an underwater cavern. When she prayed to be transformed, she was granted a mermaid's tail so that she could swim alongside the fish of the sea.

Curnan was regarded as a simpleton, and his warnings were ignored, but just as he had predicted, the maid of the well neglected her duty, and the water flooded out to create Lough Neagh.

Liban, the only survivor, was swept away into an underwater chamber, where she lived with only her lapdog for company. After a year she prayed out of loneliness to be a salmon, so that at least she might swim with other fish. She was changed, but not completely: although she got a tail, she kept a human face and upper body. Her little dog too was transformed into an otter, and swam along beside her.

For three hundred years Liban lived in the sea, and at the end of that time St Beoc of Bangor, sailing to Rome, heard angelic music coming from beneath the water. When he asked who was singing, the maiden told him she was Liban, and said that she had come expressly to meet him. She asked him to come to the mouth of the river Ollarbha (the old name of Larne Water) in a year's time, bringing boats and fishing nets. Beoc arrived punctually with other saints of Bangor, and Liban was netted by St Fergus. She was brought to land and kept in a boat half full of sea water, in which she swam about and talked to the many people who came to look at her.

The mermaid was claimed by three saints: Beoc, who had met her first; Fergus, who had caught her; and Comgall, head of the Bangor monastery. They prayed and were granted a sign from God that Liban should stay with Beoc. Offered the choice between living another three hundred years or being baptised and going at once to heaven, Liban decided for heaven, and was baptised in the name of Murgen or Muirghein, meaning 'sea birth'. She died, and was buried at Beoc's church, where many wonders and miracles were performed through her means, and she was included in a seventeenth-century Irish calendar as St Murgen.

MULROY BAY, COUNTY DONEGAL

St Columba and the salmon

On his way to IONA (Scottish Lowlands), St Colm Cillé (St Columba) stopped in Mulroy Bay, where he saw two fishermen out after salmon. The men had already caught two fine fish, but when they saw Colm Cillé coming they hid them, because they knew the saint was always asking for something. Sure enough, he wanted to know if they had any salmon, and they told him they had none. If that was true, said the saint, they would catch plenty now, but if by any chance they were lying, they would never catch salmon in the bay again – and nor they did.

The story was known in the sixteenth century, and was still current in the 1960s, when the folklore collector Horace Beck heard it from Mrs Macbride of Mulroy Bay. She went on to give an update:

> Now from that time until about twenty years ago no salmon ever was caught in this bay and then some men were out in a boat and a salmon jumped and one of the men caught him (he was a Protestant, you know) and as he caught him he said, 'St Cholm Cille was wrong.' Just then the salmon gave a great leap out of him and he was back in the sea. St Cholm Cille was right after all, for never the salmon has been caught in this bay from the day of the curse to this.

PORTACLOY, COUNTY MAYO

Fairy boats

John Henry, a storyteller of Kilgalligan in Donegal, told in 1973 how a crew from Portacloy stayed out fishing one stormy night at the insistence of a man known as 'the greedy fellow', because he had a big family to feed and was always eager for a good catch. The rest of the men remarked what a crowd there was going home, as they saw boat after boat heading past them, and they became restless and keen to make for land. At last the greedy fellow let them take in their nets, with a fine haul of mackerel, and then they hurried for home, but when they came to the beach they were amazed to see it empty of boats. What had happened to the currachs that had passed them out at sea? 'Those were the fairy currachs that came back that way, trying to get them to go home, of course,' explained Henry, and he added that the reason the greedy fellow could brave the storm was that he had been born with a caul, and was therefore safe from drowning: 'Only for the man with the caul they were done for. Not one of them would ever have returned – it would have been all up with them.'

A caul is sometimes found on a newborn child's head, being part of the membrane that has enclosed it before birth. Because a foetus does not 'drown' in the wet womb, it was thought that cauls would continue to protect their possessors from a watery death, and they were much valued by sailors until very recently, sometimes changing hands for quite large sums (*see* CAULS, p. 402).

RATHLIN ISLAND, COUNTY ANTRIM

Breccán's Cauldron

Breccán was a noble merchant of the fourth or fifth century, who traded between Ireland and Scotland with fifty boats. One day, however, he and his entire fleet were sucked into a terrible whirlpool between Rathlin Island and Antrim, and vanished utterly. Nobody knew what had become of them, until an animal skull was washed up on the beach at Drogheda. A blind poet touched the relic with his magic staff, and announced that this was the skull of Breccán's dog, drowned with its master and all his men.

The maelstrom in which they perished was at the meeting of three bodies of water, 'the sea which encompasses Ireland at the north-west, and the sea which encompasses Scotland at the north-east, and the sea to the south between Ireland and Scotland'. Where the oceans met they spun round like the paddles of a millwheel and were sucked down, leaving a gaping hole that 'would suck even the whole of Ireland into its yawning gullet', then spewing the torrent back up again with a noise like thunder and an explosive fog of spray as if a vast pot were boiling. It was named from Breccán's fate, *Coire Breccáin* ('Breccán's Cauldron').

Tale and description are taken from a ninth-century glossary, and the legend appears in medieval verses, translated:

> Breccán son of Maine, rich in graces, the Cauldron drowned with its red spray, and he lies under the heavy high-piled strand with his ship and his valiant following.
>
> Though it has buried unforgotten Breccán, his name endures in story with his bark and its burthen that lie beneath the whirlpool's stormy water.

The poem goes on to tell how a century or so later St Columba passed the whirlpool, bearing with him a handful of earth from a holy grave, which he threw into the waters and miraculously stilled them: 'he made feeble and faint its fury, so that it is now a pool right peaceful.'

The saint's calming influence has not lasted, however. The Rathlin Island channel is still very dangerous at certain tides. It is now known as *Sloc na Mara*, 'Gulp of the Sea', the name *Coire Breccáin* or CORRYVRECKAN (Scottish Lowlands) having been transferred to the still more destructive whirlpool between Jura and Scarba.

Dolphins and Porpoises

Dolphins were traditionally beloved by sailors, their habit of following ships interpreted as desire for company (although it may really be a more basic appetite, dolphins being quite intelligent enough to learn that fishing boats can lead the way to food). Classical legend tells of dolphins rescuing shipwrecked sailors, and of how they welcomed children on their backs, reports that may have arisen in part through linguistic confusion with a type of rowing boat known by the same Latin name, *delphinus*. Pliny's tale of a boy who rode to school every day on a dolphin was probably a misunderstanding of this kind, but it is true nonetheless that dolphins are companionable, and will allow human beings to swim with them. Religious art made the dolphin a symbol of Christ, bearing mankind to deliverance through the waters of baptism, and saints as well as characters of pagan myth were said to have been saved from drowning by its kindly intervention.

Porpoises are similar to dolphins, but with somewhat more swinish snouts (as indicated by their name, derived from the Latin *porcus piscis*, 'pig fish'). The twelfth-century historian Gervase of Tilbury may be referring to porpoises, although calling them dolphins, when he records that 'people say that this fish is born as a knight, and puts on its piggish appearance secretly among the waves of the sea'. He then tells a tale prefiguring many accounts of merfolk (*see* for example GALWAY, Northern Eire & Northern Ireland), in which a sailor who has speared a dolphin is taken by a strange marine horseman to a distant land, where he finds his victim, now in the shape of a knight, with the weapon still sticking in him. He removes it, restoring the dolphin-warrior to health, and is forgiven for his crime. 'That explains why, ever since then,' says Gervase, 'sailors have no longer attacked dolphins.'

Through to modern times, seamen have considered the sight of dolphins and porpoises a good omen, and held that it is unlucky to kill them. One illustration comes from the Second World War, when the steamer *Britannia* was torpedoed off the coast of Africa, and her crew took to the sea in lifeboats. After days adrift, the men were starving, and it was suggested that they should shoot one of the dolphins that

A fable tells of a dolphin that rescued a drowning monkey, supposing it to be human, but when he discovered that his passenger was only an ape, he unceremoniously ditched it.

were following them, but as noted in Lieutenant Commander Frank West's memoirs of the voyage, 'they were such beautiful and friendly creatures that I doubt if any one of us could have done so and, I cannot help thinking, that respect for superstition may have stopped us also, though none would have admitted that was so.'

The American officer William Anderson, recording his submarine voyages in the 1950s, also notes the significance he attached to appearances of porpoises:

When I saw porpoises playing around the bow as we left port, I had a good feeling, a feeling that our cruise would be a success. When I saw no porpoises, I felt just the opposite. And my predictions were usually fairly accurate.

RATHLIN O'BIRNE ISLAND, COUNTY DONEGAL

A magic island

The fishermen of Teelin used often to talk of an enchanted island, south-west of Rathlin O'Birne Island, which appeared once every seven years and would always sink again until the day that a burning spark landed on it, after which it would remain above the sea. In the twentieth century, an old man of Glencolumbkille, on the coast of Donegal, reported that it had been seen by one of his family just after sunset on a fine summer evening. The man was carrying two legs of a spinning wheel which he had just collected from the carpenter, and he stuck them in the ground as a cross to mark the spot where he had been standing when he saw the island. Then he hurried off to the nearest village, Dooey, to collect more witnesses, but when the villagers came back with him, they found a thousand crosses stuck all about the place.

At last, after much searching, the right cross was identified, and then the men set off in a boat for the magic island, taking with them a pot of smouldering embers. It was the greenest place they had ever seen, and they thought that this must be Tir na nÓg (the Land of Youth). As they neared the shore they saw a girl sitting at the foot of a tree. She was knitting stockings, and when the boat was close in, she stood up and threw her ball of wool into it, keeping hold of one end herself. Instantly the boat began to toss on the waves so that the men were almost thrown out into the sea. Terrified, they tried to cut the woollen thread with a knife, but it was too strong. Then one of the men remembered the pot of embers, and used the live coal to burn through the wool. Once they had escaped they went straight back home, and the magic island vanished. 'Often since then the same island has been seen between Rathlin and the sunset,' finished the storyteller, 'but no one has ever again tried to land on it.'

The elements of this folk tale go back a long way. In the twelfth century, Gerald of Wales reported the disenchantment by fire of a phantom island, sometimes identified as INISHBOFIN, and the magic ball of wool appears in a still earlier story, the eighth-century *Voyage of Bran* (*see* WONDER VOYAGES AND LOST LANDS, p. 458), in which an enchantress throws a ball of thread that sticks to the traveller's hand and draws him ashore.

ST MACDARA'S ISLAND, COUNTY GALWAY

Fishermen lower their sails

St MacDara's Island was held in particular reverence by the Galway fishermen, reported Roderic O'Flaherty in the seventeenth century, on account of a wooden statue of the saint that stood in the chapel there, until it was buried by order of the archbishop 'for speciall weighty reasons', no doubt to prevent superstitious practices connected with it. Even after the statue had been hidden, a custom continued for boats passing the island to dip their sails three times (as was also done at LIHOU, South-West England & Channel Islands). A captain who neglected to do so, in 1672, 'was so tossed with sea and storme, that he vowed he would never pass there again, without paying his obeysance to the saint', and a few years later a fisherman named Gill, who scorned the practice, was struck dead when his mast broke and hit him on the head.

Well into the nineteenth century, both children and boats were named 'MacDara' for luck at sea, and the ritual of lowering the sail was still observed in the 1930s.

SCURMORE, COUNTY SLIGO

The Children of the Mermaid

Between Inishcrone and Scurmore, near Killala Bay, stands a rough circle of large granite boulders, a site still marked on ordnance survey maps as the 'Children of the Mermaid'. Their legend is told in W. G. Wood-Martin's *Traces of the Elder Faiths of Ireland* (1902). One day a man walking along the seashore discovered a mermaid asleep among the rocks, and having stolen her 'magic garment' (probably an enchanted cloak, as mentioned at CAHERDANIEL, Southern Eire), he won her for his wife. When their seven children were nearly grown up, the youngest saw his father take the garment from its hiding place:

> The youth ran off to describe what he had seen to his mother, who seized with a sudden yearning to return to her native element, resumed possession of her property, and bade her children follow her to the seashore. Being now re-endowed with all the attributes of a mermaid, she touched each in succession, changed them into seven stones, and then plunged into the ocean, and has never since been seen.

There are many stories accounting for boulders or menhirs as human beings turned to stone (usually as retribution for a sin such as dancing on the Sabbath), and many more dealing with a mermaid wife's return to the sea, but it is uncommon to find the two in combination. Occasionally a mermaid mother used to take her children with her to her submarine kingdom, but more usually she abandoned them along with her husband.

SKERD ROCKS, COUNTY GALWAY

The magician's journey

'In former days,' wrote James Macpherson in 1771, 'there lived in Skerr a magician of high renown.' Although the wizard had power over the sea and the storm, he longed only for one thing: to see the green Isle of the West, where dwelt the souls of the dead. One day a tempestuous wind began to blow, low clouds covered the sky and sea, and from the mist emerged a boat, its sails spread, with a hundred oars that seemed to row.

> But it was destitute of mariners; itself seeming to live and move. An unusual terror seized the aged magician: He heard a voice though he saw no human form. 'Arise, behold the boat of the heroes – arise, and see the green Isle of those who have passed away.'

The magician entered the boat. For seven days and nights he sailed unsleeping on the stormy sea, and on the eighth the waves swelled mountain-high, darkness gathered around him and a chorus of a thousand voices cried, 'The Isle, the Isle.' Here was the land of the dead, a beautiful green land where the sun always shone, and here the magician stayed, as he thought, for one day only before sailing back to Skerr. On his return, however, he discovered that no fewer than two hundred years had passed.

Macpherson claimed this as 'One of those tales, which tradition has brought down to our times', and certainly the ideas it contains are old ones. Legends going back to the ninth century tell of an Otherworld island, destination of the dead or home of the fairies, lying in the ocean west of Ireland, and said to be visible from INISHMORE and other places on the western coast. Stories are common, too, of people who visit fairyland for a few days or hours but find that centuries have gone by in the human world.

As for the magician's point of departure, Macpherson added in a footnote that 'Skerr signifies in general a rock in the Ocean.' It may have been the Skerd or Sceirde Rocks, north-west of the Aran Islands and now the location for a wind farm, a place with a wild, uncanny quality noted by earlier writers. In 1684, Roderic O'Flaherty described how sometimes the rocks looked like a great city in the distance, full of houses and castles. At other times they would appear to be 'full of blazing flames, smoak, and people running to and fro', or like a great fleet of ships. These illusory appearances could not be an effect of sunlight on water vapour, he noted, since they could be seen just as well on dark and cloudy days.

O'Flaherty connected the visions with legends of the magical island Hy Brazil, but observed that 'the inchanted island of O'Brasil is not alwayes visible, as those rocks are, nor these rocks have allways those apparitions.' His nineteenth-century editor added a note that sea mirages were often seen around the Irish coast, for instance at YOUGHAL (Southern Eire).

TEELIN, COUNTY DONEGAL

Fairies take the drowned

Fairies were said to live on the sea as well as on land, or, to put it another way, mermaids and other sea-people were counted among the fairies, and often little or no distinction was made between the different sorts of supernatural being, except purely in terms of where they lived. The old people of Teelin in south-west Donegal, as reported in a collection of folklore made in the 1970s, believed of anyone who was drowned 'that it was the wee folk who had taken him away and because of that he could not be saved'.

Long ago in Teelin, there was a midwife who lived with her husband Paddy and his brother, an idle fellow. Late one night the woman was called out to a birth. Paddy had meant to go to the beach to catch sand eels for bait, but he was ill that evening, and so the midwife told the brother he must go instead. 'And if you don't go,' she said, 'don't be here when I return tonight!' So all alone, the man went down to the seashore to look for the eels. 'The immortal host came to the strand, and whatever they did to him, he never returned home.'

The midwife found out what had happened, because on her way home that night, when she was crossing the Glen river, she saw her brother-in-law standing on the other side. She asked him what in God's name he was doing there, and he replied that he had given her much trouble, but would never do so any more.

> 'The fairy host came on me on the strand and they have a hold on me for the rest of my life. Farewell and blessings to you now, you will never see me again! Tell Paddy to be very careful on such a day, otherwise he and the crew will have a tale to tell!'

Then he vanished, 'like a piece of mist', and was never seen again in Teelin.

TELTOWN, COUNTY MEATH

Sky-ships

According to medieval belief, above the earth there existed a world with its own sea, and vessels that to those below seemed to float among the clouds. Reports of sky-ships and their crews coming into contact with

Sirens, Sea-Nymphs, and Mermaids

Six or seven thousand years ago, a fish-tailed god was worshipped by the Akkadians of Babylonia, and the ancient Syrians and Philistines honoured the divine Atargatis or Derceto, a woman from the waist upwards, but her lower half shaped like a fish. These early marine deities were originally quite distinct from the sea-nymphs or nereids of classical legend, aquatic but all-female offspring of the god Poseidon, and from the sirens, first described by Homer as creatures half-woman, half-bird, whose song lured sailors to their doom. Over time, sirens came to be imagined as sea-dwellers, and in the first century BCE the Roman poet Horace gives them women's heads on fishes' bodies.

Thus the mermaid was born, but she continued to undergo transformation. The bird element persisted for some while, and medieval art sometimes shows complex hybrids with human faces, fish-tails, wings and claws, while as late as the nineteenth century there is an echo of these bird-fish-women in an account of a feathered mermaid seen near Exmouth, Devon, who sang 'wild melodies'.

Another idea in the mix is seal mythology. According to Greek poets from the eighth century BCE onwards, an entire Mediterranean tribe was descended from seals, via a nereid who transformed herself into that shape when pursued by Zeus's son Aeacus. Their union produced a seal-son, Phocus, whose offspring were the Phocians of Asia Minor. Intermarriage between humans and seals is a staple of British legend (*see* for example BARRA, Scottish Highlands & Islands), and similar tales about mermaids such as the Lady of GALLARUS (Southern Eire) relate how they had children by their human husbands, but always eventually returned to their natural element.

The potent female sexuality of the mermaid is an ambivalent quality, on the one hand making her vulnerable to rape and domestication by predatory males, on the other enabling her to tempt men into her power. A plaque presented to Mary Queen of Scots by her first husband the Dauphin showed her as a mermaid on a dolphin's back, combining a pun on his French title with a compliment to her beauty and charm, but later, when accused of murder and adultery, she was caricatured in the same guise, a bare-breasted mermaid, but this time in allusion to her supposedly insatiable lust.

In Christian art, the mermaid is a symbol of vanity with her looking-glass and comb, archetypal ornaments that appear at least from the Middle Ages, and which may have developed from the image of the reflective moon (connected with the sea and sea-goddesses such as Atargatis) and perhaps the comb-like plectrum with which the siren was originally supposed to pluck a lyre to accompany her singing. In legend, she can appear as vengeful or protective nature-spirit (PADSTOW, South-West England & Channel Islands, and PEEL, North-West England & Isle of Man), dangerous temptress or baby-killer (COLONSAY and KNOCKDOLIAN, both Scottish Lowlands), unusual catch (YELL, Scottish Highlands & Islands) or even saint (LARNE WATER, Northern Eire & Northern Ireland). She is as changeable and many-faced as the sea itself.

A mermaid in characteristic pose, holding up her mirror and combing her long hair.

earth-dwellers were not unlike twentieth-century accounts of UFOs. It was said that in 748, monks at Clonmacnoise saw a ship sailing through the air above them, from which an anchor was lowered and stuck in the church floor. A man swam down through the air to release it, but begged for help as he was 'drowning', and after the monks freed the anchor he returned to his vessel, which sailed away on the aerial sea. In the tenth century at Teltown, an assembly held by the high king Conghalach was interrupted when a visitor descended from a strange sky-craft to fetch an arrow that had been shot from his ship at a salmon swimming through the air, and from England there is a similar medieval account by Gervase of Tilbury (*see* BODMIN, South-West England & Channel Islands).

A remarkable aerial apparition was seen in 1679 in Tipperary, according to a pamphlet reprinted in William Hone's *Every-Day Book* (1830). Several people saw a ship sailing in the air, 'and so near to them it came, that they could distinctly perceive the masts, sails, tacklings, and men'. This was followed by two more ships careering across the sky engaged in battle, and then by a horse-drawn chariot, out of which leapt a bull and a dog that fought a monstrous serpent pursuing them. Several more ships came later, 'but the relators were so surprised and pleased with what they had seen, especially with the bull and dog, that they did not much observe them'. Hone dismissed the whole story as a deliberate fraud, but other explanations are possible. In the early nineteenth century, William Wordsworth reported a personal experience in the Lake District:

> While we were gazing around, 'Look,' I exclaimed, 'at yon ship upon the glittering sea!' 'Is it a ship?' replied our shepherd-guide. 'It can be nothing else,' interposed my companion; 'I cannot be mistaken; I am so accustomed to the appearance of ships at sea.' The Guide dropped the argument; but, before a minute was gone, he quietly said, 'Now look at your ship; it is changed into a horse.' So indeed it was, – a horse with a gallant neck and head.

Both ship and horse were clouds, and so may have been the Tipperary marvels. Alternatively, they may have been reflections. In 1798, the editor James Hardiman saw ships sailing through a calm autumn sky, and hundreds of others who witnessed the scene believed it to be supernatural. It later proved to have been a fleet on the sea, the image of which the onlookers saw cast on the clouds. Light on water vapour can produce remarkable mirages, whether of flying ships and animals, or of landscapes and houses as at YOUGHAL (Southern Eire).

TORY ISLAND, COUNTY DONEGAL

Tory earth

In the 1950s, Jimí Dixon, a fisherman, told the folklorist Seán Ó hEochaidh about what had happened to his grandfather Donnchadh, a great fisherman and a strong swimmer, one evening when he went with a friend to Port an Dúin on the north of Tory Island. As they rowed around the island they saw a rabbit sitting on the cliff, and it seemed quite unafraid of the men. Donnchadh beat his paddles against the boat, and still the rabbit did not move. 'Donnchadh, that is no natural rabbit!' said the second man.

At that moment the sea swelled mightily, and three great waves crashed against their boat. The vessel overturned, and only Donnchadh's skill in swimming got the two men to land.

> As long as they lived both men held, and I heard my grandfather speak of it a score of times, that it was a fairy rabbit they had seen on the height above them and that it was trying to drown them.

The only reason it did not succeed, Donnchadh maintained, was because they had with them 'a grain of the earth of Tory'. Tory Island earth or clay is renowned on the Donegal coast for its protective power, usually said to be effective against rats.

That there was not a single rat on Tory was reported as fact in the 1930s by Thomas Mason. He was told that 'a hard unbelieving man' from Londonderry once imported some rats as a test, but they died spontaneously within minutes of landing. Mason, whose own factory in Dublin was infested, was anxious to try the charm himself, and visited a man called Anthony Dugan, head of a family that claimed the island's first Christian convert as an ancestor. On him and his descendants St Colm Cillé (Columba) had conferred his own ability to banish vermin, and people used to travel many miles to get 'Tory clay' direct from Dugan. Mason asked in prescribed form, 'In the name of God give me some Tory clay,' and was impressed by the dignified manner in which the old man responded. He went to the ruins of a small church nearby, knelt and said a prayer, and then put a couple of handfuls of clay into a paper bag. In the event, however, Mason was unable to carry his experiment through:

> Unfortunately when I came home from my holidays the rats had disappeared, and when they returned after a couple of years I could not find the paper bag in which the clay was contained.

Atlantic Ocean

Londonderry
Antrim
Donegal
NORTHERN IRELAND
Tyrone
Fermanagh
Armagh
Down
Sligo
Monaghan
Leitrim
Cavan
Mayo
Louth
Roscommon
Longford
Meath
Rockabil
Lambay Island
Westmeath
Malahide
Galway
Dublin
Offaly
Kildare
Lahinch
IRELAND
Laois
Doonbeg Bay
Clare
Wicklow
Vicarstown
Scattery Island
Carlow
Moorestown
Tipperary
Limerick
Kilkenny
Ballyheige Bay
Brandon Creek
Wexford
Gallarus
Wexford
Kerry
Dingle
Rossbeigh Strand
Waterford
Bannow Bay
Great Blasket Island
Cork
Ballinskelligs Bay
Dunkerron Castle
Youghal
Ardmore
Cork
Caherdaniel
St. George's Channel
Trafrask
Bantry Bay
The Bull
Dursey Island
Glandore
Clear Island
Fastnet Rock
Celtic Sea

0 20 40 60 mi
0 40 80 120 km

SOUTHERN EIRE

Counties Clare, Cork, Dublin, Kerry, Waterford, Wexford

ARDMORE, COUNTY WATERFORD

St Declan's Stone

Between Youghal and Dungarvan is a rocky promontory that may once have been an island, and could become one again if the sea level rises. Tradition states that it was joined to the mainland when St Déaglán or Declan landed here in the fifth century. Upon arrival, he struck the water with his staff and the sea withdrew, creating Ardmore Bay in the curve between the headland and the cliffs to the north-east.

Two holy wells, an ancient oratory, a ruined church and a magnificent twelfth-century round tower are among the remains of Declan's 'holy city', and at the southern end of the beach below Ardmore village is a curved boulder with a narrow gap beneath. This is St Declan's Stone, visited for centuries by devotees, who used to crawl through the hole for a blessing or a cure. The rite as followed in 1833 was described by a witness as beginning at low tide with the scraping away of some of the sand that had blown under the rock. Then the pilgrims took off their shoes and stockings, the women removing their bonnets and the men their hats and coats.

> They turn up their breeches above the knee, then lying flat on the ground, put in hands, arms, and head, one shoulder more forward than the other, in order to work their way through more easily, and coming out from under the stone on the other side . . . they rise on their knees and strike their backs three times against the stone, remove beads, repeat aves, etc. They then proceed on bare knees, over a number of little rocks to the

place where they enter again under the stone, and thus proceed three times, which done, they wash their knees, dress and proceed to the well.

The legend of the stone was that it floated to this spot carrying St Declan's bell. After a visit to Wales, the saint was returning to Ireland with his disciple Runan, who left the bell behind on the Welsh shore. The bell was particularly precious, having been sent to Declan direct from heaven, and Runan was distressed, but Declan reassured him, 'Lay aside your sorrow for it is possible with God who sent that bell in the beginning to send it now again by some marvellous ship.' Sure enough, the stone on which the bell lay launched itself over the waves and overtook Declan's boat. The saint followed the stone, and where it came to rest he founded his community.

Some other rocks in the sea around Ardmore also have a legendary origin. It is said that hostile ships once approached the sacred town, but were petrified by a miracle when Declan's colleague Ultan raised his hand against them. At the time the alarm was raised, Ultan was occupied in feeding some hungry children, and only had his left hand free. Had he been able to use his stronger right hand, he said, no foreigner would ever again have invaded Ireland.

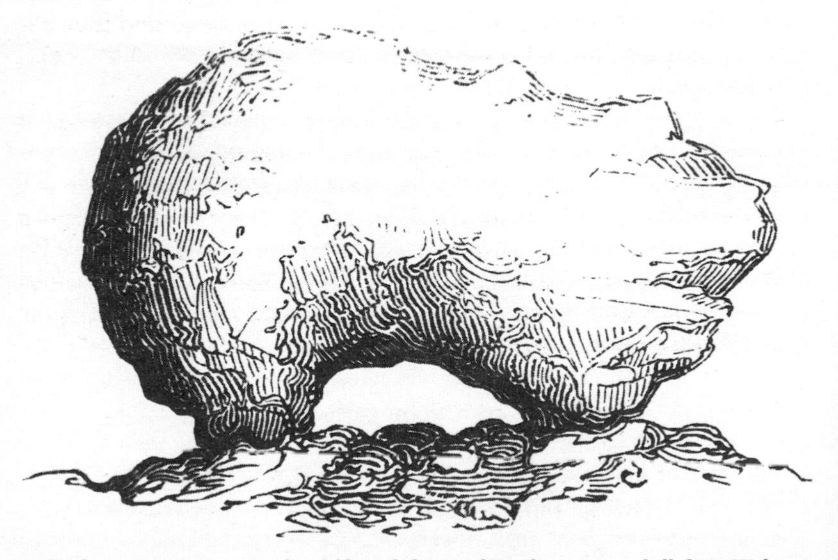

St Declan's Stone was considered blessed, having brought a saint's bell from Wales to Ireland. A holy ritual, followed with considerable difficulty, was for pilgrims to crawl under the rock.

BALLINSKELLIGS BAY, COUNTY KERRY

The piper of Ballinskelligs

Maurice Connor was the best piper in Munster. He knew every jig there was, and one of his tunes in particular had the power to set anybody or anything dancing, alive or dead though they might be. One day a great crowd of people gathered on the beach at Ballinskelligs Bay, and after Maurice had drunk a bottle of whiskey he played them his special tune. The firm, smooth sand was irresistible to dance on, and not only the men and women leapt and twirled, but the fish too jumped out of the water, and finally from the waves came a lovely young woman. She was dressed in an elegant gown white as foam, and her long green hair, just the colour of the sea, streamed down her back from under the cocked hat she wore on her head.

The sea-nymph pirouetted up to Maurice and invited him to come and be married to her. At first he refused, but after a while he followed her down to the water's edge, though his mother protested. 'Oh then, as if I was not widow enough before, there he is going away from me to be married to that scaly woman,' she cried, and lamented that Maurice's children might be hake or cod, 'and may be 'tis boiling and eating my own grand-child I'll be.' Maurice swore that every year, as a sign that he was alive and well, he would send a piece of burned wood to TRAFRASK, and then the mermaid covered him and herself up in a cloak with a big hood. The wave broke above their heads, and they disappeared.

Exactly a year later, a piece of burned wood came ashore in Trafrask, as Maurice had promised, and every year for a hundred years and more the token came from Maurice to his mother – but alas, the poor lady had died within three weeks of her son's disappearance. The wood has stopped coming now, perhaps because the anniversary of the dance has been forgotten, but Maurice still seems to be thriving under the water. Sailors off the coast of Kerry have often heard the sound of his pipes drifting up from the water, and his voice singing these words:

> Beautiful shore, with thy spreading strand,
> Thy crystal water, and diamond sand;
> Never would I have parted from thee
> But for the sake of my fair ladie.

This celebrated tale appears in Thomas Crofton Croker's *Fairy Legends and Traditions of the South of Ireland* (1825–8).

One interesting point is that although the mermaid wraps herself and

Maurice in 'a cloak with a big hood' before they enter the waves, there is an earlier reference to her 'cocked hat', also mentioned in the legend of GALLARUS, and explained there by Croker as 'the *cohuleen driuth*, or little enchanted cap, which the sea people use for diving down into the ocean'. The phrase *cohuleen driuth* may be Croker's rendering of *cochaillin draiochta*. This is a 'magic cape' or 'magic hood' that appears in several Irish legends and fairy tales, sometimes but not exclusively in connection with mermaids, conferring special gifts such as invisibility or the power to live in the sea. The Ballinskellig mermaid's hooded cloak is therefore her *cochaillin draiochta*, and the cocked hat is simply a duplication. Croker seems, however, to have heard from more than one informant that a hat rather than a cape was the magical garment, since he mentions it in several tales. There might be a link between this idea and superstitions associated with CAULS (p. 402), pieces of afterbirth on some newborn babies' heads that were widely thought to protect their owners from drowning.

BALLYHEIGE BAY, COUNTY KERRY

Cantillon family burials

The view from Ballyheige is a grand and lovely one, across the sea to the mountains of Corcaguiny. In the bay lie dangerous reefs, submerged at high tide and just visible at low water, said to be the remains of an island, which according to a story current in the twentieth century, and dating back to at least the eighteenth, was the ancestral burial ground of the Cantillon family, the ancient proprietors of Ballyheige. One of the early Cantillons, it was said, married a mermaid princess, and after her death she was interred on the island, which was then, by the power of her father, sunk beneath the waves. Her descendants too were buried there, but by no human hand. It was the custom of the Cantillons to leave their coffins on the shore, from where they were carried by the sea-people to the island.

This tradition came to an end, wrote Thomas Crofton Croker in 1828, when Connor Crowe, a Clare man related to the Cantillons by marriage, was curious to find out the truth of the old story. He took his opportunity on the death of Florence Cantillon (Florence, incidentally, was once a name used for either sex, and this Florence was a man). Connor travelled to Ardfert for the wake. 'Flory was laid out in high style, and a beautiful corpse he made.'

The coffin was carried to Ballyheigh, as the place is spelt in this version.

Prayers were said, the family departed, and Connor was left alone on the strand, where he sat drinking whiskey in the moonlight until well after twelve o'clock. Then the sound of voices rose from the sea, sweetly keening for the dead. Several strange figures emerged from the waves to stand around the coffin, saying to each other that their task of carrying the Cantillons to the island would be over when a mortal man looked on them and heard their singing.

At this moment they spied Connor and knew that the fated day had arrived. They rejoiced that their duty was done, and a great wave came from the sea to carry out the coffin, followed by the mysterious mourners.

> The sound of the lamentation died away, and at length nothing was heard but the rush of waters. The coffin and the train of sea people sank over the old church-yard, and never, since the funeral of old Flory Cantillon, have any of the family been carried to the strand of Ballyheigh, for conveyance to their rightful burial place, beneath the waves of the Atlantic.

BANNOW BAY, COUNTY WEXFORD

Barnacle geese

Now a bird sanctuary, Bannow Bay has always been full of sea fowl. Visiting in the mid nineteenth century, Mr and Mrs S. C. Hall saw it teeming with wild birds, including 'one which has been the occasion of very extraordinary opinions' – the barnacle goose. Its flesh had, they said, a delicate flavour from the sweet sea grass it fed upon.

> But the circumstance which long made it an object of the highest curiosity, was an idea that it was not produced in the usual way, from the egg of a similar parent, but that it was the preturnatural production of a shell-fish, called a barnacle. This singular absurdity is not to be charged to the Irish; it was first published to the world by Giraldus Cambrensis, who accompanied the early invaders, and saw the bird in this place.

Giraldus, or Gerald of Wales, did report the existence of the miraculous bird in the twelfth century, but the notion that certain wildfowl began life as something like barnacles attached to rotten wood is more ancient. The earliest literary reference comes from the eighth century CE, in an Anglo-Saxon riddle:

> I was locked in a narrow nest,
> My beak bound below the water
> In a dark dive; the sea surged
> Where my wings woke – my body quickened
> From the clutch of wave and wandering wood.

Belief in 'bird-fish' was widespread in medieval Europe, and in the late sixteenth century the physician John Gerard vouched personally for the truth of the wonder. Near Romney in Kent, he declared, he himself had seen a rotting tree-trunk sprouting limpet-like shells containing the birds in different stages of development, and he gave a precise description of the budding geese, starting with a sort of froth that in time turned into pale shells, which contained a silky substance fastened into the shell at one end:

> the other ende is made fast unto the belly of a rude masse or lumpe, which in time commeth to the shape and forme of a Bird: when it is perfectly formed, the shel gapeth open, & the first thing that appeereth is the foresaid lace or string; next come the legs of the Birde hanging out; and as it groweth greater, it openeth the shell by degrees, till at length it is all come foorth, and hangeth onely by the bill; in short space after it commeth to full maturitie, and falleth into the sea, where it gathereth feathers . . .

Barnacles are often seen clustered on timber, which suggested to early observers that they had grown spontaneously out of the wood. Then, it seems, the 'foot' of the barnacle was mistaken for a bird's neck and its shell for a head. Its tuft of tentacles can truly look very feathery.

The idea was persistent. In 1661, Sir Robert Moray, president of the Royal Society, wrote that he had examined a dead fir tree thrown up by the sea, on which he had found 'multitudes of little Shells; having within them little Birds, perfectly shaped'. As late as 1807 an exhibition was advertised of the 'Barnacle Tree, or tree bearing geese', found at sea by a suspiciously named 'Captain Bytheway', and until very recently, sailors maintained (perhaps for the sake of a good yarn) that geese would emerge from barnacles on a ship's bottom, or even, as reported by Calum of CROIG (Scottish Lowlands), on the soles of a man's feet. The basic idea survives to this day in the common name 'barnacle goose' and the scientific names for a shellfish – *Lepas anatifera*, 'goose-bearing barnacle' – and goose – *Branta bernicla* (although, confusingly, this is the brent goose rather than the barnacle goose, whose full name is *Branta leucopsis*).

People continued to believe in the fishy origins of the bird partly because they wanted to. If geese hatched from barnacles, they were not 'flesh' and were therefore suitable fare for church fast days, and so many otherwise devout Catholics ate the birds during Lent that the practice was forbidden by the Pope in 1215. It continued regardless, however. In the early twentieth century, it was reported that the goose was considered legitimate Lenten food in both Kerry and Ulster, under the belief that it was 'more fish than fowl'.

BANTRY BAY, COUNTY CORK

A wraith on the sea

In 1796, the French attempted to invade Ireland, landing at Bantry Bay. The venture was foiled, but it made the Irish suspicious of any strange craft seen in those waters. A year or so later, a Cork magistrate was riding home one day from Bantry, and saw a ship at anchor there. He recognised its build as foreign, and as it was nearer the shore than he liked, he stopped to inspect it carefully through his telescope. It was a still summer evening, 'so *deadly* calm, that the rays of the sun rested almost without sparkling upon the huge mirror of the bay'.

Through his glass, the magistrate was surprised to see a tall thin figure rise out of the water, under the ship's stern. His first thought was that someone was bathing, but then he saw that the figure was standing *on* the sea. Although not a superstitious man, the magistrate was terrified. He saw the thing clasp its hands, then stretch them out towards the ship, and after three or four minutes it sank slowly back into the waves.

The magistrate slept not a wink that night, and next morning before dawn he rode to Bantry and raised a force of men to board the ship. The captain, a bluff character, presented his papers, which were quite in order. The magistrate then mentioned what he had seen the day before, and at once the captain's confidence vanished. He began to tremble and had to sit down, exclaiming: 'All is in vain; the vengeance of God is everywhere. Sir, *that* has followed me from sea to sea, from harbour to harbour, in storm and calm, everywhere.' Sweating, he then poured out his confession. He had, he believed, been doomed to destruction from his birth. Under this conviction he had led a mutiny and killed his captain, and it was this man whose wraith the magistrate had seen.

This story was told by the magistrate's niece, who had been a little girl at the time of the incident. She added that some of the ship's crew had borne witness to the mutiny, and the murderer and one or two of his confederates had, she believed, been executed in Cork.

See also SEA-SONGS (p. 440).

Sea-Songs

Music was an important part of life under sail. At leisure, sailors sang ballads by landsmen – often celebrations of waves and wind, naval battles and other romantic marine images – and also songs they composed themselves, known as 'forebitters' or 'fo'c'sle songs', most of which focus on experiences ashore. Many describe a sailor's encounter with a prostitute, with some using nautical jargon for their bawdy content, as in these lines from 'Ratcliffe Highway' (a street near the London docks famous for its brothels):

> I tipped her me flipper, me towrope an' all,
> She then let her hand on me reef-tackle fall;
> She then took me up to her lily-white room,
> An' in her main riggin' I fouled me jibboom.

There are different versions, all concluding with the mariner getting robbed or catching the pox.

Jobs aboard a sailing ship often involved coordinated effort. Rhythmic shouts, helping the men synchronise their hauling on the ropes, developed into work songs called chanties (from French *chantez*, 'sing'), often largely improvised by the chantyman, whose solo lines or verses alternated with a chorus in which all hands joined. In 'Blow the Man Down', for example, the chantyman might sing, 'Oh it's where are you going to, my pretty maid?' followed by a united shout, '*Way*! Hey! *Blow* the man down!' with the stress (and the pull) on 'Way' and 'Blow'. How it continued depended on the chantyman's extempore skill, and the pretty maid's further adventures often became unprintable.

Certain chanties were exclusive to particular occasions. 'O Hurrah, My Hearties O!' was sung while pulling out a whale's teeth, while 'Shenandoah' was for weighing anchor. 'The Dead Horse Song' accompanied a ceremony after one month at sea, marking the fact that from now on the sailors would get their wages again (having received the first month's pay in advance, money invariably spent before they boarded). At the end of their time 'working off the dead horse', as they called the period when they were essentially paying a debt, a straw-stuffed canvas horse was dragged about the deck, hoisted aloft and dropped into the sea. The song that went with the ritual had the refrain 'They say my horse is dead and gone, Oh, poor old man!' and a verse that appears in different forms. One rendering begins:

> Old horse, old horse, what brought you here
> After carrying sand for many a year
> From Bantry Bay to Ballywhack
> Where you fell down and broke your back?

The mention of Bantry Bay and Ballywhack indicates that this was sung by Irish sailors. Another version, however, gives 'Botany Bay and Hammertack', and like many chanties the 'Dead Horse Song' must have been known to sailors of many nationalities.

As steam replaced sail, so chanty-singing declined, but recordings and performances by folk singers preserve this vigorous, ribald musical tradition.

An image from a nineteenth-century magazine shows the 'Dead Horse' ceremony in progress, at the moment when the horse-shaped dummy was set on fire before being thrown into the sea

BRANDON CREEK, COUNTY KERRY

St Brendan's voyage

One day St Brendan took two young brothers sailing to a small island. Brendan left the younger with the boat while he and the elder walked and discussed religion, but as the wind and the tide rose, the lad started to worry that his brother could not manage the boat alone. Brendan dismissed his concerns, but the boy became more and more uneasy. At last Brendan lost his temper and swore at his pupil. 'Begone – and be drowned to you!' he exclaimed. Taking him at his word, the boy ran to the beach, rushed into the water to help his little brother with the boat, and was swept away by the waves. Full of remorse, St Brendan asked his nurse St Itha what he should do. She told him that he should take himself out of the way of the boy's relatives, who would otherwise make a blood feud of the matter. 'Besides,' she added, 'you deserve punishment for your inconsiderate and passionate conduct. Go to sea.'

This nineteenth-century tale of how Brendan set out on his voyage is obviously legendary, but the journey itself is another matter. St Brendan was a historical character, a missionary monk of the sixth century who is known to have travelled by sea between Ireland and Wales, the Hebrides and England, and there is a strong tradition that he accomplished the

In the sixth century, St Brendan set out from Ireland on a voyage that took him and his brethren to distant lands, perhaps as far as America.

crossing of the Atlantic. The *Navigatio*, a Latin text surviving in manuscript from the tenth century but composed earlier, tells the story of how Brendan and his thirteen companions went to seek the Promised Land of the Saints, a Christian version of the Otherworld island that appears in much Celtic myth and was said to be visible from INISHMORE (Northern Eire & Northern Ireland) and other places on the western coast of Ireland. The travellers had hair-raising adventures, including landing by mistake on the back of a whale (*see* ELACHNAVE, Scottish Lowlands), and they visited many islands. On one they met a sinful but penitent monk who had been fed on fish for seven years by an otter, on another they saw Judas, tormented by demons, and on a third they were greeted by crowds of singing birds who said that they had once been angels, demoted for having considered joining Satan's mutiny, but preserved from hell because they had not finally rebelled. After seven years, having been shown the secrets of the great ocean, Brendan reached the Promised Land, and finally he returned to his monastery at Clonfert.

The religious elements are central to the *Navigatio*, understood as a parable of man's uncertain and risky voyage through life, but in another sense Brendan's journey can be, and has been, interpreted as a narrative of fact. In the 1970s the sailor Tim Severin built a leather coracle to the model specified in the *Navigatio*, and launched it from St Brendan's Creek, the very spot mentioned in the original narrative. He and his crew sailed it via the Hebrides, the Faroes, Iceland, and Greenland, eventually reaching Newfoundland more or less unscathed, as narrated in Severin's book *The Brendan Voyage* (1978).

Severin was able to link many incidents of their voyage to the tale of St Brendan. The magnificent variety of bird life they saw in the Faroes suggested a location for the Paradise of Birds, and Iceland's volcanoes brought to mind the fiery mountain seen by Brendan's crew, from which burning lumps of slag were hurled at them that made the sea boil and smoke. When Severin was weaving his way through pack ice off Labrador, he remembered a description in the *Navigatio* of Brendan's crew negotiating a silver net, harder than marble, that impeded their way. His journey, he believed, showed how Brendan's landfalls could be made in a logical progression around the North Atlantic, using the summer wind patterns.

That the voyage *could* have taken place does not, of course, prove that it actually *did*, but Severin noted that 'in every case local folklore, as well as current archaeological research, is firmly based on the tradition of the Irish visits'. On Streymoy in the Faroes, for example, it is said that Irish priests were the island's first settlers, and a creek there is called Brandarsvik,

'Brendan's Creek'. It is at least possible that the Irish indeed set foot in the New World some thousand years before Columbus and around four centuries before the Vikings.

See also RHOS-ON-SEA, Wales.

THE BULL, COUNTY KERRY

The house of the dead

Off the coast of Kerry is an islet traditionally regarded as the westernmost point of Ireland. It is sometimes known as the Bull, and one tale of its origin, told at DURSEY ISLAND, holds that it was one of three supernatural cattle transformed to stone. It has, however, another name, and a different and more uncanny legend has been attached to it from pre-Christian times.

The rock is in the form of a massive arch with a narrow aperture, resembling an ancient tomb, traditionally constructed from a large flat stone laid on two uprights, and at certain times of year the setting sun shines directly through the hole in the rock towards Derrynane beach on the mainland. In Irish mythology the land of the dead was supposed to lie in the west, towards the sunset – as were the enchanted islands reportedly seen from INISHMORE (Northern Eire & Northern Ireland) and other places – and texts from the eighth century onwards refer to the rock as the place to which dead souls travel. It was known as *Tech Duinn*, 'the house of Donn'.

Donn was the lord of the dead, his name meaning 'the dark one'. He is equivalent to Pluto, god of the underworld, in the classical pantheon, and in Norse mythology his match is Odin, who was said to lead the 'Wild Hunt' of disembodied spirits through stormy nights, a legend echoed in tales such as that told at DINGLE. Medieval historians wrote of Donn as among the early colonists of Ireland, a leader who never reached land himself, but was drowned near the rock that afterwards bore his name. A ninth-century poem claims that the rock was a memorial raised by his followers, and in it the dying Donn declares that 'to me, to my house, you shall all come after your deaths!'

CAHERDANIEL, COUNTY KERRY

Three laughs

The tale of 'A Man of the O'Sheas and the Mermaid' was collected in 1936 from a Kerry man in his seventies. One of the O'Sheas of Caherdaniel once caught a mermaid and hid her magic cloak, the traditional way for a man to persuade a sea-woman to live with him. 'That was that, and they lived comfortably and well there and got on well enough together, well enough to have children.' One Monday morning a stranger visited the couple, and O'Shea offered him a bite to eat. When he refused, the mermaid laughed, explaining afterwards that the man was a fool for rejecting food on a Monday morning (when he might not get another chance all week). She laughed again when O'Shea fell over a stone, and once more when he slaughtered a ewe, because he was killing an animal that, alive, would give both milk and clothing. The last time she laughed was when she cleared out the loft and found her cloak. Having put it on, she said she was leaving, and told O'Shea that she had laughed when he kicked the stone because under it there was a pot of gold. O'Shea dug up the pot and found the gold, and sometime later a poor scholar who saw the pot said that there was writing on the side of it to say there was another pot hidden in the same place. O'Shea and the scholar shared the second pot of gold between them and were both well off.

The narrator said that he had heard this told as a true story. The O'Sheas in that area claimed to be descended from mermaid stock, 'and they say that when they go near the sea the waves of the sea are very much given to hitting them', a report mentioned in more detail at ROSSBEIGH STRAND.

Tales of 'three laughs' go back to the Middle Ages and have remained in oral tradition, the central theme of the story being that an Otherworld being laughs three times (the Caherdaniel mermaid fitted in an extra one) at jokes only she or he can understand. In *The People of the Sea* (1954), David Thomson gives a very similar story he heard about a seal-woman who married a human husband:

> And if she stayed with him long or short all the laughs she gave were three. The first was when a stranger came and she set the potatoes before him and he would not eat. And the second was when she saw a little girleen slip on the flagstone of the door. And the third was when she found her cloak that he had hidden. For the man of the house was one day above on the roof, putting a new thatch to it, and while he was working there above, the cloak fell down at her foot. She snatched it up and out with her to the sea.

Before she swam away her husband asked why she had laughed those three times. The first was because a man had lost his chance of food for the day. The second was because under the flagstone gold was buried, but the girl who slipped did not know it. 'And the third laugh she gave was with the cloak, for by that she knew she could go back to the sea.'

CLEAR ISLAND, COUNTY CORK

A forgone rescue

A Clear Island man was out one day in his boat, scavenging for any good pickings cast overboard from the transatlantic liners that sailed around Cape Clear. He found a bone that looked promising, but it had no meat on it and so he threw it back in the sea, saying, 'To heck with you.' No sooner had he done this than a horseman came out of the waves, and asked the islander to come down into the water with him. The bone had speared the old man of the sea, and had to be removed by the one who had thrown it. The rider advised him not to shake hands, and so when the islander had pulled the bone out and the old sea-king put out his hand, the islander offered him the bone instead of his own hand. Sure enough, so strong was the old man's grip that the bone was crushed to powder.

The old man of the sea, too short-sighted to notice that he had clutched a bone instead of a living hand, was impressed when the islander did not scream. The Clear islanders were as tough as ever, he said, and since the man had been kind enough to come and relieve him, he could have anything in the house that he liked.

The man looked around and saw a young girl there, but paid her no attention. His eye fell on a length of rope, and he said that would be useful for mooring his boat. The girl was most upset. 'Bad cess to you,' she said, 'that you did not ask for me and a poor day would never come near you, because the last chance has been given me and I will be here for ever.'

This story was collected from Clear Island in 1941. Several Irish and Scottish tales feature a man who loses the chance to redeem somebody from a magic island, and such narratives of 'forgone rescues' usually have a comic or cynical flavour, since the man is apt to choose some trivial object to take away instead of the woman he could save. Sometimes it is the entire island whose enchantment he could break. In a version recorded in 1938 in west Cork, a fisherman lost in the fog found himself on an unfamiliar shore, where he met a woman combing her red hair. He asked her for a light for his pipe, and she asked whether he would not prefer a woman. No, he said, a light would do nicely. 'Aw, shame on you! Why did you not say a woman?' she exclaimed. If he had done so,

A sailor from Clear Island was taken underwater to meet the old man of the sea, a supernatural being of surprising strength.

said the storyteller, the spell on the island would have been broken - a variation on tales where a burning spark could disenchant an island, as for instancc at RATHLIN O'BIRNE ISLAND (Northern Eire & Northern Ireland).

CORK, COUNTY CORK

Flowers from the waves

Barra or Finbar, patron saint of Cork, was once sailing across the Irish Channel when he met St Scothine walking on the water. 'What is the cause of thy walking on the sea?' asked Barra. 'It is not sea at all, but a plain full of clover-blossom,' said Scothine, and to prove his point he picked a flower and threw it to Barra in his boat. 'What is the cause of a vessel swimming on the plain?' he asked in his turn. For answer, Barra put his hand into the sea and caught a salmon, which he flung to Scothine.

This fifteenth-century legend is reminiscent of an episode in the eighth-century Irish poem *Voyage of Bran* (see WONDER VOYAGES AND LOST LANDS, p. 458), when the sea-god Manannán meets the sailor Bran on the ocean and says that what Bran perceives as waves full of fish, he himself sees as flowery pastures. Like Manannán, St Barra was in other accounts said to ride on horseback across the sea between Wales and Ireland, while Scothine, or Scoithín, is sometimes linked with another sea legend, the sunken city off LAHINCH.

DINGLE, COUNTY KERRY

The Wild Hunt at sea

Ghost stories of the 'Wild Hunt' were once told throughout northern Europe. The basic tradition, which goes back at least to the twelfth century, is of a group of spectral horsemen pursuing the souls of the damned along deserted roads or through the night sky, led sometimes by the Devil, sometimes by a pagan deity such as Artemis, Greek goddess of the hunt, or the Norse god Odin. Some Irish tales describe the hunt as taking place over or through the sea, in line with the idea that the land of the dead was an island (often identified as THE BULL).

One such folk tale, printed in the Irish folklore journal *Béaloideas* in 1930, told of a sea captain who arrived in Dingle with a cargo of wood from Canada. The strangest thing he had seen on his travels, he said, was a hare that had run across the sea one fine night and jumped aboard his ship. Soon a pack of hounds and huntsmen followed, and one of the riders asked the captain if he had seen a hare. The captain said he had seen nothing, and the hunt rode away over the waves, but when the hounds could not find the scent they came back to the ship, and the rider said that the hare must be aboard. He demanded that it should be sent out immediately. The crew, frightened of the strange rider, forced the hare out from the cabin where it had been hiding. 'When it came up on deck it jumped into the sea and off it went followed by the hounds and horses and nobody saw them ever after.'

It was well known that witches often appeared in the form of hares (*see* PORTREE, Scottish Highlands & Islands), and although the narrator does not say so, this animal was clearly a witch, and therefore a damned soul.

DOONBEG BAY, COUNTY CLARE

Soul cages

Thomas Crofton Croker's *Fairy Legends and Traditions of the South of Ireland* (1825–8) includes a tale of drowned souls kept in cages by a sea-spirit. This uncanny practice was discovered by Jack Dogherty, a fisherman, who one day spied a Merrow, or sea-fairy, in Doonbeg Bay. It was a strange-looking creature with a fish's tail, scaly legs, short arms like fins, green hair, long green teeth, a red nose and little piggy eyes. Although it wore no clothes, it held a cocked hat under its arm; this was one of the enchanted caps that allowed Merrows to live underwater (*see* BALLINSKELLIGS BAY).

The Merrow was called Coomara, which Croker tells us is *cú-mara*, 'the sea-hound', and the origin of the name MacNamara, 'the son of the sea-hound' (*see* ERRIS HEAD, Northern Eire & Northern Ireland). Coomara said that he and Jack's grandfather had been good friends, and had often had a glass of brandy together. He invited Jack to visit him, and on the appointed day he gave Jack a spare cap and took him down to the seabed. In the Merrow's cellar the two of them drank brandy out of seashells, and ate fish for supper.

Things that looked like lobster pots stood around the walls, and Coomara told Jack that these were soul cages, in which he kept the souls of drowned sailors. Jack was horrified, and after he got home he could not get the soul cages out of his head. He invited Coomara to visit the land, and tried to get him drunk. The first evening they drank brandy, but that had little effect on the Merrow. The next time, however, Jack tried him on poteen, and that did the trick. While Coomara slept it off, Jack took his magic cap and went quickly down to the undersea cellar. There he inspected the cages, and although he could see nothing (souls being invisible), he heard a little chirp or whistle coming from each. He opened them all and then returned to the surface.

Coomara never noticed his souls were gone. He and Jack stayed friends, and every so often Jack visited his cellar in order to release any new souls. One day, however, he threw a stone into the sea and nobody came, so he thought that Coomara must have died or gone away.

Croker notes that 'Merrows are said to be as fond of wine as snakes are of milk,' and the idea that fairies and sea-folk liked a drink features in several tales, including that told at HORN HEAD (Northern Eire & Northern Ireland). The folklore collector Patrick Kennedy, writing in 1866, was probably thinking of Croker's story when he wrote that 'mermen do not seem on the whole to be an attractive or interesting class':

> Their hair and teeth are green, their noses invariably red, and their eyes resemble those of a pig. Moreover, they have a penchant for brandy, and keep a look-out for cases of that article that go astray in shipwrecks. Some naturalists attribute the hue of their noses to extra indulgence in that liquor.

Referring to this sort of opinion, P. W. Joyce commented in *Old Celtic Romances* (1894) that many of the ancient Gaelic tales were presented in an unfavourable and vulgar manner by more modern writers, 'distorted to make them look *funny*, and their characters debased to the mere modern conventional stage Irishman'. Part of the Irish stereotype, as seen by the English, was that they were a nation of drunkards.

DUNKERRON CASTLE, COUNTY KERRY

Blood in the water

About two miles west of Kenmare, off the Sneem Road, is Dunkerron Castle. Now a ruin, perched on a rock in thick woodland, it is not easy to find, but the sight of the tower, wreathed in green branches, is worth the effort. Good walking boots and local guidance are recommended.

The castle was once the seat of the O'Sullivans, and legend records that long ago the chief of the family fell in love with a mermaid. Many times he begged her to marry him, and at last, overcome by passion, he dived into the bay and clasped her in his arms. She admitted that she returned his love, but said she had to get permission from the undersea king:

> Once more must I visit the chief of my race,
> His sanction to gain ere I meet thy embrace.
> In a moment I dive to the chambers beneath:
> One cause can detain me – one only – 'tis death!

Waiting for his sweetheart, to his horror O'Sullivan saw a gush of blood stain the water. The waves boiled as if in fury, and carried him to shore.

In *Fairy Legends and Traditions of the South of Ireland* (1825–8), Thomas Crofton Croker gives this story (partly as a ballad), which he heard from an old man one night when they were out in a boat on Kenmare Bay. A very similar legend was told by the brothers Grimm, of a Magdeburg butcher who fell in love with a mermaid and waited for her beside the Elbe, only to see blood spots in the water that showed she had been put to death by her jealous family.

DURSEY ISLAND, COUNTY CORK

Sea-cattle

Many centuries ago, a family who lived on Dursey Island found near the beach a beautiful coal-black bull and cow, and kept them for their own. They were well satisfied with the animals, which produced a calf the next year. The cow gave plenty of good milk, but one day a servant girl cursed her and struck her with a rope. Mooing in protest, the cow walked towards the sea, followed by her mate and child. All three plunged into the waves, and became three rocks, afterwards known as the Bull, the Cow, and the Calf.

This folk tale, from Patrick Kennedy's *Legendary Fictions of the Irish Celts*

A special breed of cattle was said to live in the sea, and a marine bull, cow, and calf lived for a while on Dursey Island, before they were driven away by a cruel maidservant.

(1866), and quite different from the legend that identifies THE BULL as an Isle of the Dead, is one of many reports of sea-cattle that came ashore and could be domesticated. Such stories are widespread in the lore of Scotland, Wales, and the Isle of Man, as well as Ireland. Sometimes the animals were said to be seals that transformed themselves into the shape of cows, and other tales attribute a sacred symbolism to the beasts. Included in Lady Wilde's *Ancient Legends* (1887) is the tale of a mermaid who appeared on the western coast and said that she had come to announce the arrival of *Bo-Finn*, *Bo-Ruadh*, and *Bo-Dhu* – the White, Red, and Black Cows. She returned to the sea, and exactly a year later three cows emerged from the waves. The Black Cow went south, the Red Cow went north and the White Cow went to the centre of Ireland, where she bore twins, male and female, from which descended a great race of cattle. The White Cow then went into a cave by the sea, 'the entrance to which no man knows', to sleep until the arrival of Eire's true king – or, according to an alternative version, she was struck dead by a red-haired woman, an allegory for the subjugation of Ireland by Elizabeth I of England.

FASTNET ROCK, COUNTY CORK

A haunted U-boat

A submarine is one of the most claustrophobic places conceivable. Visiting one, even on the surface, you can all too easily imagine life underwater for the men crowded into rooms no wider than passages, the miles of sea pressing down above their heads, unseen but inescapable. The idea of a haunted or jinxed submarine is horribly obvious, the oppressive atmosphere tightened that extra turn by fear of a supernatural curse.

German U-boat 65 was unlucky from the beginning. During her building, a steel girder fell on two labourers, killing both, and then when the submarine was nearly finished, the engine-room door jammed, trapping three more workmen in a narrow chamber filling with poisonous fumes. When entry was forced, all three were dead. On UB-65's maiden voyage in 1916, a storm swept one of the crew overboard, and when the ship's torpedoes were being installed, one of them exploded, killing five men, badly injuring several others and causing extensive damage. The submarine was repaired and ready to set out when reports began to circulate that the second officer, one of those who had died in the explosion, had been seen as a ghost. One of the men who saw him deserted before the submarine could sail again.

Rumours spread of the phantom's appearances, and one night UB-65 was cruising in the Channel when a lookout and the captain both saw the dead officer on deck. The captain died a few weeks later in a British air raid, and the crew became increasingly demoralised, unreassured even when a pastor visited the submarine to hold a service of exorcism. The new captain reportedly declared his ship to be 'haunted by devils', and a petty officer wrote later of his own conviction that the submarine was haunted, his account quoted in Hector Bywater's *Their Secret Purposes* (1932):

> One night at sea I saw an officer standing on deck. He was not one of us. I caught only a glimpse of him, but a shipmate who was nearer swore that he recognized our former second officer, who had been killed long before by a torpedo explosion. On other nights, while lying in my bunk, I saw a strange officer walk through the ship. He always went into the forward torpedo-room, but never came out again.

A gunner went mad, jumped into the sea and drowned. Soon afterwards the chief engineer fell and broke a leg, one of the gun crew was washed overboard, and the coxswain died when UB-65 and some other submarines were attacked. 'The men were so depressed that they went about like sleep-walkers, performing their duties automatically and starting at every unusual sound,' reported the petty officer, who survived UB-65's final voyage because he was in hospital at the time.

The end came on 10 July 1918, when the submarine exploded for reasons unknown, perhaps because one of her own warheads malfunctioned. There were no survivors. According to both US and German naval records, this happened near the Fastnet Rock off the southern tip of Clear Island, but recent investigations by marine archaeologists suggest that in fact the remains of UB-65 lie on the seabed much further to the south-west, off the Cornish coast. Her fate, therefore, poses one final mystery.

GALLARUS, COUNTY KERRY

The Lady of Gollerus

On the shore of Smerwick harbour one summer morning just at daybreak, Dick Fitzgerald saw at the foot of a rock a beautiful young creature combing her long green hair. He guessed at once that she was a Merrow or sea-fairy, for he spied near her on the strand the enchanted cap that enabled her to live and breathe under the waves (*see* BALLINSKELLIGS BAY). Knowing that she would not be able to leave the land without her cap, he crept closer and seized it. When she saw it had gone she began to cry, so to comfort her Dick took her hand (which had a thin white web between the fingers), and she stopped crying and asked if he meant to eat her. Dick said he would rather marry her, and the Merrow was agreeable to that, but first she bent her head and whispered some words into the water at the foot of her rock. Dick saw her words travelling out across the ripples like a breath of wind, and the Merrow explained that she was sending a message to her father, the king of the waves.

Dick and the Merrow were married by the priest, and they settled down together at Gollerus (Gallarus), a little village to the east of Smerwick harbour. Three years passed, each year bringing a child, and the Merrow made the best of wives and mothers. One day, however, Dick had to go to Tralee, leaving the Merrow at home minding the children. When she was cleaning the house, what should she find in a hole in the wall, hidden by a fishing net, but her own magic cap. Looking at it, she thought of her father and mother and brothers and sisters, and she longed to see them. She could see no harm in going home just for a visit, and so she took her cap and slipped down to the shore. She thought she heard a faint sweet singing inviting her to come down under the sea, and at once she forgot Dick and the children and dived into the water.

When Dick came back and found the cap gone, he knew at once what had happened. Year after year he waited for her in vain, and would never believe that she stayed away of her own free will, but maintained that her father the king of the waves had kept her by force. In the neighbourhood, concludes Thomas Crofton Croker in *Fairy Legends and Traditions of the South of Ireland* (1825–8), she was always spoken of as a model wife, and referred to as the Lady of Gollerus.

GLANDORE, COUNTY CORK

Clíona's Wave

Sea and wind together can make some very odd noises. A visitor to Portnatraghan, on the Antrim coast, described his experience:

> He heard, suddenly, a heavy long-drawn sigh quite close beside him as he imagined, though no human being was in sight, and as he continued to listen intently, the sigh was repeated at regular intervals. When he regained self-possession he investigated the matter, and discovered that the sound, which had so startled him, issued from a fissure in the rocks over which he was standing. Close to this he found a second fissure from which unearthly groans proceeded, so like those of a human being, that it was distressing to listen to them.

Such moans were often interpreted as the cry of a spirit. At Glandore in Cork, a 'very peculiar, deep, hollow, and melancholy roar' was sometimes heard among the cliffs, wrote P. W. Joyce in 1870, and was associated with the banshee Clíona (pronounced Cleena). The tide in Glandore harbour was known as Clíona's Wave, one of the 'great waves' of Ireland supposed to foretell a king's death, and the plaintive sound from the sea caverns was likewise taken as a fatal omen.

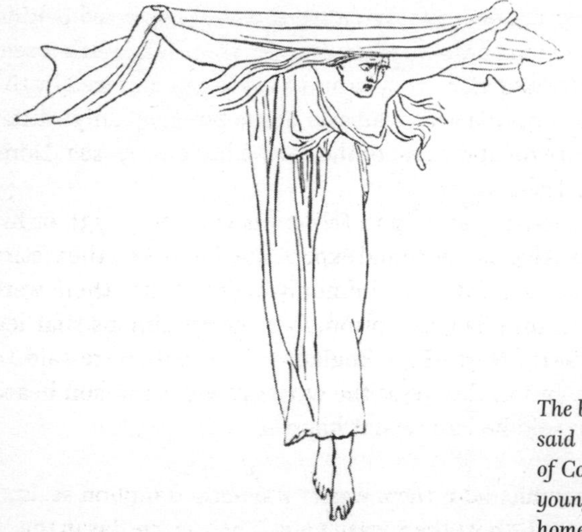

The banshee Clíona was said to haunt the coast of Cork, where she lured young men to her rocky home.

Clíona first appears in medieval legend as a maiden who went in a boat to meet her lover but was drowned at Glandore. By the nineteenth century, she featured in folk tales as a powerful hag who enticed young men to her 'house' on the seashore, a large square rock, from which they never returned, and local tales current in Cork, reported in Dáithí Ó hÓgáin's *Lore of Ireland* (2006), held that she could be seen dancing with the fairies around her rock, sometimes in the form of a big white rabbit.

See also ROSSBEIGH STRAND.

GREAT BLASKET ISLAND, COUNTY KERRY

The king of the seals

Some fishermen were out near Great Blasket Island one day, and as they had caught nothing they decided to kill one of the crowds of seals on the rocks. When the boat approached almost all the animals swam away, leaving only one enormous old bull seal, covered in barnacles and limpets. The men whacked him with an oar and killed him, but then found the carcase was too big to go into their boat, so they cut off the head and tail. Even so, when the body was across the boat, the two ends trailed in the water, and as they rowed along the seal's blood flowed into the sea, dyeing it red. The other seals swam along beside the boat like mourners following a coffin, crying and pushing against the body of the huge dead seal.

As the men drew near to Great Blasket, a sea-monster appeared behind the boat. Appalled, the men vowed that they would never kill another seal if their lives were spared, and when they came to land, still followed by the grieving seals, they did not take the dead seal home but took only a little of the fat for oil and jettisoned the rest of the corpse back in the sea. None of the men ever killed a seal again.

Either Horace Beck, writing this up in *Folklore and the Sea* (1973), or his informant, who claimed this as a personal experience, identified the victim as 'the dead king of the seals'. It was commonly believed that there were kings among both seals and fish, exceptionally large specimens that led the rest (*see* CULLERCOATS, North-East England). The seals were said to elect a chief annually, and in the 1940s the writer David Thomson heard from a Mayo man who said he had seen it happen:

> And I was one day fishing and if there wasn't a thousand million seals there in the sea around the boat there wasn't one. There is one day in the

> year, you understand, when they send the seals in thousands from along the coast to choose their king. And they disperse to their own places after. I saw them. And this king. I remember well his face, for 'twas like the face of an old man.

LAHINCH, COUNTY CLARE

'The Phantom City'

A lost land called Cill Stuifín or Cill Stuithín is said to lie in the sea near Lahinch, and to have formed the town's south-western district until it was drowned in a flood. It is supposed to surface once every seven years, and it is ill-omened: those who see it will die, it is claimed, within a year, a tradition referred to in a nineteenth-century poem by Gerald Griffin, 'The Phantom City':

> A story I heard on the cliffs of the west,
> That oft, through the breakers dividing,
> A city is seen on the ocean's wild breast
> In turretted majesty riding.
> But brief is the glimpse of that phantom so bright,
> Soon close the white waters to screen it,
> And the bodement, they say, of the wonderful sight,
> Is death to the eyes that have seen it.

Folk tales collected in the area say that a Druid or chieftain once lived in Cill Stuifín, and before he went away to war, long ago, he cast a spell over his land to sink it in the sea until his return. He was killed in battle, or lost the key to disenchant his home, and so it stays submerged.

A related tradition, reported in *Notes and Queries* (1853), was that a great city had been swallowed up by the waves as punishment for the crimes of its inhabitants, and could be seen once every seven years from the cliffs of Moher, which overlook the sea beyond Lahinch. Far from being a signal of doom, the apparition would bring enormous wealth to one who saw it, if only he could keep his eyes fixed on it until he reached the spot, but this was almost impossible to do. A labourer once saw the city emerge in all its splendour from the deep:

> He called to his companions to look at it; but though they were close to him, he could not attract their attention: at last, he turned round to see

> why they would not come; but on looking back, when he had succeeded in attracting their attention, the city had disappeared.

This is reminiscent of the island that could be seen only from a particular spot on CEMAES HEAD (Wales), and the idea that the city was drowned as a punishment is also like the Welsh tales associated with CARDIGAN BAY (Wales). Irish legends of phantom islands more usually describe them as Otherworld realms like Hy Brazil (*see* INISHMORE, Northern Eire & Northern Ireland), which was also often said to be visible every seven years.

The name Cill Stuifín may indicate a connection with St Stephen (some tales claim that the underwater reef that causes white foaming breakers even on a calm day is in fact a church built by the saint) or with the more obscure St Scoithín or Scothine. A legend told at CORK indicates that Scoithín was seen as a patron of the sea, and the sunken island off Clare may have been identified as his *cill* (cell, or hermitage).

LAMBAY ISLAND, COUNTY DUBLIN

Manx shearwaters

In the early twentieth century, Dublin workmen repairing Lord Revelstoke's house on Lambay Island spent 'two nights of terror' there, and then insisted on sleeping on the mainland. They said that they had heard terrible noises and were sure the place was haunted.

The sounds they heard may well have been made by Manx shearwaters, which are among the many seabirds that roost on the island, and which make a horrible racket. In *The Islands of Ireland* (1936), Thomas Mason writes that he sympathises with the workmen, 'as I remember the shock I received when I first heard the noise at close quarters on a pitch dark night'. Shearwaters nest in colonies, and the mating pairs of birds take turns to sit on their eggs:

> One of the birds is fishing all day and returns late at night, heralding its return in a most riotous manner with blood-curdling cries which could be best described as a cross between a donkey's bray and a cock's crow.

In the eastern Mediterranean it is said that they are possessed by the souls of the damned.

Wonder Voyages and Lost Lands

The first Wonder Voyage was the Odyssey. Its route passed the sirens, birds with women's heads who lured sailors to their death, but the ever-resourceful Odysseus ordered his crew to plug their ears, and had himself tied to the mast so that he could safely hear their irresistible song.

In ancient times, the final frontier was the sea, where marvels and terrors were possible and nothing was known for certain. The modern role of science fiction was for centuries filled by magical seafaring epics known as Wonder Voyages, from the Odyssey in classical legend, through the Arabian adventures of Sinbad, to the *imrama* (voyages, or literally 'rowings') of medieval Celtic myth.

Among the oldest of the *imrama*, dating from around the eighth century, is the tale of Maelduin, a prince of the Aran Islands who sets off with his crew to avenge his father's murder. They wander for years among strange islands, including one where any man who lands forgets his companions – echoing Odysseus' visit to the Lotus-Eaters – and another ruled by an amorous queen who is reluctant to let the strangers go, again like Odysseus' encounters with Circe and Calypso. More characteristic of Irish legend is the detail that when Maelduin sails away, the queen throws a thread that sticks to the fingers of the man who catches it, and eventually they only escape by cutting off his hand (*see* RATHLIN O'BIRNE ISLAND, Northern Eire & Northern Ireland). Their most fateful encounter is with a hermit, so old that his white hair covers him to his feet, who tells Maelduin that he must forgive his father's assassins as God has forgiven his own sins. When this Christian command has been fulfilled, the travellers are allowed to return to Ireland.

A similar story is told in *The Voyage of Bran*, an Irish poem dating from the twelfth century in manuscript but composed considerably earlier, and another is St Brendan's journey from BRANDON CREEK (Southern Eire), widely supposed to be based on a real expedition, but closely allied to fabulous adventures like those of Bran and Maelduin. The saint eventually finds a blessed paradise, but does not stay there (a common theme among most of these tales is that the wanderers, like Dorothy in *The Wizard of Oz* – a twentieth-century take on the Wonder Voyage tradition – ultimately go home). St Brendan's Isle or Isles were marked on maps from the thirteenth century onwards, in places from the Canaries to the Azores. They were also known as the Fortunate Isles, and sometimes equated with Hy Brazil, the legendary island said to be visible from INISHMORE (Northern Eire & Northern Ireland) and other places on the west coast of Ireland. Another questionable island that was shown on sea charts until the twentieth century was Mayda, located west of Brittany. Unusually, Mayda has no legendary or literary sources, appearing only on maps, and it has been proposed that this was once a genuine reef, later submerged.

Among the analogues of these lost realms is Avalon, the island where King Arthur was taken after his final battle. Sometimes identified as Glastonbury – which was never an island, but was once surrounded by swamps – it was alternatively said to lie in the Atlantic, an offshoot of the ancient Celtic tradition of an Isle of the Dead far to the west, towards the setting sun.

See also THE BULL (Southern Eire).

MALAHIDE, COUNTY DUBLIN

Roth Ramhach *the giant ship*

An old prophecy was attributed to St Colm Cillé (St Columba) and some other ancients, dealing with a giant ship, the *Roth Ramhach* or 'Rowing Wheel', which was to contain a thousand beds, with a thousand men in each bed, and would sail with equal ease over sea and land. It would be wrecked by the pillar-stone of Cnámchoill, a wood that once grew near the present town of Tipperary, and then a fleet of one thousand hostile ships would land at Inbher Domhnann (Malahide Bay). The enemy would rampage through the land, capturing cattle and women, and would eventually be defeated in a great battle at Tara, the seat of the High King. All this may be understood as part allegory, part description of Danish invasions in the tenth and eleventh centuries, and the scholar Eugene O'Curry states positively that the 'prophecy' can have been written no earlier than the twelfth century, well after the events it purported to foretell.

The vast ship was supposed to be built with the help of Simon Magus, a first-century sorcerer mentioned in the New Testament (Acts 8:9–24). Its legend pre-dates, and may have partly inspired, tales in many countries of an impossibly enormous vessel (*see* DOVER, South-East England).

MOORESTOWN, COUNTY KERRY

The Son of the Sea

In the late nineteenth century, the Scottish folklorist John Francis Campbell heard a story about three Irish princesses who went bathing in a sea loch:

> One of them went to sleep, and when she awoke, she felt a most amazing sensation. She saw a seal making off a little way from her. She became heavy. And when the child was born, he was as hairy as a goat.

The princess called her son MacCuain, the Son of the Sea. When he grew up, he was a strange and dangerous man, who would sometimes be seized with a frenzy of anger, when his face became horrible and his eyes grew as big as those of a whale. He fought his grandfather and took over his kingdom, but in the end was vanquished in his turn by the hero Fionn Mac Cumhaill.

A local version of this tale, or its beginning, was told about a young man named Moore who lived at Moorestown on the Dingle peninsula, and suffered terribly from insomnia. One night a travelling scholar was staying at his house, and as they sat up late, Moore mentioned his affliction. The stranger remarked that he had read that the children of supernatural beings could not sleep. His suspicions aroused, Moore asked his mother how he had been conceived, and she confessed that her husband was not his real father. One day, she told her son, she had gone swimming in the sea and fainted in the water, and after that she had become pregnant. Moore realised from this that he was the son of a sea-spirit. He mounted his horse and rode into the sea, and was never seen again.

ROCKABILL, COUNTY DUBLIN

The goddess of the Boyne

In the west of Ireland there was once a secret well, forbidden for anyone to visit except the High King Nechtan. His wife Bóand declared that the well had no power over her, and she dared to walk around it three times, but was punished for her pride when three waves broke from the spring, blinding her and shattering her hand and thigh. Bóand fled, but the waters of the well pursued her, creating the river Boyne. Eventually she came to the sea, and rushed into the waves followed by her faithful lapdog Dabilla. Bóand was afterwards worshipped as a mighty river-goddess, and Dabilla was not forgotten either. Swept out to sea, the dog was divided in two pieces, and became the Rockabill islands, two small stony outcrops about two and a half miles off the coast midway between Dublin and Drogheda.

This was the ancient explanation for their origin, but a different story was given in a nineteenth-century Monaghan folk tale, which said that the islands were a cow and calf turned to stone by the poisoned gaze of Balar. He was the leader of a tribe of legendary pirates called the Fomhóire, and he had just one eye, so fearsome that it destroyed all it looked on and had to be shielded by a heavy lid. This tyrant stole a magical cow and calf and drove them to the Dublin coast, but when the cow halted and tried to turn back, Balar raised his eyelid to see what the trouble was. At once the cow and calf were changed into the Rockabill islands, echoing the story told at DURSEY ISLAND.

ROSSBEIGH STRAND, COUNTY KERRY

The avenging wave

On the north shore of the Iveragh peninsula is an unusual beach, Rossbeigh Strand, which points sharply into Dingle Bay, giving it one calm, sheltered side, while the other is exposed to the Atlantic blast. All kinds of cross-currents are active in the bay, and the tides create what is known as the Wave of Tóim, one of the four 'great waves' that in early Irish literature were said to roar at the death of a true king (the others being at Ballintoy in Antrim, Dundrum Bay in Down, and GLANDORE BAY in Cork).

A folk tale from Dáithí Ó hÓgáin's *Lore of Ireland* (2006) relates that a daring young man named O'Shea once saw Tír na nÓg (the Land of Youth) appear above the waters of Dingle Bay, raced towards it on his horse, and seized a cloak that hung drying on a wall there. At once the island began to sink, and as he galloped away with his prize he was pursued by the Wave of Tóim, so fast and destructive that it cut away the rear of his horse. A variant version, current in the nineteenth century, talked of the 'avenging wave' that rushed along 'crested with lightnings', to the terror of the local fishermen. They said it was caused by an act of cruelty long ago, when a man called Shea once killed a mermaid even though she begged for mercy. The next time he ventured out on the sea, the wave rushed towards him, and as fast as he could flee towards land it pursued him faster, overtaking him before he could reach safety, and drowning not only Shea himself but all the rest of his crew. Even long afterwards, it still chased his descendants (*see* CAHERDANIEL).

SCATTERY ISLAND, COUNTY CLARE

St Senan defeats a monster

A monastery was founded on Scattery Island in the sixth century by St Senan (which means 'old man', so it cannot have been his original name, but that is how he became known). An account of his miracles, said to have been written not long after his death, and translated from Gaelic in the late seventeenth century, tells how an angel summoned him to the south shore of the Shannon, overlooking Scattery, and told him how the island had been kept pure and sinless since the creation of the world.

> 'A Sea Monster from the beginning of all ages guarded it and suffered no pagans or infidels to enter therein. The western world contains not a more hallowed island.'

Only one thing remained to be done before Senan could build his church and settle down, and that was to expel the sea-monster. The saint sat down upon a stone, which instantly transported him to a hill in the middle of the island, and he and the angel went forth in search of the monster,

Scattery Island was inhabited by a scaly, bristled beast with great tusks and steely talons, until the monster was banished by St Senan.

which did not take them long. The dragon smelt them, and rushed out to attack:

> His fore parts with huge bristles standing on end, like those of a boar; his mouth gaping wide open, with a double row of crooked sharp tusks, and with such openings that his entrails may be seen; his back like a round island, full of scales and shells; his legs short and many, with such steely talons that the pebble-stones where he ran among them sparkled.

He gave off such heat that he scorched the ground as he passed, and boiled the sea when he dived into it.

St Senan made the sign of the cross, and commanded the beast in the name of Jesus to go and sink himself in a lake on Mount Callan, County Clare. When the island was monster-free, the saint and the angel blessed it, and the angel assured Senan that none of his monks should drown in the rough sea around the island, nor should anyone buried in its earth be damned:

> A high rough tide goes round this land,
> It hath this privilege from God's own hand;
> No hell shall after death torment
> True Christians that are buried in it.

Scattery continued to be regarded as divinely protected. At least until the mid twentieth century, it was a custom among fishermen sailing down the Shannon to take their boats sunwise (clockwise) around the island, and a pebble from it was often carried as a lucky charm.

TRAFRASK, COUNTY CORK

Cluasach and the sea-maiden

A tale current in south-west Ireland tells of a great sailor, Cluasach Ó Fáilbhe, who travelled with a sea captain known as the Fair-haired Merchant. Together they set out to find where the sea ends, and after seeing many marvels they came at last to a great wall of brass across the ocean. Here they cast anchor, but when they tried to pull it up again they found it was stuck on the seabed. Diving to free the ship, Cluasach found himself in a splendid city, where he met a beautiful sea-maiden who paid court to him. He said he must go home, but swore to return to her as soon as he had set his affairs in order.

Cluasach and the Merchant set sail, but found that a great wave was following them, and in it the sea-maiden was swimming. Cluasach threw a knife at the wave, knocking out one of the maiden's eyes, and at this a horseman came riding over the sea to them, saying that he was the maiden's brother, and warning Cluasach to keep his promise. Shortly after his return to Ireland, Cluasach went down to the sea, where a wave swept him away to rejoin the mermaid, but first he vowed to send a burned sod of earth every May morning to the strand of Trafrask, to show he was still alive, and the peat duly arrived for many years.

This is linked to the nineteenth-century story told at BALLINSKELLIGS BAY, and also holds echoes of a much earlier text, a thirteenth-century verse about an Irish pilot who sailed to Norway and found his ship stuck on an unseen obstacle. Diving in, he found some beautiful sea-nymphs were responsible, 'nine female forms, fair and firm'. They seduced him, and he spent nine happy nights underwater with them. One of them became pregnant, and at this he fled, though vowing to return. Having spent seven years in Norway, however, he went home to Ireland, and the angry mermaids pursued him to the mouth of the Delvin river, where the mother of his child threw it upon the rocks, killing it.

VICARSTOWN, COUNTY KERRY

The prince of Spain

Sailing around the west of Ireland in 1968, Horace Beck put in at Dingle and went to Vicarstown, a settlement near Dunquin. Here, behind the house of a local storyteller and poet, he was shown 'a curious ancient stone' with a crudely carved image of a galleon. Beneath the memorial, said the bard, lay the body of a Spanish prince, and he told Beck the tale of how the prince's fleet, part of the great Armada, was blockaded by the English in Smerwick harbour, and how the vessels fought their way out into Blasket Sound in a hurricane, 'And the seethe and the surge and their sails were torn to rags until they came to the Blasket Strand.'

All six ships went down, and every man on board perished except an eight-year-old boy who tied himself to a barrel and floated to land. He spoke only Spanish, and it was not until he had been taken to Cork that an interpreter could be found. Then the boy explained what had happened, and some of the drowned, including the prince who had captained the fleet, were retrieved from the Sound and buried. Later, according to the storyteller, some Spaniards came to visit the prince's grave, intending

A prince of Spain, fleeing from the English after the defeat of the Armada, was shipwrecked in a dreadful hurricane off Blasket Strand, near Vicarstown.

to take his body home, but when they saw the spot 'they said he was buried in the wildest and the most beautiful place that he could find,' and so they left him where he was.

The stone can still be seen in a field just north of Dunquin school. To interpret the linear markings on it as a picture of a ship requires a little imagination, but the old burial ground is still a wild and lovely spot.

WEXFORD, COUNTY WEXFORD

St Martin's warning

The gruesome legend of St Martin's martyrdom was that he was ground to death in a mill, and folk tradition forbade anything to be done on St Martin's Day (11 November) that involved turning or twisting. This prohibition covered not only the grinding of corn and the spinning of wool, but also fishing (because the boats would have to turn in the water), and it was said that 'the sea belongs to Martin on that night'.

According to a tale printed in 1867, all the Wexford fishermen were once warned on 11 November to get to shore by an apparition of the saint himself, pacing along the waves, and everyone who ignored the omen died that day. A ballad by the nineteenth-century poet John Boyle O'Reilly, 'The Fishermen of Wexford', tells how the tradition was strictly observed in Wexford, until one year a fabulous shoal of herring came to the bay on St Martin's Eve. Fishermen go out in the evening, and for them to do so on this occasion would mean breaking the taboo, since the boats would still be at sea after midnight. The temptation, however, was too great, and although their wives and parents begged them to stay ashore, the fishermen were adamant:

> And scoffingly they said, 'To-night our nets shall sweep the Bay,
> And take the Saint who guards it, should he come across our way!'
> The keels have touched the water, and the crews are in each boat;
> And on St Martin's Eve the Wexford fishers are afloat!

They cast their nets, but no sooner had they done so than a human shape rose from the waves. In the moonlight they saw the figure point to the land three times, and then sink again, one white hand clenched above its head. As it disappeared from sight they felt the rush of the herring underneath the boats:

Defying the dread warning, every face was sternly set,
And wildly did they ply the oar, and wildly haul the net.
But two boats' crews obeyed the sign, – God-fearing men were they, –
They cut their lines and left their nets, and homeward sped away;
But darkly rising sternwards did God's wrath in tempest sweep,
And they, of all the fishermen, that night escaped the deep.
Oh, wives and mothers, sweethearts, sires! well might ye mourn next day;
For seventy fishers' corpses strewed the shores of Wexford Bay!

YOUGHAL, COUNTY CORK
Visions and mirages

Around the turn of the eighteenth century, several sightings were reported in southern Ireland of the phenomena known as 'fata Morgana', mirages popularly attributed to the spells of the fairy (*fata* in Italian) Morgan le Fay, known from tales of King Arthur. One of the most striking, described in J. N. Brewer's *Beauties of Ireland* (1825–6), occurred at about 5 a.m. on a day in June 1801:

> All the coast opposite the river of Youghal, on the Waterford side, being covered with a dense vapour, presented on the right, next the sea, the objects of an alpine region; in the back-ground were snow-capped mountains, while woods and a cultivated country appeared in front.

As the sun rose, the snow seemed to roll down the illusory mountains, and then around the land to the left of the river appeared splendid parks surrounding mansion houses, some of which could be seen in such detail that even their door-knockers were visible, and whether their windows were open or shut. All this vanished after about half an hour.

Seemingly miraculous visions like these have been seen from many places, including RAMSGATE (South-East England). In southern Ireland, the folklorist Thomas Westropp witnessed one on a clear evening in 1872, when just as the sun was setting, a dark island suddenly appeared far out to sea. He could clearly see two hills, one covered in woods, and in between a plain from which towers rose, and curls of smoke as if fires were lit in the houses. Several other people saw the same thing, and one of them cried out that he could see New York.

What Westropp and his friends saw might have been a fog bank, reflecting the light of the setting sun, or perhaps a real island below the horizon.

Things normally out of sight can be visible at sea when the sun's rays are at a certain angle, because water bends light. This is easily demonstrated: put a coin in a bowl with a rim high enough that you cannot see the coin, then pour water into the bowl and the coin will come into view, although it has not moved.

Even with scientific knowledge it is hard not to believe the evidence of our eyes, and wonders like those described at Youghal and other places may well have inspired legends of phantom islands to the west of INISHMORE (Northern Eire & Northern Ireland) and other places, where souls travelled after death or where the fairy people lived.

BIBLIOGRAPHY

AA Road Book of Ireland (London, 1932)
An Account of the Dreadful Battle of Whales, or Sea Monsters, in the Firth of Edinburgh the 25 of April Instant 1707 (Edinburgh, 1707)
Adamnan, *Life of Saint Columba*, ed. William Reeves (Edinburgh, 1874)
Adams, Morley, *In the Footsteps of Borrow & Fitzgerald* (London, n. d.)
Adams, W. H. Davenport, *The History, Topography, and Antiquities of the Isle of Wight* (London & Ryde, 1856)
Ahier, Philip, *Stories of Jersey Seas*, 3 parts (continuous numbering) (Huddersfield, c. 1955)
L'Amy, John H., *Jersey Folk Lore* (Jersey, 1927)
Anderson, M. D., *History and Imagery in British Churches* (London, 1971)
Anderson, Commander William R., *Nautilus 90 North* (London, 1959)
Andrews, Elizabeth, *Ulster Folklore* (Dumfries, 2003) [1913]
[Anon.,] *Angels, Cherubim, and Gods* (London, 1861)
Annals of the Kingdom of Ireland, by the Four Masters, trans. John O'Donovan, 7 vols (Dublin, 1851)
Anson, Peter F., *Fisher Folk-Lore* (London, 1965)
—, *Fishing Boats and Fisher Folk on the East Coast of Scotland* (London, etc., 1930)
Armstrong, Edward A., *The Folklore of Birds* (New York, 1970) [1958]
Armstrong, Warren, *Sea Phantoms* (London, 1963)
Ashton, William, *The Battle of Land and Sea* (Southport, etc., 1909)

Baker, Margaret, *Folklore of the Sea* (Newton Abbot & London, 1979)
Banks, John, *Reminiscences of Smugglers and Smuggling* (London, c .1873)
Barham, Richard Harris, *The Ingoldsby Legends*, 3 vols (London, 1894) [1840–47]
Barham, Richard Harris Dalton, *The Life and Letters of the Rev. Richard Harris Barham*, 2 vols (London, 1870)
Baring-Gould, Sabine, *A Book of the West*, 2 vols (London, 1899)
—, *The Vicar of Morwenstow: A Life of Robert Stephen Hawker*, 3rd edn (London, 1876)
Barron, W. R. J., & Burgess, Glyn S., eds, *The Voyage of Saint Brendan: Representative Versions of the Legend in English Translation* (Exeter, 2002)
[John Barrow,] *The Eventful History of the Mutiny and Piratical Seizure of H. M. S. Bounty* (London, 1831)
Bassett, Fletcher S., *Legends and Superstitions of the Sea and of Sailors in all Lands and at all Times* (London, 1885)
Bassett, Michael G., *'Formed Stones', Folklore and Fossils* (Cardiff, 1982)
Bassett, Wilbur, *Wander-Ships: Folk-Stories of the Sea with Notes upon their Origin* (Chicago, 1917)
Bathurst, Rev. William Hiley, *Roman Antiquities at Lydney Park, Gloucestershire* (London, 1879)

Bauman, Richard, 'Three Legends from the Ayrshire Coast', in *Scottish Studies*, vol. 8, part 1 (Edinburgh, 1964) 33–44

Baxter, Richard, *The Certainty of the World of Spirits* (London, 1834) [1691]

Beck, Horace, *Folklore and the Sea* (Middletown, Connecticut, 1973)

–, 'The Prince of Spain', in *Journal of the Folklore Institute*, vol. 8 (1971) 48–56

Beckett, Commander W. N. T., *A Few Naval Customs, Expressions, Traditions and Superstitions* (Portsmouth, c. 1932)

Beddington, Winifred G., & Christy, Elsa B., *It Happened in Hampshire* (Winchester, 1937)

Belloc, Hilaire, *The Four Men* (London, etc., 1912)

Bence-Jones, Mark, *A Guide to Irish Country Houses* (London, 1988)

Bennett, Alfred Gordon, *Focus on the Unknown* (London, etc., 1953)

Benwell, Gwen, & Waugh, Arthur, *Sea Enchantress: The Tale of the Mermaid and her Kin* (London, 1961)

Beresford, Admiral Lord Charles, *Memoirs*, 2 vols (London, 1914)

Bett, Henry, *English Legends* (London, etc., 1950)

Beveridge, David, *Between the Ochils and Forth* (Edinburgh & London, 1888)

Bignell, Alan, *Kent Lore* (London, 1983)

Bird, James, *Dunwich; A Tale of the Splendid City* (London, 1828)

Bisson, Sidney, *Jersey: Our Island* (London, 1950)

Black, G. F., *County Folk-Lore 3: Orkney and Shetland* (London, 1903)

Blakeborough, Richard, *Wit, Character, Folklore & Customs of the North Riding of Yorkshire* (London, 1898)

Bligh, William, 'The Mutiny of the Bounty', in *Voyages and Travels* (London, 1853) 537–608 [1792]

Blind, Karl, 'Scottish, Shetlandic, and Germanic Water-Tales', in *The Contemporary Review*, vol. 40 (London, July–December 1881) 186–208, 399–423, 534–63

Blundell, William, *A History of the Isle of Man*, 2 vols (Douglas, 1876–7) [1648–56]

Boase, Wendy, *The Folklore of Hampshire and the Isle of Wight* (London, 1976)

Boece, Hector, *The Chronicles of Scotland*, trans. John Bellenden (1531), 2 vols., ed. R. W. Chambers & Edith C. Batho (Edinburgh & London, 1938–41)

Bois, G. J. C., *Jersey Maritime Folklore* (texts and research notes prepared for the Jersey Maritime Museum, 1996)

Bone, David W., *Capstan Bars* (Edinburgh & New York, 1931)

Boswell, James, *The Journal of a Tour to the Hebrides* (London, 1785)

Bottrell, William, *Traditions and Hearthside Stories of West Cornwall*, 3 vols (Penzance, 1870–80)

Bowen, Frank C., *Sea Slang* (London, 1929)

Brainard, John, *The Poems of John G. C. Brainard* (Hartford, 1841)

Bramwell, James, *Lost Atlantis* (London, 1937)

Brand, John, *The History and Antiquities of the Town and County of the Town of Newcastle upon Tyne*, 2 vols (London, 1789)

Bray, Anna Eliza, *Traditions, Legends, Superstitions, and Sketches of Devonshire on the Borders of the Tamar and the Tavy*, 3 vols (London, 1838)

Brewer, J. N., *The Beauties of Ireland*, 2 vols (London, 1825–6)

Brewer's Dictionary of Phrase and Fable, 2nd rev. edn, Ivor H. Evans (London, 1981)

Bridges, T. C., *The Book of the Sea* (London, Bombay & Sydney, 1927)

Briggs, Katharine M., & Tongue, Ruth L., *Folktales of England* (London, 1965)

Broadley, A. M., *The Wessex Mumming Play* (Bridport, 1908)

Broadley, A. M., & Bartelot, R. G., *Nelson's Hardy: His Life, Letters and Friends* (London, 1909)

Brockie, William, *Legends and Superstitions of the County of Durham* (Sunderland, 1886)

Broome, Dora, *Fairy Tales from the Isle of Man* (Douglas, 1963) [1951]

Brown, Raymond Lamont, *Phantoms, Legends, Customs and Superstitions of the Sea* (London, 1972)

Browne, Thomas, *Pseudodoxia Epidemica: or, Enquiries into Very many Received Tenents, And commonly Presumed Truths*, 2nd edn, corrected and enlarged (London, 1650)

Browning, Robert, *The Poetical Works of Robert Browning*, 9 vols (Oxford, 1983–2004)

Bruce-Mitford, R. L. S., 'The Sutton Hoo Ship Burial', in R. H. Hodgkin, *A History of the Anglo-Saxons*, 2 vols (3rd edn, London, etc., 1952) vol. 2, 696–734

Bruford, Alan, & MacDonald, Donald A., *Scottish Traditional Tales* (Edinburgh, 1994)

Bryan, George S., *Mystery Ship: The Mary Celeste In Fancy and In Fact* (Philadephia

& New York, 1942)
Bryant, Arthur, *Jackets of Green* (London, 1972)
Buchan, Peter, *Annals of Peterhead* (Peterhead, 1819)
Buckland, Francis T., *Curiosities of Natural History*, second series (London, 1860)
Bullock, H. A., *History of the Isle of Man* (London, 1816)
Burns, Robert, *An Address to the Deil*, illustrated by Thomas Landseer (London, 1830) [1795]
Burton, Maurice & Robert, *Encyclopedia of Fish* (London, 1975)
Byatt, Andrew, & Fothergill, Alastair, & Holmes, Martha, *The Blue Planet* (London, 2001)
Byron, Lord, *The Island, or Christian and his Comrades*, 3rd edn (London, 1823)
Bywater, Hector C., *Their Secret Purposes* (London, 1932)

Camden, William, *Britannia*, trans. Edmund Gibson with additions and improvements (London, 1695) [1586 in Latin]
Campbell, Commander A. B., *Customs and Traditions of the Royal Navy* (Aldershot, 1956)
Campbell, Lord Archibald, *Records of Argyll* (Edinburgh and London, 1885)
Campbell, Rear-Admiral Gordon, *My Mystery Ships* (London, 1928)
Campbell, John Francis, *Popular Tales of the West Highlands*, 2 vols (Edinburgh, 1994) [1860–62]
Campbell, John Gregorson, *Superstitions of the Highlands & Islands of Scotland* (Glasgow, 1900)
—, *Witchcraft & Second Sight in the Highlands & Islands of Scotland* (Glasgow, 1902)
Campbell, John L., & Hall, Trevor H., *Strange Things* (London, 1968)
Candlin, Lillian, *Tales of Old Sussex* (Newbury, 1985)
Carew, Richard, *The Survey of Cornwall* (London, 1602)
Carmichael, Alexander, *Carmina Gadelica*, 5 vols (Edinburgh & London, 1928–54)
Carrington, Richard, *Mermaids and Mastodons* (London, 1961)
Carter, Captain Harry, *The Autobiography of a Cornish Smuggler*, intro. & notes by John B. Cornish (Truro, 1894)
'de Castre, William' (W. B. Gerish), *Norfolk Folk-Lore Collection*, 6 vols (Great Yarmouth Public Libraries, 1916–18)
Chambercombe Olde Manor Farm (no author) (Ilfracombe, c. 1998)
Chambers, Robert, *The Book of Days*, 2 vols (London & Edinburgh, 1863–4)
—, *The Picture of Scotland*, 2 vols (Edinburgh, 1827)
—, *Popular Rhymes of Scotland* (Edinburgh & London, 1826);
—, Popular *Rhymes of Scotland* new edn (4th) (London & Edinburgh, 1870)
Chandler, Allen, *Chichester Harbour* (Haslemere, 1968)
Charlton, Lionel, *The History of Whitby* (York, 1779)
The Cheshire Sheaf, 2 vols (Chester, 1878–83)
Child, Francis James, *English and Scottish Popular Ballads*, 5 vols (New York, 1957) [1882–98]
Choice Notes from 'Notes and Queries': Folk Lore (London, 1859)
Churchyarde, Thomas, *A Prayse, and Reporte of Maister Martyne Forboishers Voyage to Meta Incognita* (London, c. 1578)
Clark Hall, John R. (trans.), *Beowulf and the Finnsburg Fragment: A Translation into Modern English Prose* (London, 1911)
Coleridge, Samuel Taylor, *The Poetical Works*, ed. James Dykes Campbell (London, 1925)
Colgrave, Bertram (trans. & notes), *Two Lives of Saint Cuthbert* (Cambridge, 1940)
Collinson, John, *The History and Antiquities of the County of Somerset*, 3 vols (Bath, 1791)
Collman, M., *Hants and Dorset's Legends & Folklore* (St Ives, 1975)
Conan Doyle, Arthur, *The Captain of the Polestar and Other Tales*, 3rd edn (London & New York, 1892)
—, *The Case-Book of Sherlock Holmes* (London, 1927)
Condry, William, *Exploring Wales* (London, 1972) [1970]
Cook, W. Victor, *The Story of Sussex* (Hove, 1920)
Cooper, Ernest, *Mardles from Suffolk* (London, 1932)
Cordingly, David, 'Introduction', in Captain Charles Johnson, *A General History of the Pirates* (London, 2002) [1998] vii–xiv
Cormac's Glossary, trans. John O'Donovan, ed. & notes by Whitley Stokes (Calcutta, 1868)
Costello, Louisa Stuart, *The Falls, Lakes, and Mountains of North Wales* (London, 1845)
Courtney, Miss M. A., *Cornish Feasts and Folk-Lore* (Penzance, 1890)
Cowdell, Paul, 'Cannibal Ballads', in *Folk*

Music Journal, vol. 10, no. 5 (2011) 723–47
Coxe, Antony D. Hippisley, *A Book about Smuggling in the West Country, 1700–1850* (Padstow, 1984)
Coxhead, J. R. W., *The Devil in Devon* (Bracknell, 1967)
Croker, Thomas Crofton, *Fairy Legends and Traditions of the South of Ireland*, 3 vols (London, 1825–8)
Cromek, R. H., *Remains of Nithsdale and Galloway Song* (London, 1810)
Cruikshank, E. A., *The Life of Sir Henry Morgan* (Toronto, 1935)
Cunningham, Allan, *Traditional Tales of the English and Scottish Peasantry*, 2 vols (London, 1822)
Cunningham, Andrew S., *Upper Largo, Lower Largo, Lundin Links, and Newburn* (Leven, 1907)

Dacombe, Marianne R., *Dorset Up Along and Down Along* (Gillingham, 1935)
Dalton, John Neil (ed.), *The Cruise of Her Majesty's Ship* 'Bacchante' *1879–1882, compiled from the private journals, letters, and note-books of Prince Albert Victor and Prince George of Wales*, 2 vols (London, 1886)
Dampier, William, *A Voyage to New Holland, &c. In the Year, 1699* (London, 1703)
Dana, Richard Henry, *Two Years Before the Mast* (London, etc., 1981) [1840]
Dance, Peter, *Animal Fakes and Frauds* (Maidenhead, 1976)
Darton, F. J. Harvey, *The Marches of Wessex* (London, 1922)
Dasent, Sir G. W. (trans.), *The Orkneyingers' Saga* (London, 1894)
Davidson, Hilda R. Ellis, 'The Chariot of the Sun', in *Folklore*, vol. 80 (London, Autumn 1969) 174–80
–, 'Scandinavian Folklore in Britain', in *Journal of the Folklore Institute*, vol. 7 (1970) 177–86
Davidson, Thomas, *Rowan Tree and Red Thread* (Edinburgh & London, 1949)
Davies, John, *A History of Wales* (London, New York, etc., 1993)
Davies, Jonathan Ceredig, *Folk-Lore of West and Mid-Wales* (Aberystwyth, 1911)
Davis, Edward B., 'A Whale of a Tale: Fundamentalist Fish Stories', in *Perspectives on Science and Christian Faith*, vol. 43, no. 4 (Ipswich, Mass., December 1991) 224–37
Defoe, Daniel, *Robinson Crusoe* (Edinburgh, 1847) [1719]
–, *A Tour Thro' the Whole Island of Great Britain . . .* 2nd edn, 3 vols (London, 1738)
Denham, Michael Aislabie, *The Denham Tracts*, ed. Dr James Hardy, 2 vols (London, 1892–5) [1846–59]
Denys de Montfort, Pierre, *Histoire Naturelle, Générale et Particuliere des Mollusques*, 6 vols (Paris, 1802–5)
Devonshire Association, *Report and Transactions*, vol. 30 (Plymouth, 1898)
Dickens, Charles, *David Copperfield* (London, etc., 1985) [1849–50]
–, *The Pickwick Papers* (Oxford, 1988) [1837]
Dinsdale, Tim, *The Leviathans* (London, 1966)
Ditmas, E. M. R., *The Legend of Drake's Drum* (St Peter Port, 1973)
Dixon, John H., *Gairloch* (Edinburgh, 1886)
Doel, Fran & Geoff, *Folklore of Dorset* (Stroud, 2007)
–, *Folklore of Kent* (Stroud, 2009) [2003]
Donnelly, Ignatius, *Atlantis: The Antediluvian World* (New York, 1882)
Dorson, Richard M., 'Sources for the Traditional History of the Scottish Highlands and Western Islands', in *Journal of the Folklore Institute*, vol. 8 (The Hague, 1971) 147–84
Douglas, Norman, *Birds & Beasts of the Greek Anthology* (London, 1974) [1928]
Drayton, Michael, *Poly-Olbion*, 2 vols (London, 1612–22)
Dunton, John, *Athenianism* (London, 1710)
Dyer, Thomas Firminger Thiselton, *English Folk-Lore* (London, 1880)
–, *The Folk-Lore of Plants* (London, 1889)

Eardley-Wilmot, Sir Sydney M., *An Admiral's Memories* (London, 1927)
Edwards, J. M., *Cambridge County Geographies: Flintshire* (Cambridge, 1914)
Egerton, John Coker, *Sussex Folk and Sussex Ways* (London, 1924) [1892]
Elder, Abraham, *Tales and Legends of the Isle of Wight* (London, 1839)
Ellis, Richard, *The Book of Whales* (New York, 1982) [1980]
Emerson, P. H., *English Idyls*, 2nd edn (London, 1889)
Emlyn, Ivor, *The Smalls: A Sketch of the Old Light-House* (Solva, 1858)
The Encyclopaedia Britannica, 11th edn, 29 vols (New York, 1910–11)
English's Reminiscences of Old Folkestone Smugglers (Folkestone, 1885)
Epstein, Nicky, *Knitting on Top of the World*

(New York, 2008)
Evans, Andrew, 'The Levitating Altar of Saint Illtud', in *Folklore*, vol. 122 (April 2011) 55–75
Evans, Angela Care, *The Sutton Hoo Ship Burial* (London, 1994) [1986]
Evans, Gwladys, *Hampshire Haunts and Legends* (Colden Common, 1988)
Evans, Roger, *Somerset in the Footsteps of Coleridge and Wordsworth* (Bridgewater, 2000)
Exquemelin, Alexander O., *The Buccaneers of America*, trans. from the Dutch by Alexis Brown, intro. by Jack Beeching (London, etc., 1969) [1678]

Falle, Philip, *An Account of the Isle of Jersey* (London, 1694)
Farmer, David, *Oxford Dictionary of Saints*, 5th rev. edn (Oxford, 2011) [1978]
Farre, Rowena, *Seal Morning* (London, etc., 1960) [1957]
Fea, Allan, *The Real Captain Cleveland* (London, 1912)
Fenland Notes & Queries, vol. 1 (Peterborough & London, 1889–91)
Fenton, Richard, *A Historical Tour through Pembrokeshire* (London, 1811)
Fielding, Henry, *The Journal of a Voyage to Lisbon* (London, 1755)
Fitz-Geffrey, Charles, *Sir Francis Drake, His Honorable lifes commendation, and his Tragicall Deathes lamentation* (Oxford, 1596)
Foley, Helen, *A Handful of Time* (London, 1961)
Forbes, Athol, *The Romance of Smuggling* (London, 1909)
Forbes, Thomas R., 'The Social History of the Caul', in *Yale Journal of Biology and Medicine*, vol. 25, no. 6 (New Haven, 1953) 495–508
Forby, Rev. Robert, *The Vocabulary of East Anglia*, 2 vols (London, 1830)
Fordyce, T., *Local Records*, 2 vols numbered as 3 and 4 (Newcastle, 1876)
Forester, C. S., *The Commodore* (London, 1945)
Foster, Jeanne Cooper, *Ulster Folklore* (Belfast, 1951)
Foster, John Wilson (ed.), *Titanic* (London, etc., 1999)
Fournier, Laurent Sébastien, 'The Embodiment of Social Life: Bodylore and the Kirkwall Ba' Game', in *Folklore*, vol. 120 (August 2009) 194–212
Friedrich, Walter L., *Fire in the Sea: The Santorini Volcano* (Cambridge, 2000)
Fritze, Ronald H., *Travel Legend and Lore* (Santa Barbara, Denver & Oxford, 1998)
Fuller, Thomas, *The Profane State* (Cambridge, 1648)
—, *The History of the Worthies of England*, 4 vols in 1 (London, 1662)
Funk & Wagnalls Standard Dictionary of Folklore Mythology and Legend (New York, 1972) [1949]

Gardner, Thomas, *An Historical Account of Dunwich* (London, 1754)
Garle, Hubert, *A Driving Tour in the Isle of Wight* (Newport, 1905)
Garrett, Richard, *Voyage into Mystery* (London, 1987)
Gaskell, Elizabeth, *Sylvia's Lovers* (London, 1880) [1863]
Geoffrey of Monmouth, *British History*, ed. J. A. Giles, in *Old English Chronicles* (London, 1908) 89–292
Gerald of Wales, *The History and Topography of Ireland*, trans. John J. O'Meara (Harmondsworth, etc., 1982)
—, *The Itinerary of Archbishop Baldwin through Wales, A.D. MCLXXXVIII*, trans. Sir Richard Colt Hoare, 2 vols (London, 1806)
—, *The Journey through Wales/The Description of Wales*, trans. Lewis Thorpe (London, etc., 1978)
Gerard, John, *The Herball or Generall Historie of Plantes*, 2 vols (London, 1597)
Gervase of Tilbury, *Otia Imperialia*, trans. S. E. Banks & J. W. Binns (Oxford, 2002) [c. 1200 in Latin]
Gibbings, Robert (ed.), *Narratives of the Wreck of the Whale-Ship Essex of Nantucket which was destroyed by a whale in the Pacific Ocean in the year 1819, told by Owen Chase First Mate Thomas Chappel Second Mate and George Pollard Captain of the said vessel* (London, 1935) [1821]
Gibson, Wilfrid, *Collected Poems 1905–1925* (London, 1926)
Gilbert, C. S., *An Historical Survey of the County of Cornwall*, 2 vols (Plymouth & London, 1817–20)
Gilbert, Davies, *The Parochial History of Cornwall*, 4 vols (London, 1838)
Gilbert, W. S., *The 'Bab' Ballads* (London, 1869)
Gill, W. Walter, *A Manx Scrapbook* (London & Bristol, 1929)
—, *A Second Manx Scrapbook* (London &

Bristol, 1932)
Glyde, John, *The Norfolk Garland* (London & Norwich, 1872)
Godden, Rumer, *In This House of Brede* (London, 1969)
Gordon, R. K. (trans. & notes), *Anglo-Saxon Poetry* (London & New York, 1976) [1926]
Gould, R. T., *The Case for the Sea-Serpent* (London, 1930)
Graham, Dougal, 'The Ancient and Modern History of Buck-Haven', in *The Collected Writings of Dougal Graham*, ed. George MacGregor, 2 vols (Glasgow, 1883) vol. 2, 217–37
Granville, Wilfred, *Sea Slang of the Twentieth Century* (New York, 1950)
Greenoak, Francesca, *All the Birds of the Air* (London, 1979)
Gregor, Revd Walter, *Notes on the Folk-Lore of the North-East of Scotland* (London, 1881)
—, 'Some Folk-Lore of the Sea', in *Folk-Lore Journal*, vol. 3 (London, 1885) 52–6, 180–5, 305–11
—, 'Some Folk-Lore of the Sea', in *Folk-Lore Journal*, vol. 4 (London, 1886) 7–17
Gregory, Augusta, *Visions and Beliefs in the West of Ireland*, 2 vols (New York & London, 1920)
Griffin, Gerald, *Poetical Works* (Dublin, 1926)
Grose, Francis, *A Provincial Glossary*, new edn (London, 1811)
Gunn, John, *The Orkney Book* (London, etc., 1909)
Gunn, Rev. John, 'Proverbs, Adages and Popular Superstitions, still preserved in the Parish of Irstead', in *Norfolk Archaeology*, vol. 2 (Norwich, 1849) 291–308
Gutch, Eliza, *County Folk-Lore Vol. II: North Riding of Yorkshire, York and the Ainsty* (London, 1901)
Gutch, Eliza, & Peacock, Mabel, *County Folk-Lore Vol. V: Lincolnshire* (London, 1908)
Guthrie, Ellen, 'Superstitions of the Scottish Fishermen', in *Folk-Lore Journal*, vol. 7 (London, 1889) 44–7
Gwynn, Edward, *The Metrical Dindsenchas*, 5 vols (Dublin, 1903–35)

Hadfield, R. L., *The Phantom Ship* (London, 1937)
Hakluyt, Richard, *Voyages and Discoveries*, abridged edn, ed. Jack Beeching (London, etc., 1972) [1598–1600]
Hall, Rev. James, *Travels in Scotland, by an Unusual Route*, 2 vols (London, 1807)
Hall, Mr & Mrs S. C., *The Book of South Wales* (London, 1861)
—, *Ireland: its Scenery, Character, &c.*, 3 vols (London, 1841–3)
Halliwell, James Orchard, *The Norfolk Anthology* (London, 1852)
—, *Rambles in Western Cornwall by the Footsteps of the Giants* (London, 1861)
—, *The Yorkshire Anthology* (London, 1851)
Hallywell, Henry, *Melampronoea: or a Discourse of the Polity and Kingdom of Darkness* (London, 1681)
Hamilton, Robert, *The Natural History of the Amphibious Carnivora* (Edinburgh, London & Dublin, 1839)
Hammond, Peter, *Food & Feast in Medieval England* (Stroud, 2005) [1993]
Hancock, F., *A History of Minehead* (Taunton, 1903)
Hardie, R. P., *The Tobermory Argosy: A Problem of the Spanish Armada* (Edinburgh & London, 1912)
Hardwick, Michael & Molly, *The World's Greatest Sea Mysteries* (London, 1967)
Hardy, Florence Emily, *The Life of Thomas Hardy* (London, New York, etc., 1965)
Hardy, Thomas, *Life's Little Ironies* (London, 1894)
Hargrove, Ethel, *Wanderings in the Isle of Wight* (London, 1913)
Harland, John, & Wilkinson, T. T., *Lancashire Legends, Traditions, Pageants, Sports, &c.*, (Manchester & London, 1882)
Harper, Charles G., *The Ingoldsby Country* (London, 1904)
Harrison, Paul, *Sea Serpents and Lake Monsters of the British Isles* (London, 2001)
Harrison, William, *Mona Miscellany*, second series (Douglas, 1873)
Hasted, Edward, *The History and Topographical Survey of the County of Kent*, 2 vols (Canterbury, 1778–82)
Hawker, R. S., *The Cornish Ballads and Other Poems* (Oxford & London, 1869)
—, *Footprints of Former Men in Far Cornwall* (London, 1908) [1870]
Hawkes, Ken, *Sark* (Vale, 1995) [1987]
Healy, James N., *Irish Ballads and Songs of the Sea* (Cork, 1976) [1967]
Heath, Edward, *Sailing* (London, 1975)
Heaume, Doris O., *The Book of Guernsey Songs and Dances* (Guernsey, 1976)
Hemsley, T. W., *St. Nicholas Church, Brighton* (Brighton, 1896)
Henderson, George, *The Norse Influence on Celtic Scotland* (Glasgow, 1910)
Henry, John *see* Ó Catháin (1983)

Herodotus, *Works*, trans. Canon Rawlinson, notes by A. J. Grant, 2 vols (London, 1897)
Heron-Allen, Edward, *Barnacles in Nature and in Myth* (London, 1928)
Heuvelmans, Bernard, *In the Wake of the Sea-Serpents*, trans. Richard Garnett (London, 1968)
Hewett, Sarah, *Nummits and Crummits: Devonshire Customs, Characteristics, and Folk-lore* (East Ardsley, 1976) [1900]
Hibbert, Samuel, *A Description of the Shetland Islands* (Edinburgh, 1822)
Hill, Douglas, *Magic and Superstition* (London, etc., 1968)
The History and Legend of Chambercombe Manor (no author) (St Ives, c. 2001)
Hobbs, Sandy & Cornwell, David, 'Hanging the Monkey', in *Dear Mr Thoms . . .*, vol. 35 (1994) 17–20
Hogan, Edmund, *Onomasticon Goedelicum* (Dublin, 1993) [1910]
Hogg, Garry, *The Far-Flung Isles: Orkney and Shetland* (London, 1961)
Hogg, James, *The Poems of James Hogg*, ed. & intro. by William Wallace (London, 1903)
Holdsworth, Edmund W. H., *Deep-Sea Fishing and Fishing Boats* (London, 1874)
Hole, Christina, *British Folk Customs* (London, 1976)
—, *Traditions and Customs of Cheshire* (London, 1937)
Holinshed, Raphael, *Chronicles*, 3 vols (London, 1586)
Holweck, F. G., *A Biographical Dictionary of the Saints* (Detroit, 1969) [1924]
Hone, William, *The Every-Day Book and Table Book*, 3 vols (London, 1830)
Hornibrook, J. L., 'The Case of the *Marie Celeste*', in *Chambers's Journal*, sixth series, vol. 7, no. 355 (London & Edinburgh, 1904) 671–2
Howells, W., *Cambrian Superstitions* (London, 1831)
Hughes, Meirion, & Evans, Wayne, *Rumours and Oddities from North Wales* (Llanrwst, 1995) [1986]
Hughes, Thomas, *The Stranger's Handbook to Chester* (Chester & London, 1856)
Hugill, Stan, *Shanties and Sailors' Songs* (London, 1969)
Hugo, Victor, *Toilers of the Sea*, trans. Mary W. Artois, 4 vols (London, 1895) [1866]
Hunt, Robert, *Popular Romances of the West of England*, new edn (London, 1903) [1865]
Hutchins, Jane, *Discovering Mermaids and Sea Monsters* (Tring, 1968)
Hutchins, John, *The History and Antiquities of the County of Dorset*, 3rd edn corrected, augmented and improved by William Shipp & James Whitworth Hodson, 4 vols (London, 1861)
Hutchinson, William, *A View of Northumberland*, 2 vols (Newcastle, 1778)
Hutton, J. Bernard, *Frogman Extraordinary: 'The Commander Crabb Case'* (London, 1960)
Hutton, Ronald, *The Stations of the Sun: A History of the Ritual Year in Britain* (Oxford & New York, 1996)

Igglesden, Charles, *Those Superstitions* (London, 1932)
Ingersoll, Ernest, *Birds in Legend Fable and Folklore* (New York, etc., 1923)
Ingledew, C. J. Davison, *The Ballads and Songs of Yorkshire* (London, 1860)

Jack, John, *An Historical Account of St Monance, Fife-shire* (Edinburgh, London & Cupar, 1844)
Jal, Auguste, *Scènes de la Vie Maritime*, 2 vols (Paris, 1832)
James, M. R., *Suffolk and Norfolk* (London & Toronto, 1930)
Jeans, Peter D., *Seafaring Lore and Legend* (New York, London, etc., 2004)
Jenkins, D. E., *Bedd Gelert: Its Facts, Fairies, & Folk-Lore* (Portmadoc, 1899)
Jobson, Allan, *Dunwich Story* (Lowestoft, 1961)
—, *An Hour-Glass on the Run* (London, 1959)
Johnson, Captain Charles, *A General History of the Pyrates* (London, 1972) [1724]
Johnson, W. H., *The Superstitions and Curious Beliefs of Old Sussex* (Lewes, 2009)
Jones, T. Gwynn, *Welsh Folklore and Folk-Custom* (Cambridge, 1979) [1930]
Jones, William, *The Broad, Broad Ocean* (London, 1871)
—, *Credulities Past and Present* (London, 1880)
Jonsson, Lars, *Birds of Sea and Coast*, trans. Roger Tanner (Harmondsworth, etc., 1978)
Joyce, P. W., *Old Celtic Romances*, 2nd edn (London, 1894)
—, *The Origin and History of Irish Names of Places*, 3 vols (Dublin, London & Edinburgh, 1870–1913)

Kemp, Peter (ed.), *The Oxford Companion to Ships and the Sea* (Oxford, New York & Melbourne, 1988) [1976]
Kennedy, Patrick, *The Banks of the Boro:*

A Chronicle of the County of Wexford (London & Dublin, 1867)
—, *Legendary Fictions of the Irish Celts* (London, 1866)
Kennedy-Fraser, Marjory, *A Life of Song* (London, 1929)
— & Macleod, Kenneth, *Songs of the Hebrides*, 3 vols (London & New York, 1909–21)
Kenney, James F., *The Sources for the Early History of Ireland: Ecclesiastical* (Dublin, 1993) [1929]
Kerényi, C., *The Heroes of the Greeks* (London, 1959)
Killip, Margaret, *The Folklore of the Isle of Man* (London & Sydney, 1975)
'Kilrounie', *The Kingdom: A Descriptive and Historical Hand-book to Fife* (Cupar & Edinburgh, 1882)
King-Hall, Admiral Sir Herbert, *Naval Memories and Traditions* (London, 1926)
Kingshill, Sophia, 'The Tobermory Treasure', in *Folklore*, vol. 121, no. 3 (London, 2010) 334–45
Kipling, Rudyard, *Puck of Pook's Hill* (London, etc., 1994) [1906]
Kircher, Athanasius, *Mundus Subterraneus* (Amsterdam, 1678)
Krapp, George Philip, & Dobbie, Elliott van Kirk (eds), *The Exeter Book* (London & New York, 1936)

Lambarde, William, *A Perambulation of Kent* (Bath, 1970) [1570]
Lambert, M., & Marx, Enid, *English Popular Art* (London & New York, 1951)
Lancashire Federation of Women's Institutes, *Lancashire Lore* (Kendal, 1971)
Lane, Louisa, *Recollections and Legends of Serk* (Guernsey, c. 1880)
Lane, Margaret, *The Tale of Beatrix Potter* (London, 1970) [1946]
Lane, William G., *Richard Harris Barham* (Columbia, 1967)
Lang, Andrew, 'The Mystery of the Tobermory Galleon Revealed', in *Blackwood's Magazine*, vol. 191, no. 1157 (March 1912) 422–36
Langmaid, Captain Kenneth, *The Sea, Thine Enemy* (London, 1966)
Lanier, Sidney, *Poems of Sidney Lanier* (London, 1892)
Laughton, L. G. Carr, *Old Ship Figure-heads & Sterns* (London & New York, 1925)
Lawrence, Berta, *Coleridge and Wordsworth in Somerset* (Newton Abbot, 1970)
Legg, Penny, *Folklore of Hampshire* (Stroud, 2010)
Leland, John, *The Itinerary of John Leland in or about the years 1535–1543*, ed. Lucy Toulmin Smith, 5 vols (London, 1907–10)
Lewis, C. S., *The Voyage of the* Dawn Treader (London, 1968) [1952]
[Lewkenor, John,] *Metellus his Dialogues* (London, 1693)
Linder, Leslie (compiler), *The Beatrix Potter Papers at Hill Top* (Stroud, 1987)
van Linschoten, Jan Huygen, *Discours of Voyages into the East & West Indies* (Amsterdam & Norwood, New Jersey, 1974) [1598]
Lockhart, John Gibson, *Memoirs of the Life of Sir Walter Scott*, 4 vols (Paris, 1837–8)
Lockhart, John Gilbert, *Curses, Lucks and Talismans* (London, 1938)
—, *Mysteries of the Sea* (London, 1924)
Logan, W. H., *A Pedlar's Pack of Ballads and songs* (Edinburgh, 1869)
Lovett, Edward, *Folk-Lore & Legend of the Surrey Hills and of the Sussex Downs & Forests* (Caterham, 1928)
—, *Magic in Modern London* (Croydon, 1925)
Lowes, John Livingston, *The Road to Xanadu*, 2nd rev. edn (London, 1951) [1927]
Lucas, E. V., *Highways and Byways in Sussex* (London, 1935) [1903]
Lucian of Samosata, trans. William Tooke, 2 vols (London, 1820)
Lysaght, Patricia, & Ó Catháin, Séamas, & Ó hÓgáin, Dáithí (eds), *Islanders and Water-Dwellers* (Dublin, 1999)

MacCodrum, John, *The Songs of John MacCodrum: Bard to Sir James MacDonald of Sleat*, ed. William Matheson (Edinburgh, 1938)
McCormick, Donald, *The Mystery of Lord Kitchener's Death* (London, 1959)
MacCulloch, Edgar, *Guernsey Folk Lore* (London & Guernsey, 1903)
McDermid, Val, *The Grave Tattoo* (London, 2006)
MacDonald, Rev. Archibald (ed.), *The Uist Collection: The Poems and Songs of John Mac Codrum, Archibald MacDonald, and some of the minor Uist bards* (Glasgow, Edinburgh & Oban, 1894)
MacDonald, Donald, *Tales and Traditions of the Lews* (Stornoway, 1967)
McDonald, Patrick, *A Collection of Highland*

Vocal Airs (Edinburgh, 1784)
MacDougall, Philip, *Phantoms of the High Seas* (Newton Abbot & London, 1991)
MacGregor, Rev. Alexander, *The Life of Flora Macdonald, and her Adventures with Prince Charles* (Inverness, 1882)
McIntire, Walter T., *Lakeland and the Borders of Long Ago* (Carlisle, 1948)
Mackay, Aeneas E. G., *A History of Fife and Kinross* (Edinburgh & London, 1896)
Mackay, Charles, *The Poetical Works of Charles Mackay* (London & New York, 1876)
McKay, John G., *More West Highland Tales*, 2 vols (Edinburgh & London, 1940–60)
MacLean, J. P., *A History of the Clan MacLean* (Cincinnati, 1889)
McLeay, Alison, *The Tobermory Treasure* (London, 1986)
McPherson, J. M., *Primitive Beliefs in the North-East of Scotland* (London, New York & Toronto, 1929)
Macpherson, James, *An Introduction to the History of Great Britain and Ireland* (Dublin, 1771)
MacPherson, John, *Tales of Barra* (Edinburgh, 1960)
Macquoid, Thomas and Katharine, *About Yorkshire* (London, 1883)
MacTaggart, John, *The Scottish Gallovidian Encyclopedia* (London, 1824)
The Mabinogion, trans. Jeffrey Gantz (London, etc., 1987) [1976 trans.]
Mair, Craig, *A Star for Seamen: The Stevenson Family of Engineers* (London, 1978)
Malory, Thomas, *Le Morte d'Arthur* (London, 1933) [1485]
Mandeville, Sir John, *The Travels of Sir John Mandeville*, trans. & intro. by C. W. R. D. Moseley (London, etc., 1983)
Maple, Eric, 'Ship', in *Man, Myth and Magic* (London, 1970) 2565–71
March, Jenny, *Cassell Dictionary of Classical Mythology* (London, 1998)
Marples, Morris, 'Old English Football at Workington', in *Cumbria*, vol. 19, no. 12 (March 1970) 647–8
Marryat, Captain C. B., *Poor Jack* (Paris, 1841)
–, *Snarleyyow* (London, 1897) [1837]
Marsden, John, *Sea-Road of the Saints* (Edinburgh, 1995)
Marshall, Michael, *Herm: Its Mysteries and its Charm* (Guernsey, 1978) [1958]
Martin, Martin, *A Description of the Western Isles of Scotland* (London, 1703)
Martyn, Myles, *Somerset Tales, Fact and Legend* (South Molton, 1975)
Masefield, John, *The Collected Poems of John Masefield* (London, 1932)
Mason, Thomas H., *The Islands of Ireland* (London, 1936)
Melville, Herman, *Moby-Dick, or The Whale* (London, 1992) [1851]
Metford, J. C. J., *Dictionary of Christian Lore and Legend* (London, 1983)
Meyer, Kuno, & Nutt, Alfred, *The Voyage of Bran*, 2 vols (London, 1895–7)
Meyrick, Samuel Rush, *The History and Antiquities of the County of Cardigan* (London, 1810)
Miller, Hugh, *My Schools and Schoolmasters* (Edinburgh, 1854)
–, *Scenes and Legends of the North of Scotland* (Edinburgh & London, 1835)
Milton, John, *The History of Britain* (London, 1670)
–, *Paradise Lost* (Oxford, 2005) [1667]
Mitchell, Elizabeth Harcourt, *Her Majesty's Bear* (London, 1884)
Monger, George P., *Marriage Customs of the World* (Oxford, etc., 2004)
Moore, A. W., *The Folk-Lore of the Isle of Man* (Douglas & London, 1891)
Moore, J., *Sussex Legends and Folklore* (St Ives, 1976)
Morgan, Alun, *Legends of Porthcawl and the Glamorgan Coast* (Cowbridge & Bridgend, 1974)
Morgan, Mary, *A Tour to Milford Haven, in the year 1791* (London, 1795)
Morganwg, Iolo, *The Triads of Britain*, trans. W. Probert (London, 1977) [1823]
Morris, Jan, *Wales* (London, New York, etc., 1998) [1984]
Morvan, F., *Legends of the Sea*, trans. David Macrae (Geneva, 1980)
Müller, F. Max, *Lectures on the Science of Language*, 6th edn, 2 vols (London, 1871)
Munnelly, Tom, 'Songs of the Sea', in *Béaloideas*, vols 48–9 (Dublin, 1980–81) 30–58
Murray's Handbook for Essex, Suffolk, Norfolk and Cambridgeshire (London, 1892)
Murray's Handbook for Travellers in Devon and Cornwall, 3rd rev. edn (London, 1856)
Murray's Handbook for Travellers in Durham and Northumberland (London, 1864)

Nelson, Esther, *Island Minstrelsy* (London & Liverpool, 1839)
Newall, Venetia, 'Up-Helly-Aa: A Shetland Fire Festival', in Theresa Buckland & Juliette Wood, *Aspects of British Calendar*

Customs (Sheffield, 1993) 57–73
Newbolt, Henry, *Admirals All* (London, 1897)
Newland, Robert J., & North, Mark J., *Dark Dorset* (Weston-super-Mare, 2002)
Nicholas, Thomas, *Annals and Antiquities of the Counties and County Families of Wales*, rev. edn, 2 vols (London, 1875)
Nichols, John Gough (ed.), *The Topographer and Genealogist*, 3 vols (London, 1846–58)
Nicholson, Norman, *Selected Poems* (London, 1966)
Norden, John, *Speculi Britanniae Pars: A Topographical and Historical Description of Cornwall* (London, 1728) [1610]
Norman, J. R., *A History of Fishes* (London, 1931)
North, F. J., *Sunken Cities* (Cardiff, 1957)
Noyes, Alfred, *Collected Poems* (London, 1963) [1950]

Ó Catháin, Séamas, *Irish Life and Lore* (Dublin & Cork, 1982)
Ó Catháin, Séamas (trans.), *Stories of Sea & Shore, told by John Henry* (Dublin, 1983)
O'Curry, Eugene, *Lectures on the Manuscript Materials of Ancient Irish History* (Dublin & London, 1861)
O'Flaherty, Roderic, *A Chorographical Description of West or h-Iar Connaught, written A.D. 1684*, ed. James Hardiman (Dublin, 1846)
O'Grady, Standish H. (ed. & trans.), *Silva Gadelica*, 2 vols (London & Edinburgh, 1892)
Ó hEochaidh, Seán, & Mac Neill, Máire, & Ó Catháin, Séamas, *Fairy Legends from Donegal* (Dublin, 1977)
Ó hÓgáin, Dáithí, *The Hero in Irish Folk History* (Dublin & New York, 1985)
–, *The Lore of Ireland* (Cork, 2006)
–, *The Sacred Isle* (Woodbridge & Doughcloyne, 1999)
O'Keeffe, C. M., 'Persecution of the Wren', in *Ulster Journal of Archaeology*, vol. 4 (Belfast, 1856) 171–2
O'Kelleher, A., & Schoepperle, G., *Life of Columcille compiled by Manus O'Donnell in 1532* (Illinois, 1918)
O'Reilly, John Boyle, *Songs, Legends, and Ballads* (Boston, 1878)
Ó Súilleabháin, Seán, *Irish Folk Custom and Belief* (Dublin, 1967)
O'Sullivan, Sean, *Folktales of Ireland* (London, 1966)
Official Guide to Whitby, 32nd edn (Whitby, 1931)
Olaus Magnus, *History of the Goths, Swedes, & Vandals* (London, 1658) [1555]
Opie, Iona, & Tatem, Moira, *A Dictionary of Superstitions* (Oxford, 1992) [1989]
Owen, Elias, *Welsh Folk-Lore* (Oswestry & Wrexham, 1896)
Owen, William, *The Cambrian Biography* (London, 1803)
Oxford English Dictionary, 2nd edn (Oxford, 2001) [1989]

Page, John Lloyd Warden, *The Coasts of Devon and Lundy Island* (London, 1895)
Page, Michael, & Ingpen, Robert, *Encyclopaedia of Things that Never Were* (Limpsfield, 1985)
Paine, Lincoln P., *Ships of the World* (New York, 1997)
Paine, Ralph D., *The Book of Buried Treasure* (London, 1911)
Partridge, Eric, *The Routledge Dictionary of Historical Slang*, abridged by Jacqueline Simpson (London, 1973)
Pastoureau, Michel, *Le Cochon* (Paris, 2009)
Paterson, James, *History of the Counties of Ayr and Wigton*, 3 vols (Edinburgh 1863–6)
Pembrokeshire Antiquities (Solva, 1897)
Pennant, Thomas, *British Zoology*, 4 vols (London, 1776)
–, *A Tour in Wales*, 2 vols (London, 1778–83)
Pepys, Samuel, *Diary and Correspondence*, 5 vols (London, 1848–9)
Perkins, Thomas, & Pentin, Herbert, *Memorials of Old Dorset* (London & Derby, 1907)
Peterson, Roger, & Mountfort, Guy, & Hollom, P. A. D., *Collins Field Guide: Birds of Britain & Europe*, 5th rev. edn (London, 2004)
Philipot, Thomas, *Villare Cantianum: or Kent Surveyed and Illustrated* (London, 1659)
Phillips, Ken, *Shipwrecks of the Isle of Wight* (Newton Abbot, etc., 1988)
Plato, *Timaeus* and *Critias*, trans. H. D. P. Lee (London, 1971)
Pliny, *Natural History*, trans. & notes by John Bostock & H. T. Riley, 6 vols (London, 1855–7)
Poe, Edgar Allan, *A Descent into the Maelström and Other Stories* (London, 1948) [1841]
–, *The Narrative of Arthur Gordon Pym of Nantucket* (New York, 1838)
Poingdestre, Jean, *Caesarea or A Discourse of*

the Island of Jersey, ed. William Nicolle (St Helier, 1889)
Pontoppidan, Revd Erich, *The Natural History of Norway*, 2 vols in 1 (London, 1755)
Poole, Charles Henry, *The Customs, Superstitions, and Legends of the County of Somerset* (London, 1877)
Pope, Dudley, *Harry Morgan's Way* (London, 1977)
Porter, Enid, *The Folklore of East Anglia* (London, 1974)
Porter, John, *History of the Fylde of Lancashire* (Fleetwood & Blackpool, 1876)
Potter, Beatrix, *The Fairy Caravan* (London & New York, 1952) [1929]
Power, Frank, *The Kitchener Mystery* (London, 1926)
Power, Rev. P. (ed.), *Life of St. Declan of Ardmore and Life of St. Mochuda of Lismore* (London, 1914)
Prichard, T. Jeffery Llewelyn, *Welsh Minstrelsy* (London, 1824)
Pugh, David T., *Tides, Surges and Mean Sea-Level* (Chichester, etc., 1987)
Putley, John, *The Isles of Scilly in the English Civil Wars* (Bristol, 2003)

Quinn, Tom, *Smugglers' Tales* (Newton Abbott, 1999)

Radford, E. & M. A., *Encyclopaedia of Superstitions*, ed. & revised by Christina Hole (London, 1961)
Radford, Ken, *Tales of North Wales* (Letchworth, 1982)
Raine, James, *Saint Cuthbert* (Durham, 1828)
Ramsay, Raymond H., *No Longer on the Map* (New York, 1973)
Reader's Digest, *Folklore, Myths and Legends of Britain* (London, 1977) [1973]
Rees, Rice, *An Essay on the Welsh Saints* (London, 1836)
Rees, W. J., *Lives of the Cambro British Saints* (Llandovery, 1853)
Reeves, James, *The Everlasting Circle* (London, etc., 1960)
Reid, Hugh G., *Past and Present* (Edinburgh, 1871)
Rhys, John, *Celtic Folklore: Welsh and Manx*, 2 vols (Oxford, 1901)
Roberts, George, *The Social History of the People of the Southern Counties of England* (London, 1856)
Roberts, W. Langley, *Legends and Folk Lore of Lancashire* (London & Glasgow, 1931)
Robertson, Morgan, *Futility* (New York, 1900) [1898]
Robertson, R. Macdonald, *Wade the River, Drift the Loch* (Edinburgh & London, 1948)
Robinson, James, *The Lewis Chessmen* (London, 2004)
Robson, Michael, *Rona: The Distant Island* (Stornoway, 1991)
Robson, Peter, 'Dorset Garland Days on the Chesil Coast', in Theresa Buckland & Juliette Wood (eds), *Aspects of British Calendar Customs* (Sheffield, 1993) 155–66
Roeder, C. (ed.), *Manx Notes and Queries* (Douglas, 1904)
Rogers, Nicholas, *The Press Gang* (London & New York, 2007)
Rogers, Stanley, *Sea-Lore* (London, Bombay & Sydney, 1929)
Rogers, Captain Woodes, *A Cruising Voyage round the World* (London, 1718)
Rogozinski, Jan, *The Wordsworth Dictionary of Pirates* (Ware, 1997) [1995]
Rollinson, William, *Cumbrian Dictionary of Dialect, Tradition and Folklore* (Otley, 1997)
Rondeletius, Gulielmus, *Libri de Piscibus Marinis* (Lugdunum, 1554)
Rorie, David, 'Some Superstitions of the Fifeshire Fisher-Folk', in *Folk-Lore*, vol. 15 (London, 1904) 95–8
Ross, Anne, *The Folklore of the Scottish Highlands* (London, 1976)
Roud, Steve, *The Penguin Guide to the Superstitions of Britain and Ireland* (London, 2003)
Rowling, Marjorie, *The Folklore of the Lake District* (London, 1976)
Rudkin, Ethel H., *Lincolnshire Folklore* (Wakefield, 1973) [1936]
Rye, Walter, *Norfolk Songs, Stories and Sayings* (Norwich, 1897)
—, *Tourist's Guide to the County of Norfolk* (London, 1892)

Sandford, Mrs Henry, *Thomas Poole and his Friends*, 2 vols (London, 1888)
Sandys, George, *Ovid's Metamorphosis Englished* (Oxford, 1632)
Saul, A., 'The Herring Industry at Great Yarmouth c. 1280–c. 1400', in *Norfolk Archaeology*, vol. 38, part 1 (Norwich, 1981) 33–43
Savile, Rev. Bourchier Wrey, *Apparitions: A Narrative of Facts* (London, 1874)
Saxby, Jessie M. E., *Shetland Traditional Lore* (Edinburgh, 1932)
Schofield, James, *An Historical and Descriptive*

Guide to Scarbrough and its Environs (York, c. 1787)
Scot, Reginald, *The Discoverie of Witchcraft* (n. p., 1584)
Scott, Walter, *Letters on Demonology & Witchcraft* (Ware, 2001) [1830]
—, *Minstrelsy of the Scottish Border* (London, etc., 1931) [1801–3]
—, *Peveril of the Peak*, 4 vols (Edinburgh, 1823)
—, *The Pirate*, 3rd edn, 3 vols (Edinburgh, 1822) [1821]
—, *Rokeby*, 5th edn (Edinburgh, 1813)
Searle, Adrian, *Isle of Wight Folklore* (Wimborne, 1998)
Sebag-Montefiore, Hugh, *Dunkirk* (London, etc., 2006)
'A Seneachie', in *An Historical and Genealogical Account of the Clan Maclean* (London & Edinburgh, 1838)
Service, James, *Metrical Legends of Northumberland* (Alnwick, 1834)
Severin, Tim, *The Brendan Voyage* (London, 1996) [1978]
Sharpe, Sir Cuthbert, *The Bishoprick Garland* (London, 1834)
Shelvocke, George, *A Voyage Round the World* (London, 1971) [1726]
Shewan, Andrew, *The Great Days of Sail* (London, 1927)
Sikes, Wirt, *British Goblins* (London, 1880)
Simeon of Durham, *The Historical Works of Simeon of Durham*, trans. Rev. Joseph Stephenson (London, 1855)
Simmons, James C., *Castaway in Paradise: The Incredible Adventures of True-Life Robinson Crusoes* (Dobbs Ferry, NY, 1993)
Simpkins, John Ewart, *County Folk-Lore 7: Fife* (London, 1914)
Simpson, A. W. Brian, *Cannibalism and the Common Law: The Story of the Tragic Last Voyage of the* Mignonette *and the Strange Legal Proceedings to Which It Gave Rise* (Chicago & London, 1984)
Simpson, David A., *Shore of the Saints: The people, history and folklore of the North East Coast, Whitby to Berwick-upon-Tweed* (Durham, 1991)
Simpson, Eve Blantyre, *Folk Lore in Lowland Scotland* (East Ardsley, 1976) [1908]
Simpson, Jacqueline, *Folklore of Sussex* (Stroud, 2009) [1973]
—, *Icelandic Folktales & Legends* (Stroud, 2004) [1972]
— & Roud, Steve, *A Dictionary of English Folklore* (Oxford, 2000)
Sinclair, Sir John (ed.), *The Statistical Account of Scotland, 1791–1799*, 21 vols (Wakefield, 1978) [1791–9]
Skene, William F., *The Four Ancient Books of Wales*, 2 vols (Edinburgh, 1868)
Smith, Alan, 'The Image of Cromwell in Folklore and Tradition', in *Folklore* vol. 79, (London, Spring 1968) 17–39
Smith, Charles, *The Ancient and Present State of the County of Kerry* (Dublin, 1774)
Smollett, Tobias, *The Adventures of Peregrine Pickle*, 2 vols (London, 1831) [1751]
Smyth, Admiral William Henry, *The Sailor's Word-Book* (London, 2005) [1867]
Snell, D. J., *The Wreckers of Dunraven* (Swansea, 1894)
Society of Antiquaries of Scotland, *Proceedings* (Edinburgh, 1852–)
Southey, Robert, *Madoc*, 2 vols (London, 1815) [1805]
Spelman, Henry, *The History and Fate of Sacrilege* (London, 1698) [1632]
Spencer, Marianne Robertson, *Annals of South Glamorgan* (Carmarthen, 1913)
Stammers, Michael, *Figureheads and Ship Carving* (London, 2005)
Stanley, Louis T., *Journey through Cornwall* (London, 1958)
Steele, Richard, *The Englishman*, 2 vols (London, 1714–37)
Stevens, Todd, & Cumming, Edward, *Ghosts of Rosevear and The Wreck of the Nancy Packet* (Glasgow, 2009) [2008]
Stevenson, Robert Louis, *The English Admirals* (Glasgow, 1923) [1881]
—, *Treasure Island* (London, New York, etc., 1883)
Stjerna, Knut, *Essays on Questions connected with the Old English poem of Beowulf*, trans. John R. Clark Hall (Coventry & London, 1912)
Stokes, Whitley, *The Martyrology of Oengus the Culdee* (London, 1905)
Stone, Percy G., *Legends and Lays of the Wight* (London, 1912)
Story, Alfred T., 'The Sea-Serpent', in *Strand Magazine*, vol. 10 (London, July–December 1895) 161–71
Stow, John, *The Chronicles of England* (London, 1580)
Strachey, John St Loe, *The Adventure of Living* (London, 1922)
Stratford, Neil, *The Lewis Chessmen* (London, 1997)
Suffling, Ernest R., *History and Legends of the*

Broad District (London, c. 1891)
Swinburne, Henry, *Travels in the Two Sicilies*, 2 vols (Dublin, 1783–6)
Swinson, Arthur, *Scotch on the Rocks: The True Story of the 'Whisky Galore' Ship* (London, 1963)

Taylor, A. B., 'The Name 'St. Kilda', in *Scottish Studies*, vol. 13, part 1 (Edinburgh, 1969) 145–58
Taylor, Revd J. W., *Historical Antiquities of Fife*, 2 vols (Cupar, 1868)
Tennyson, Alfred, *The Poems of Tennyson*, ed. Christopher Ricks (London & New York, 1972)
Thackeray, William Makepeace, *Ballads and The Rose and the Ring* (London, 1885)
Thom, Valerie M., *Fair Isle: An Island Saga* (Edinburgh, 1989)
Thomson, David, *The People of the Sea: An Enthralling Journey in Search of the Celtic Seal Legends* (Paladin, 1980) [1954]
Thornber, William, *The History of Blackpool and its Neighbourhood* (Blackpool, 1985) [1837]
Thorpe, Benjamin, *Northern Mythology*, 3 vols (London, 1851–2)
Tongue, Ruth, *Somerset Folklore* (London, 1965)
Tonkin, J. C. & Row, Prescott, *Lyonesse: A Handbook for the Isles of Scilly* (St Mary's & London, 1898)
Train, Joseph, *An Historical and Statistical Account of the Isle of Man*, 2 vols (Douglas, 1845)
Trevelyan, Marie, *Folk-Lore and Folk-Stories of Wales* (London, 1909)
Trubshaw, Bob (ed.), *Explore Phantom Black Dogs* (Loughborough, 2005)
Tugwell, George, *The North-Devon Scenery-Book* (London & Ilfracombe, 1863)
Turnbull, Ronald, *Sandstone and Sea Stacks* (London, 2011)

Udal, John Symonds, *Dorsetshire Folk-Lore* (Exeter, 1989) [1922]

Vaughan, Robert, *British Antiquities revived* (Oxford, 1662)
Vaughan Williams, R., & Lloyd, A. L. (eds), *The Penguin Book of English Folk Songs* (Harmondsworth, etc., 1959)
Verne, Jules, *The Green Ray*, trans. Margaret de Hauteville (London, 1883) [1882]
'Verstegan, Richard' [Rowlands, Richard], *A Restitution of Decayed Intelligence* (Antwerp, 1605)
Vining, Elizabeth Gray, *Flora MacDonald: Her Life in the Highlands and America* (London, 1967)
Virgil, *The Georgics*, trans. L. P. Wilkinson (London, etc., 1982)
Vox Piscis: Or, The Bookfish Contayning Three Treatises which were found in the belly of a Cod-fish in Cambridge Market, on Midsummer Eve last, Anno Domini 1626 (London, 1627)

Wade-Evans, A. W., *Vitae Sanctorum Britanniae et Genealogiae* (Cardiff, 1944)
Walcott, Mackenzie E. C., *The East Coast of England from the Thames to the Tweed* (London, 1861)
Waldron, George, *A Description of the Isle of Man* (Douglas, 1865) [1726]
Wales, Tony, *A Sussex Garland* (London, 1979)
Wallace, James, *An Account of the Islands of Orkney* (London, 1700)
Walsh, John Evangelist, *Unraveling Piltdown* (New York & Toronto, 1996)
Waring, Edward, *Ghosts and Legends of the Dorset Countryside* (Tisbury, 1977)
Warwick, C. H., *The Hidden Room: The Legend of Chambercombe Manor* (Ilfracombe, 1987)
Waterton, Charles, *Wanderings in South America*, 4th edn (London, 1839)
Watson, George, *John Gow the Orkney Pirate* (Wick, 1978)
Watson, William J., *The History of the Celtic Place-Names of Scotland* (Edinburgh, 2005) [1926]
de Watteville, Alastair, *The Island of Staffa* (Romsey, 1993)
[Welch, John,] *A Six Day's Tour through the Isle of Man; by a Stranger* (Douglas, 1836)
Welham, M. G. & J. A., *Frogman Spy* (London, 1990)
Wells, Gerard, *Naval Customs and Traditions* (London, 1930)
Wendover, Roger of, *Flowers of History*, trans. J. A. Giles (Felinfach, 1993) [13th century]
Wernerian Natural History Society, *Memoirs*, vol. 1, 1808–10 (Edinburgh, 1811)
West, Lieutenant-Commander Frank, *Lifeboat Number Seven* (London, 1960)
Westropp, Thomas J., 'Brasil and the Legendary Islands of the North Atlantic', in *Proceedings of the Royal Irish Academy*,

vol. 30 (1912–13) section C, 223–60
–, 'A Study of the Folklore on the Coast of Connacht, Ireland', in *Folk-Lore*, vol. 29 (London, 1918) 305–19
Westwood, Jennifer, *Albion* (London, 1985)
–, *Gothick Norfolk* (Princes Risborough, 1989)
–, 'Lost Atlantis', in Judith Flanders (ed.), *Mysteries of the Ancient World* (London, 1999) 128–37
Westwood, Jennifer, & Kingshill, Sophia, *The Lore of Scotland* (London, 2009)
Westwood, Jennifer, & Simpson, Jacqueline, *The Lore of the Land* (London, 2005)
Whall, W. B., *Sea Songs and Shanties* (Glasgow, 1927)
Wheeler, R. E. M. & T. V., *Report on the Excavation of the Prehistoric, Roman, and Post-Roman Site in Lydney Park, Gloucestershire* (Oxford, 1932)
Whitfield, H. J., *Scilly and Its Legends* (Penzance & London, 1852)
Wild, Robert, *Iter Boreale* (London, 1674)
Wilde, Lady, *Ancient Legends, Mystic Charms, and Superstitions of Ireland*, 2 vols (London, 1887)
Wilkie, James, *Bygone Fife* (Edinburgh & London, 1931)
Williams, David, *John Evans and the Legend of Madoc* (Cardiff, 1963)
Williams, John, *An Enquiry into the Truth of the Tradition, concerning the Discovery of America, by Prince Madog ab Owen Gwynedd* (London, 1791)
Williams, Thomas, *Life of Sir James Nicholas Douglass* (London, etc., 1900)
Williams, Wendy, *Kraken* (New York, 2011)
Williamson, Craig (trans & notes), *A Feast of Creatures: Anglo-Saxon Riddle-Songs* (London, 1983) [1982]
Wilson, Daniel, *The Archaeology and Prehistoric Annals of Scotland* (Edinburgh, 1851)
Wood, J. G., *Out of Doors* (London, 1874)
Wood-Martin, W. G., *Pagan Ireland: An Archaeological Sketch* (London & New York, 1895)
–, *Traces of the Elder Faiths of Ireland*, 2 vols (London, New York & Bombay, 1902)
Woodford, Cecile, *Sussex Ways and Byways* (Willingdon, 1968)
Woodman, Pardoe, & Stead, Estelle, 'Communicated by W. T. Stead', in *The Blue Island: Experiences of a New Arrival beyond the Veil* (London, 1922)
Wordsworth, William, *A Guide through the District of the Lakes in the North of England*, 5th edn (Kendal, 1853)
Worsley, Richard, *The History of the Isle of Wight* (London, 1781)
Worsley, Roger, *Open Secrets* (Llandysul, 1987)
The Wreckers' Haunt: A Legend of Chambercombe (no author) (Ilfracombe, 1898)
Wright, Elizabeth Mary, *Rustic Speech and Folk-Lore* (Oxford, etc., 1913)
Wright, G. N., *Scenes in North Wales* (London, 1833)
Wright, Louis B., (ed.), *A Voyage to Virginia in 1609* (Charlottesville, 1965), incorporating Silvester Jourdain, *A Discovery of the Bermudas, Otherwise Called the Isle of Devils* [1610] and William Strachey, *A True Reportory of the Wreck and Redemption of Sir Thomas Gates* . . . [1610]
Wyndham-Quin, W. H., *Dunraven Castle* (London, 1926)

Newspapers and periodicals:

Aberdeen Chronicle (Aberdeen)
Athenaeum (London)
Béaloideas (Dublin)
Colchester Gazette (Colchester)
Daily Express (London)
Daily Telegraph (London)
Dear Mr Thoms . . . (Stockport & Glasgow, 1986–2002) [privately produced newsletter of the British Folk Studies Forum, later as *Letters to Ambrose Merton*]
East Anglian Daily Times (Ipswich)
Eastern Daily Press (Norwich)
FLS News (London) [newsletter of the Folklore Society]
Folk-lore Journal (London)
Gentleman's Magazine (London)
Illustrated London News (London)
Land and Water (London)
Leisure Hour (London)
Liverpool Echo (Liverpool)
Manchester Guardian (Manchester)
Mirror of Literature, Amusement, and Instruction (London)
Notes and Queries (London)
Pembroke County Guardian (Haverfordwest)
Shields Hustler (North Shields)
The Times (London)
Articles from the jounal *Folklore* can be accessed via the website *www.tandfonline.com*

REFERENCES

County systems have changed so often that it can be hard to keep track. In the case of England, we have used the traditional county names rather than the up-to-date administrative areas.

For the Republic of Ireland, I have used the name 'Eire' in order to avoid any wordy formulation distinguishing between Northern Ireland and the northern counties of the Republic.

Spellings of place-names are a minefield. Where there is an anglicised name, I have generally used that without adding the Gaelic or Welsh, and I hope this will give no serious offence. A few personal names too are problematic: St Brighid of Ireland, for example, appears in some Scottish sources as Brigid. As she is an Irish saint, I have gone for the Irish spelling.

Finally, neither Shakespeare nor any other playwright is cited in bibliography or references, but where quotes from plays appear, the scene and act are given in the text, which I hope will be enough for anyone who wants to track them down.

Abbreviations:

EB	*Encyclopaedia Britannica*
LFWI	Lancashire Federation of Women's Institutes
N&Q	*Notes and Queries*
OED	*Oxford English Dictionary*
PSAS	*Proceedings of the Society of Antiquaries of Scotland*
WNHS	Wernerian Natural History Society

SOUTH-WEST ENGLAND & CHANNEL ISLANDS

ABBOTSBURY, DORSET Doel (2007) 29–32; Hole (1976) 77; Hutchins (1861) vol. 2, 724–5; Hutton (1996) 241; Robson (1993) 155–66; Udal (1989) 104. Information from Diana Leake, Bridport Library

BIDEFORD, DEVON Baring-Gould (1899) vol. 1, 142ff; Hakluyt (1972) 355–60; Stevenson (1923) *passim*; Tennyson (1972) 1241–5

BODMIN, CORNWALL Bassett (1885) 66, 487–8; Gervase of Tilbury (2002) 78–83; Halliwell (1852) 163; Jones (1880) 2–4; Lowes (1951) 272; Stow (1580) 1211

BOSCASTLE, CORNWALL Courtney (1890) 106–7; Gilbert (1817–20) vol. 2, 576–7; Hawker (1869) 3–7; Hunt (1903) 438–9; Jones (1880) 101–4; N&Q series 1, vol. 11 (1855) 176. Information from Christina Carson, St Just Library; Cornish Studies Library; Rebecca David, Boscastle Tourist Information Centre; Rev. Robert Thewsey, Forrabury church

BOULEY BAY, JERSEY Ahier (c. 1955) part 2, 152–3; L'Amy (1927) 116–17; Bois (1996) 38

BRECQHOU, CHANNEL ISLANDS Baker (1979) 126–7; Bryan (1942) 189; Carrington (1961) 55–8; Hugo (1895) vol. 4, 12–33; Denys de Montfort (1802–5) vol. 2, 256–385

BRISTOL van Linschoten (1974) 89–90; Lowes (1951) 272, 512; Norman (1931) 70; Pliny (1855–7) vol. 6, 2–3; Sandys (1632) 109–10

BRIXHAM, DEVON Baker (1979) 36; Bassett (1885) 433, 463; Beckett (c. 1932) 23–4; Gutch (1901) 47; Lovett (1925) 54–5; Opie & Tatem (1992) 92; Porter (1974) 34–5
BROCKWEIR, GLOUCESTERSHIRE Benwell & Waugh (1961) 260–1
BUDE, CORNWALL Baring-Gould (1876) 17–21; Hawker (1908) 214–15
CAWSAND, DEVON Baring-Gould (1899) vol. 1, 366–7
CHAMBERCOMBE MANOR, ILFRACOMBE, DEVON *Chambercombe Olde Manor Farm* (c. 1998) *passim*; *The History . . . of Chambercombe Manor* (c. 2001) *passim*; *Leisure Hour* (1865) 657–61, 673–7, 689–94, 705–8, 721–5, 737–42, 753–6, 769–74; Page (1895) 81–3; Tugwell (1863) 174–85; Warwick (1987) *passim*; *The Wreckers' Haunt* (1898) *passim*; *www.ancestry.com/mbexec/message/an/surnames.oatway/4*
FOWEY, CORNWALL Carew (1602) 134–5; Child (1957) vol. 5, 131–2; [Lewkenor] (1693) 57; Norden (1728) 54–5; Norman (1931) 432; Pennant (1776) vol. 3, 193–4
GWENVOR SANDS, CORNWALL Bottrell (1870–80) vol. 1, 274–7, vol. 3, 121, 168; Courtney (1890) 73–5; Hunt (1903) 132–46; Wright (1913) 196
KEYNSHAM, SOMERSET Bassett (1982) *passim*; Collinson (1791) vol. 2, 400–401
KYNANCE COVE, CORNWALL Hunt (1903) 326–7; Murray's (1856) 226
LAND'S END, CORNWALL Carew (1602) 3; Courtney (1890) 66–7; North (1957) 179–80; Westropp (1912–13) 238; Westwood & Simpson (2005) 114–15; Whitfield (1852) 16–24
LIHOU, CHANNEL ISLANDS Bassett (1885) 82; Hugo (1895) vol. 1, 8; MacCulloch (1903) 143–7
LULWORTH COVE, DORSET Dacombe (1935) 19–20; Hardy (1894) 205–16; Hardy (1965) 391–2; Perkins & Pentin (1907) 16–17, 180–81, 189–90; Waring (1977) 79–80
LUNDY, DEVON Murray's (1856) 135; Page (1895) 229, 236–7
LYDNEY, GLOUCESTERSHIRE Bathurst (1879) *passim*; Ó hÓgáin (1999) 136–7; Rhys (1901) vol. 2, 445–8; Wheeler (1932) 1–44, 132–7
LYME REGIS, DORSET Jones (1880) 159–60; Paine (1997) 466; Wright (1965) *passim*
MINEHEAD, SOMERSET Dunton (1710) 351–60; Hancock (1903) 394–402; Jones (1880) 68–9; Poole (1877) 101–2; Scott (1813) 63, 320–31; Tongue (1965) 84
MORWENSTOW, CORNWALL Baring-Gould (1876) *passim*; Hawker (1908) 32–45; Murray's (1856) 252–3
MOUNT'S BAY, CORNWALL Bottrell (1870–80) vol. 1, 209–10; Brown (1972) 80–81; Halliwell (1861) 58–9; Hunt (1903) 256–8, 288–90
MUDEFORD, DORSET Boase (1976) 141; Evans (1988) 32; Hole (1976) 28–9, 167–8. Information from Rev. Helen Griffiss, All Saints church
PADSTOW, CORNWALL Gilbert (1838) vol. 3, 280; Halliwell (1861) 201–2; Hunt (1903) 151–2; Norden (1728) 68
PATERNOSTERS, JERSEY Ahier (c. 1955) part 1, 11, 65–80; L'Amy (1927) 43–56; Bisson (1950) 48; Bois (1996) 6, 21; Hawkes (1995) 56
LA PERELLE BAY, GUERNSEY MacCulloch (1903) 380–2; Thorpe (1851–2) vol. 1, 179–80
PLYMOUTH, DEVON Baker (1979) 71; Bray (1838) vol. 2, 171–5; Briggs & Tongue (1965) 94–5; Fitz-Geffrey (1596) [n. p.]; Murray's (1856) 99
PORLOCK, SOMERSET Armstrong (1970) 211–24; Emerson (1889) 67–71, 154; Greenoak (1979) 29–30, 160; Marryat (1841) 230–5; Opie & Tatem (1992) 345–6; Roeder (1904) 108; Tongue (1965) 48
LE PORT DU CREUX, SARK Falle (1694) 24–7; Hawkes (1995) 91; Lane (c. 1880) 24–7; MacCulloch (1903) 467–73; Poingdestre (1889) 84
PORTLAND, DORSET [Anon] (1861) 396–7; Doel (2007) 72–3; Holinshed (1586) vol. 3, part 1, 645; Hutchins (1861) vol. 2, 829; Newland & North (2002) 21; Waring (1977) 2–3, 76–8
PORTLEMOUTH, DEVON Page (1895) 319–20
PRUSSIA COVE, CORNWALL Carter (1894) *passim*; Courtney (1890) 87–9; Hunt (1903) 452; Murray's (1856) 232
ROCQUAINE BAY, GUERNSEY Chris Denton, personal communication (14.11.2010); Gill (1932) 46–8; Verne (1883) *passim*
LA ROCQUE, JERSEY Bois (1996) 5; Marshall (1978) 11; Poingdestre (1889) 75
ST AGNES, SCILLY ISLES Leland (1907–10) vol. 1, 190; Tonkin & Row (1898) 97–8; Westwood & Simpson (2005) 98–9; Whitfield (1852) 173–5
ST JUST, CORNWALL Bottrell (1870–80) vol. 1, 247–9
ST MARTIN'S POINT, GUERNSEY Ahier (c. 1955) part 2, 201–4; Bisson (1950) 12–13; Bois (1996) 36–9; Falle (1694) 77;

MacCulloch (1903) 427, 429–30; Poingdestre (1889) 76
ST OUEN'S BAY, JERSEY Ahier (c. 1955) part 3, 233–6; L'Amy (1927) 42–3; Bois (1996) 16; Jones (1880) 104–5; Spelman (1698) 159–60
SHEEPSTOR, DEVON Coxhead (1967) 33–4; Devonshire Association (1898) 92–3; Wells (1930) 122–3
STEART, SOMERSET Briggs & Tongue (1965) 54–5
STUDLAND HEATH, DORSET Collman (1975) 15–21; Darton (1922) 18–19; Malory (1933) 388; Waring (1977) 78–9
TORQUAY, DEVON Jones (1979) 40; Savile (1874) 199–212
TOTNES, DEVON Drayton, vol. 2 (1612–22) 10–11, 20–21; *Folk-Lore Journal* (1885) 79; Halliwell (1861) 3–11; Milton (1670) 4–14
TRESCO, SCILLY ISLES Halliwell (1861) 243; Murray's (1856) 190; Putley (2003) *passim*; Westwood & Kingshill (2009) 13; Whitfield (1852) 119–33
VAZON BAY, GUERNSEY Falle (1694) 19; Hugo (1895) vol. 1, 21–3; MacCulloch (1903) 204–7, 221–2, 450–66, 550–7; *Reader's Digest* (1977) 192
WATCHET, SOMERSET Armstrong (1970) 214–5; Bassett (1885) 449; Coleridge (1925) 95–110, Evans (2000) 20; *Funk & Wagnalls* (1972) 33; Jones (1880) 12; Lawrence (1970) 106; Lowes (1951) *passim*; Martyn (1975) 33; Melville (1992) 165; Poole (1877) 122–3; Sandford (1888) vol. 1, 247–9; Shelvocke (1971) 72–3; Tongue (1965) 212–13
WELCOMBE, DEVON Baring-Gould (1876) 107–13; Hawker (1908) 125–38, 301–4
WESTERN ROCKS, SCILLY ISLES Bottrell (1870–80) vol. 2, 229–3; Stevens & Cumming (2009) 12; Williams (1900) 6–17
WESTON-SUPER-MARE, SOMERSET Anson (1965) 111; Brockie (1886) 121; Campbell (1956) 44; *Choice Notes* (1859) 175, 270–71; Gill (1932) 137; Gregor (1885) 180–85, 199; Jones (1880) 117; Tongue (1965) 149–50
WEYMOUTH, DORSET Baker (1979) 36–8, 67–9; Beckett (c. 1932) 60; MacDougall (1991) 176–80; Maple (1970) 2571
WIMBORNE MINSTER, DORSET Coxe (1984) 61–71; Doel (2007) 64–5; Forbes (1909) 55–6; Roberts (1856) 374–5
ZENNOR, CORNWALL Benwell & Waugh (1961) 127–39; Bottrell (1870–80) vol. 2, 288–9; Stanley (1958) 129–31

SOUTH-EAST ENGLAND

ALTON, HAMPSHIRE Baker (1979) 166–7; Boase (1976) 172–3; Bryant (1972) 168–9; Campbell (1956) 147; Evans (1988) 36; King-Hall (1926) 25
ARUNDEL, WEST SUSSEX Bassett (1885) 270–71; Browne (1650) 104–5; Douglas (1974) 77; Greenoak (1979) 195–7; Ingersoll (1923) 22; Lovett (1928) 26; Opie & Tatem (1992) 216–7; Pliny (1855–7) vol. 1, 76, vol. 2, 512–13; Radford (1961) 209; Wild (1674) 19
BERMONDSEY, LONDON Savile (1874) 246–52
BOSHAM, WEST SUSSEX Chandler (1968) 40; Cook (1920) 33–4; Moore (1976) 11; Simpson (2009) 18–20
BRADING, ISLE OF WIGHT Adams (1856) 93–4; Boase (1976) 103; Elder (1839) 111–155; Garle (1905) 17–18; Hargrove (1913) 30, 167; Searle (1998) 32–4; Stone (1912) 4; Worsley (1781) 195–7
CHATHAM, KENT MacDougall (1991) 163–9; Pepys (1848–9) vol. 1, 206
CHICHESTER, WEST SUSSEX Hardwick (1967) 13–28; Hutton (1960) *passim*; *The Times* 1.1.1987, 2; Welham (1990) *passim*
COWES, ISLE OF WIGHT Baker (1979) 39–44; Campbell (1956) 34; Heath (1975) 141–58; Shewan (1927) 78–83; Wells (1930) 93
CUCKMERE HAVEN, EAST SUSSEX Candlin (1985) 32; Cook (1920) 170–75; Lucas (1935) 308; Moore (1976) 38–9; Simpson (2009) 162; Westwood & Simpson (2005) 783
DEVIL'S DYKE, WEST SUSSEX Belloc (1912) 31–42; Candlin (1985) 36–7; Moore (1976) 12–16; Simpson (2009) 59–61
DOVER, KENT Bassett (1917) 349–53; *Funk & Wagnalls* (1972) 1083; Jal (1832) vol. 2, 106–11; Mitchell (1884) 209; Thorpe (1851–2) vol. 1, 38–9
DUNNOSE POINT, ISLE OF WIGHT Beresford (1914) vol. 1, 155–6; MacDougall (1991) 7–8, 121–9; Phillips (1988) 93–117
FOLKESTONE, KENT Anson (1965) 109–10; Armstrong (1970) 217–24; Bassett (1885) 126; Buckland (1860) 285–7; Greenoak (1979) 133–6
GOODWIN SANDS, KENT Armstrong (1963) 43–9; Bignell (1983) 61–4; Brown (1972) 69–75; Doel (2009) 55–6; Grose (1811) 74; Kemp (1988) 348–9; Lambarde (1970) 94–7; Westwood & Simpson (2005) 380, 382–3, 387–8
HAWKHURST, KENT Banks (c. 1873) 96–123, Doel (2009) 49–50; *English's* (1885) 64–7; Forbes (1909) 47–54; MacDougall (1991) 134–7; Quinn (1999) 79–98

EAST ANGLIA

REEDHAM, NORFOLK Baring-Gould (1899) vol. 1, 135–8; 'Verstegan' (1605) 158–60; Walcott (1861) 313; Wendover (1993) 193–7; Westwood (1989) 31–2

SCROBY SANDS, NORFOLK Suffling (c. 1891) 191–2

SHERINGHAM, NORFOLK N&Q series 3, vol. 5 (1864) 236–7; Rye (1892) 88–9; Rye (1897) 66–7; Westwood (1985) 145–9; Westwood (1989) 34–5; Westwood & Simpson (2005) 500–501, 511–14

SUTTON HOO, SUFFOLK Bruce-Mitford (1952) *passim*; Clark Hall (1911) 9–11; Evans (1994) *passim*; Stjerna (1912) 97–132

THORPE-LE-SOKEN, ESSEX Quinn (1999) 169–70; Walcott (1861) 23

THORPENESS, SUFFOLK Dinsdale (1966) 87–9; *Eastern Daily Press* 24.7.1912, 5–6, 25.7.1912, 5; Harrison (2001) 63–9; Heuvelmans (1968) 388, 435–6

NORTH-EAST ENGLAND

BAMBURGH, NORTHUMBERLAND Child (1957) vol. 1, 306–13; Denham (1892–5) vol. 2, 331–2; Hutchinson (1778) vol. 2, 162–4; Murray's (1864) 224–5; Service (1834) 147–60; Walcott (1861) 358–9; Westwood & Simpson (2005) 547

BISHOPWEARMOUTH, TYNE AND WEAR Brockie (1886) 214

BOSTON, LINCOLNSHIRE Douglas (1974) 69–70; Dyer (1889) 305; *Fenland Notes & Queries* (1889–91) 206; Greenoak (1979) 32–4; Gutch & Peacock (1908) 148–50; Jonsson (1978) 34–5; Milton (2005) 111; *OED* (2001) vol. 3, 936; Virgil (1982) 68

CRIMDON DENE, COUNTY DURHAM Forbes (1909) 32–4; Gill (1932) 87

CULLERCOATS, TYNE AND WEAR Anson (1965) 108; Burton (1975) 143; Fordyce (1876) vol. 3, 243; Harrison (2001) 51; Heuvelmans (1968) 84–5, 235; *Illustrated London News* vol. 14 (1849) 326; Martin (1703) 143; Thomson (1980) 26, 118–19

DUNSTANBURGH CASTLE, NORTHUMBERLAND Denham (1892–5) vol. 2, 260–61; Service (1834) vi, 12, 13–26; Walcott (1861) 362

FARNE ISLANDS, NORTHUMBERLAND Colgrave (1940) 94–7, 214–15; Murray's (1864) 214; Raine (1828) 21–5; Service (1834) 53; Simpson (1991) 42; Walcott (1861) 375

FILEY BRIG, NORTH YORKSHIRE Anson (1965) 24; Bassett (1885) 260; *Brewer's* (1981) 518–9; Brockie (1886) 137–8; *Funk & Wagnalls* (1972) 471; Gutch (1901) 18, 73–4; Hone (1830) vol. 3, part 2, 637–8; Jones (1880) 52–3; [Lewkenor] (1693) 57; N&Q series 4, vol. 6 (1870) 394–5; Pennant (1776) vol. 3, 193–4; Walcott (1861) 178

GUNTHORPE, LINCOLNSHIRE Bassett (1885) 52; *Brewer's* (1981) 788; Forby (1830) vol. 2, 233; Gunn (1849) 297; Gutch & Peacock (1908) 17; Rudkin (1973) 14–15

HARTLEPOOL, COUNTY DURHAM Forbes (1909) 59–63; Hobbs & Cornwell (1994) *passim*; Simpson (1991) 11; Westwood & Kingshill (2009) 302

HAUXLEY, NORTHUMBERLAND Beck 1973) 360–61; Denham (1892–5) vol. 2, 360–61; Gregor (1885) 53; Holdsworth (1874) 114

HESLEDEN, COUNTY DURHAM Bassett (1885) 29; Brockie (1886) 200–201; *Choice Notes* (1859) 164; Dickens (1985) chap. 30, 506–7; *LFWI* (1971) 49–50; Pliny (1855–7) vol. 1, 128; Sharpe (1834) 12; Thomson (1980) 101

HULL, EAST YORKSHIRE Davis (1991) *passim*; Ellis (1982) 110; Hardwick (1967) 223–6; Hone (1830) vol. 3, part 1, 629–31; Jones (1871) 70–88; Metford (1983) 262; Rogers (1929) 134–58;

LINDISFARNE, NORTHUMBERLAND Brockie (1886) 157–8; Simeon of Durham (1855) 654–87; Simpson (1991) 50–52

LONGSTONE LIGHTHOUSE, NORTHUMBERLAND Fordyce (1876) vol. 3, 97–9, 172–3; Simpson (1991) 43–4

NORTH SHIELDS, TYNE AND WEAR Gaskell (1880) 218–19; Hugill (1969) 13; Macquoid (1883) 318–23; Rogers (2007) 1–4; *Shields Hustler*, no. 83 (1929) 19; Simpson (1991) 23–4

ROBIN HOOD'S BAY, NORTH YORKSHIRE Charlton (1779) 146–7; Child (1957) vol. 3, 211–13; Gutch (1901) 446; Ingledew (1860) 48–51; *Official Guide to Whitby* (1931) 132–4

SALTBURN-BY-THE-SEA, NORTH YORKSHIRE Gutch (1901) 44–5; Jones (1880) 64; Nichols (1846–58) vol. 2, 415–16

SCARBOROUGH, NORTH YORKSHIRE Gutch (1901) 52; Schofield (c. 1787) 62–3

SKINNINGROVE, NORTH YORKSHIRE Gutch (1901) 45; Nichols (1846–58) vol. 2, 416; Walcott (1861) 252

STAITHES, NORTH YORKSHIRE Anson (1965) 12; Gutch (1901) 49, 103–4; Kemp (1988) 199–203; Opie & Tatem (1992) 35; Simpson (1991) 7; Walcott (1861) 238, 245

WHITBY, NORTH YORKSHIRE *Athenaeum*, no. 3766 (1899) 899; Bassett (1982) 3–7; Camden (1695) 751; Charlton (1779) 32, 353–4; Gutch (1901) 71; Opie & Tatem (1992) 363–4

YORK, NORTH YORKSHIRE Blakeborough (1898) 269–73; Gutch (1901) 367–74; Halliwell (1851) 129–38; Ingledew (1860) 193–202; Macquoid (1883) 22–7

NORTH-WEST ENGLAND & ISLE OF MAN

BLACKPOOL, LANCASHIRE Bassett (1885) 478; Thornber (1985) 342–3

BOWNESS-ON-SOLWAY, CUMBRIA Rollinson (1997) 18; Rowling (1976) 83. Information from Reverend Peter Blackett, St Michael's church; Jessie McKerrow, Middlebie church; Reverend Hugh Steele, Dornock parish church; Reverend Ronald Seaman, formerly Dornock parish church, and Jeanette Seaman

BRIGHAM, CUMBRIA [Barrow] (1831) *passim*; Bligh (1853) *passim*; Byron (1823) 14; Rollinson (1997) 28–9

CHESTER, CHESHIRE Hole (1937) 188–9; Hughes (1856) 31–2

RIVER DEE, CHESHIRE Armstrong (1963) 65–73; Davies (1911) 169–73; MacDougall (1991) 54–62

DRIGG, CUMBRIA Bett (1950) 142; Lane (1970) 161–7; Linder (1987) 25–7; Potter (1952) 85–96; Rollinson (1997) 79; Rowling (1976) 141; Westwood & Simpson (2005) 131

KING WILLIAM'S BANKS, ISLE OF MAN Blundell (1876–7) vol. 2, 24–5; Gill (1932) 448–50; MacTaggart (1824) 307–8; N&Q series 10, vol. 5 (1906) 126; Waldron (1865) 76; Welch (1836) 128–9

KIRK MICHAEL, ISLE OF MAN Blundell (1876–7) vol. 1, 52, vol. 2, 3; Bullock (1816) 368; Gill (1932) 158, 192–4, 465; Harrison (1873) 15; Rhys (1901) vol. 1, 298–301; Roeder (1904) 38

LAXEY, ISLE OF MAN Broome (1963) 53–7; Gill (1929) 351–2; Moore (1891) 50–51; Train (1845) 147–8; Waldron (1865) 67–8

LIVERPOOL, MERSEYSIDE Armstrong (1963) 151; Baker (1979) 149–50; *Daily Telegraph* 7.7.1959; Godden (1969) 275; *Liverpool Echo* 7–8.7.1959; Maple (1970) 2571; Radford (1961) 14–15

MAUGHOLD HEAD, ISLE OF MAN Gill (1929) 385–6; Nelson (1839) 73–83

MILLOM, CUMBRIA Nicholson (1966) 33; Rollinson (1997) 24, 181–2

MORECAMBE, LANCASHIRE Jobson (1959) 147; *Manchester Guardian* 13.10.1956, 5; Morris (1998) 159; Opie & Tatem (1992) 407; Porter (1974) 49; Trevelyan (1909) 320; Wilde (1887) vol. 2, 106

NETHERTON, CHESHIRE *Cheshire Sheaf* (1878–83) vol. 1, 4–5, 13, 23; Dickens (1988) 545

NIARBYL, ISLE OF MAN Gill (1929) 432–3; Waldron (1865) 65–6

OVER, CHESHIRE Hole (1937) 193

PEEL, ISLE OF MAN Gill (1929) 240–44, 323–4; Gill (1932) 96; Scott (1823) vol. 2, 184–5; Train (1845) vol. 1, 278–9, 300–301; Waldron (1865) 12–13

PORT ERIN, ISLE OF MAN Waldron (1865) 52–7

PRIEST SKEAR, LANCASHIRE *LFWI* (1971) 12

RAVENGLASS, CUMBRIA Camden (1695) 820; Hewett (1976) 13; Opie & Tatem (1992) 301; Pliny (1855–7) vol. 2, 431; Roud (2003) 355

ROSSALL POINT, LANCASHIRE Ashton (1909) 35–42; Porter (1876) 23–5; Roberts (1931) 47–50; Thornber (1985) 54–7

SPANISH HEAD, ISLE OF MAN Gill (1929) 497–500; Harrison (1873) 250–51; Moore (1891) 37; Train (1845) vol. 2, 79, 143–4

WALLASEY, MERSEYSIDE Hole (1937) 56; Milton (1670) 272–3

WALNEY ISLAND, CUMBRIA Rollinson (1997) 189–90

WORKINGTON, CUMBRIA Dasent (1894) 6; Davidson (1970) 179–80; *EB* (1910–11) vol. 10, 617–18; Fournier (2009) *passim*; McIntire (1948) 213; Marples (1970) *passim*; Rowling (1976) 115

WALES

ABERYSTWYTH, CEREDIGION Davies (1911) 145–7

BARDSEY ISLAND, GWYNEDD Fuller (1662) vol. 4, 29; Gerald of Wales (1806) vol. 2, 83; Morris (1998) 89; Pennant (1778–83) vol. 2, 196–7. Information from Tristan Gray Hulse

BARRY ISLAND, VALE OF GLAMORGAN Jones (1979) 233–4; Rees (1853) 357–8; Sikes (1880) 352–3; Wade-Evans (1944) 91–3

CAER ARIANRHOD, GWYNEDD Condry (1972) 45; Croker (1825–8) vol. 3, 175–6; *Mabinogion* (1987) 97–117; North (1957) 213–33; Rhys (1901) vol. 1, 207–19

CARDIGAN BAY, CEREDIGION Davies (1993) 4–5; Gerald of Wales (1806) vol. 1, 217; Meyrick (1810) 50–51, 72–4; Morganwg (1977) 34–5; Morris (1998) 26–8; North (1957) *passim*; Owen (1803) 314–15; Pennant (1778–83) vol. 2, 335; Prichard (1824) vi, 1–139; Rees (1836) 234–5; Rhys (1901) vol. 1, 382–9, vol. 2, 412–17; Skene (1868) vol. 1, 302, vol. 2, 352; Vaughan (1662) 41

CEMAES HEAD, PEMBROKESHIRE Davies (1911) 90–91, 130–32; Jones (1979) 53; Morris (1998) 87–8;

CONWY, CONWY Hughes & Evans (1995) 53–4; Rhys (1901) vol. 1, 198–9; Trevelyan (1909) 331–2

CONWY BAY, CONWY Ashton (1909) 172–89; Costello (1845) 54–5; Edwards (1914) 48; North (1957) *passim*; Rhys (1901) vol. 2, 414–15; Trevelyan (1909) 119

FISHGUARD, PEMBROKESHIRE Davies (1911) 218–19; Miller (1835) 108–9; *Pembrokeshire Antiquities* (1897) 59–60

GRASSHOLM ISLAND, PEMBROKESHIRE Gill (1932) 449; *Pembroke County Guardian* 1.11.1896, 5; Rhys (1901) vol. 1, 171–3; Turnbull (2011) 233; Wordsworth (1853) 108–10

LINNEY HEAD, PEMBROKESHIRE Benwell & Waugh (1961) 108–10; Morgan (1795) 302–6; N&Q series 2, vol. 9 (1860) 360–61

LLANDDWYN ISLAND, ISLE OF ANGLESEY Rees (1853) 299; Rhys (1901) vol. 1, 219

LLANGWYRYFON, CEREDIGION Geoffrey of Monmouth (1908) 171–2; Holweck (1969) 1000; Meyrick (1810) 330

LLANRUMNEY, CARDIFF Beck (1973) 327; *Captain Morgan Jamaica Rum*, label; Cruikshank (1935) *passim*; Exquemelin (1969) 119–215; Pope (1977) *passim*; Rogozinski (1997) 227–9

LLIGWY, ISLE OF ANGLESEY Trevelyan (1909) 131–2

MENAI STRAIT, ISLE OF ANGLESEY Gerald of Wales (1978) 187–8; Radford (1982) 24–5; Wright (1833) 3

MILFORD HAVEN, PEMBROKESHIRE Howells (1831) 119; Morganwg (1977) 18; Sikes (1880) 8–10; Southey (1815) vol. 1, 111–12

NEWPORT Benwell & Waugh (1961) 216; *Daily Express* 12.9.1957, 1; *www.newport.gov.uk*

OYSTERMOUTH, SWANSEA Evans (2011) *passim*; Hall (1861) 343–5

PORTHCAWL, BRIDGEND Bassett (1917) 98; Morgan (1974) 102–4; Trevelyan (1909) 3–4

RHOOSE, VALE OF GLAMORGAN Trevelyan (1909) 298–9

RHOS-ON-SEA, CONWY Davies (1993) 255, 336; Hughes & Evans (1995) 90–91; Jenkins (1899) 273–7; Kemp (1988) 509; Morris (1998) 343–5; Southey (1815) *passim*, esp. vol. 1, vii–ix; Williams (1791) *passim*; Williams (1963) *passim*. Information from Katherine Fetherstonhaugh, Colwyn Bay Library

ST BRIDES, PEMBROKESHIRE Trevelyan (1909) 59–65

ST GOVAN'S HEAD, PEMBROKESHIRE N&Q series 1, vol. 6 (1852) 96–7, vol. 12 (1855) 201; Trevelyan (1909) 332–3

THE SMALLS, PEMBROKESHIRE Emlyn (1858) *passim*; Fenton (1811) 127–31; Langmaid (1966) 45–8; Worsley (1987) 77–80. Information from St David's church, Whitchurch

TENBY, PEMBROKESHIRE Davies (1911) 324; Jones (1880) 113–14

TUSKER ROCK, VALE OF GLAMORGAN Armstrong (1963) 52–5; Nicholas (1875) vol. 2, 570; Snell (1894) *passim*; Spencer (1913) 48–55; Trevelyan (1909) 119; Wyndham Quin (1926) 34–7

THE WOLVES, CARDIFF Rees (1853) 358; Wade-Evans (1944) 93

SCOTTISH LOWLANDS

ABERDEEN, ABERDEENSHIRE Baker (1979) 30–1; Jeans (2004) 97–9; Shewan (1927) 82–3

AILSA CRAIG, SOUTH AYRSHIRE Bauman (1964) 33–6; Beck (1973) 38

ARBIGLAND, DUMFRIES AND GALLOWAY Beck (1973) 320–27; Forbes (1909) 152–7; Logan (1869) 32–42; MacTaggart (1824) 373–6; Rollinson (1997) 87

BUCKHAVEN, FIFE Graham (1883) vol. 2, 217–37; Taylor (1868) vol. 2, 155–7; Wilkie (1931) 142–3

BUCKIE, MORAY N&Q series 1, vol. 11 (1855) 142; Gregor (1881) 145; McPherson (1929) 34–5; Simpson (1976) 128–9

BULLERS OF BUCHAN, ABERDEENSHIRE Buchan (1819) 32–5; McPherson (1929) 72–3; Reid (1871) 258–9

BURGHEAD, MORAY Anson (1965) 76–8; Guthrie (1889) 46; McPherson (1929) 19–22

COLDINGHAM, SCOTTISH BORDERS Colgrave (1940) 79–83, 189–91, 319–20

COLL, ARGYLL AND BUTE Hamilton (1839) 315–17; Heuvelmans (1968) 121–2; *Land and Water* vol. 14 (1872) 158; *WNHS*, vol. 1 (1811) 442–4
COLLIESTON, ABERDEENSHIRE Anson (1965) 97–8; Gregor (1881) 198; Ó hEochaidh, Mac Neill & Ó Catháin (1977) 383; Ó hÓgáin (2006) 195
COLLIN, DUMFRIES AND GALLOWAY MacTaggart (1824) 241–3; Westwood & Kingshill (2009) 142–3
COLONSAY, ARGYLL AND BUTE Benwell & Waugh (1961) 17; Campbell (1900) 116–19; Scott (1931) 650–60
CORRYVRECKAN, ARGYLL AND BUTE Beck (1973) 46–50; Martin (1703) 236–8; Watson (2005) 94, 187
CRAIL, FIFE Anson (1965) 24–5, 55; Hall (1807) vol. 1, 98–9; Schofield (*c.* 1787) 151–2; Simpkins (1914) 411
CROIG, ARGYLL AND BUTE Campbell (1900) 221; Campbell (1956) 164; Rogers (1929) 155; Southey (1815) vol. 1, 234
RIVER DEE, ABERDEENSHIRE Benwell & Waugh (1961) 100–101; Chambers (1864) vol. 2, 613; *N&Q* series 2, vol. 6 (1858) 371
DUART CASTLE, MULL Campbell (1900) 222–3; Lysaght *et al.* (1999) 359–69: 366–7
DUNFERMLINE, FIFE Scott (1931) 97–102
EDINBURGH *An Account of the Dreadful Battle . . .* (1707) *passim*; Jones (1880) 117; Smith (1968) 35–6
ELACHNAVE, ARGYLL AND BUTE Barron & Burgess (2002) 34–5, 38, 46; Buckland (1860) 289 [whales quite common to 14th century]; Gordon (1976) 252–5; Heuvelmans (1968) 49–51; Krapp & Dobbie (1936) 316; Marsden (1995) 119–20; Milton (2005) 23
EYEMOUTH, SCOTTISH BORDERS Anson (1930) 38
FIRTH OF FORTH Beveridge (1888) 200–201; Pugh (1987) 231–3
GIGHA, ARGYLL AND BUTE Martin (1703) 219–20; Wood-Martin (1902) vol. 2, 105; *http://megalithix.wordpress.com/category/scotland/argyll-bute/page/2/* (accessed 14.12.2011)
GREENOCK, INVERCLYDE Beck (1973) 327–38; Jeans (2004) 236–40; Johnson (1972) 440–51, 685–6; MacDougall (1991) 24–8; Rogozinski (1997) 57–8, 179–181
INCHKEITH, FIFE 'Kilrounie' (1882) 52–3
INVERARAY, ARGYLL AND BUTE Anson (1965) 87; Gill (1932) 136; Mason (1936) 48–51, 176–7; Moore (1891) 153; Rhys (1901) vol. 1, 344–5; Wood-Martin (1895) 110–13
IONA, ARGYLL AND BUTE Adamnan (1874) 17–18, 47–8; Boswell (1785) 419; Marsden (1995) 129–41; Severin (1996) 140–43
IRVINE, NORTH AYRSHIRE Fuller (1648) 351; Olaus Magnus (1658) 45–7; Scott (2001) 187–93
ISLAY, ARGYLL AND BUTE Campbell (1900) 116–19; Meyer & Nutt (1895–7) vol. 1, 30
KNOCKDOLIAN, SOUTH AYRSHIRE Bauman (1964) 36–9; Chambers (1870) 331–2; Cromek (1810) 229–48; Paterson (1863–6) vol. 2, 159–64
LADY'S ROCK, ARGYLL AND BUTE Beck (1973) 43–5
LISMORE, ARGYLL AND BUTE Beck (1973) 44–5; Westwood & Kingshill (2009) 12–13; Westwood & Simpson (2005) 808–9
LITTLE ROSS, DUMFRIES AND GALLOWAY MacTaggart (1824) 468–9. Information from Mairi Hunter and Ruth Airley, Ewart Library, Dumfries
LOCHAR MOSS, DUMFRIES AND GALLOWAY Chambers (1826) 8; Cromek (1810) 292–3; Davidson (1949) 8–9
LOWER LARGO, FIFE Cunningham (1907) 45–55; Defoe (1847) *passim*; Rogers (1718) 124–31; Simmons (1993) 1–29; Steele (1714–37) vol. 1, 121–4. Information from Dr David Caldwell, Edinburgh Museum
MORAY FIRTH Martin (1703) 143; Miller (1835) 282–4
NORTH BERWICK, EAST LOTHIAN Davidson (1949) 147–59; Fielding (1755) 102–4; *N&Q* series 1, vol. 10 (1854) 26; Opie & Tatem (1992) 61; Scot (1584) *passim*, esp. 57–60; Westwood & Kingshill (2009) 269–72
PETERHEAD, ABERDEENSHIRE Beck (1973) 295–6; Brockie (1886) 208
PORTESSIE, MORAY Anson (1965) 79; Gregor (1886) 12–13
PORTGORDON, MORAY *Aberdeen Chronicle* 20.8.1814, 4; McPherson (1929) 72–3
PRESTONPANS, EAST LOTHIAN Anson (1965) 111–13; Dana (1981) 260; *N&Q* series 1, vol. 5 (1852) 5
RATTRAY HEAD, ABERDEENSHIRE Bridges (1927) 244–5; Jeans (2004) 284–5
ROSEHEARTY, ABERDEENSHIRE Gregor (1885) *passim*
ROTHESAY, ARGYLL AND BUTE Byatt, Fothergill & Holmes (2001) 340–41; Carrington (1961) 47–55; Gould (1930) 22;

Heuvelmans (1968) 54–7; Pliny (1855–7) vol. 2, 420–22; Pontoppidan (1755) vol. 2, 210–13; Williams (2011) 72
ST MONANS, FIFE Jack (1844) 5–7, 35–7; Jones (1880) 98–9; Pastoureau (2009) *passim*; Rorie (1904) 96–7; Simpkins (1914) 21–5, 124–5, 216
SOLWAY FIRTH Burns (1830) 14; Cunningham (1822) vol. 2, 258–95
STAFFA, ARGYLL AND BUTE Boece (1938–41) vol. 1, 300; de Watteville (1993) *passim*
TIREE, ARGYLL AND BUTE Campbell (1900) 223; Moore (1891) 151; Severin (1996) 99
TOBERMORY, ARGYLL AND BUTE Campbell (1885) 329–33, 345–7; Campbell (1900) 242; Campbell (1902) 28; Campbell (1994) vol. 1, 24–5; Chambers (1827) vol. 2, 404–5; Hardie (1912) *passim*; Kingshill (2010) *passim*; Lang (1912) *passim*; Mackay (1876) 88–90; MacLean (1889) 109, 330–31; McLeay (1986) *passim*; Paine (1911) 192–217; 'A Seneachie' (1838) 73; *www.uisge.com/ud/tobermory*; *www.undiscoveredscotland.co.uk/mull/tobermory*
TRESHNISH ISLES, ARGYLL AND BUTE Campbell (1902) 42

SCOTTISH HIGHLANDS & ISLANDS

BARRA, WESTERN ISLES Farre (1960) 28–42; Kennedy-Fraser (1929) 189–90; Kennedy-Fraser & Macleod (1909–21) vol. 2, 146–8, Geoff Sample, personal communication (7.12.2010); Thomson (1980) 211–23
BURRA, SHETLAND Carrington (1961) 49; Hibbert (1822) 564–5; Olaus Magnus (1658) 231; Pontoppidan (1755) vol. 2, 210–13; Tennyson (1972) 246–7
CANNA, HIGHLAND Kennedy-Fraser & Macleod (1909–21) vol. 1, 15; MacCodrum (1938) xxxiv–xliv, 240; McDonald (1784) 7; MacDonald (1894) iv–vii; Henderson (1910) 260–62; Thomson (1980) 211–23
CROMARTY, HIGHLAND Miller (1854) 22–4
EDAY, ORKNEY Beck (1973) 220–21; Blind (1881) 403–9; Dana (1981) 80–81; Forester (1945) 45; Gill (1932) 284; Saxby (1932) 91–6; Wallace (1700) 60–61
EILEAN FLADDAY, HIGHLAND Campbell (1902) 215–16
EILEAN LEATHAN, WESTERN ISLES Dorson (1971) 174; McKay (1940–60) vol. 1, 394–409; MacPherson (1960) 81–91
ERISKAY, WESTERN ISLES MacPherson (1960) 155–63; Swinson (1963) *passim*
EYNHALLOW, ORKNEY Gunn (1909) 391–4; Hogg (1961) 27
FAIR ISLE, SHETLAND Anson (1965) 31; Epstein (2008) 47–8; Lambert & Marx (1951) 61; MacCulloch (1903) 30–33; Mackay (1896) 100; Thom (1989) 84–8
FLANNAN ISLES, WESTERN ISLES Gibson (1926) 171–4; Hardwick (1967) 9–10; Mair (1978) 231–2
FOULA, SHETLAND Greenoak (1979) 150–52; Hibbert (1822) 586–8; Martin (1703) 283–4; Pennant (1776) vol. 2, 446–8; Peterson *et al.* (2004) 122–3
GIZZEN BRIGGS, HIGHLAND Hutchins (1968) 5, 26–7; Westwood & Kingshill (2009) 356–7
GRIMSAY, WESTERN ISLES Bruford & MacDonald (1994) 387–90, 479; Campbell (1902) 15–16; MacDonald (1894) xviii–xix
HARRIS, WESTERN ISLES Campbell (1902) 21–2; McPherson (1929) 206
LERWICK, SHETLAND Davidson (1969) 176–8; Davidson (1970) 178–9; Hole (1976) 201–2; Hutton (1996) 43; Newall (1993) 57–73. Information from Shetland Library
LOCHBOISDALE, WESTERN ISLES Beck (1973) 238–40; Child (1957) vol. 4, 360–69
LYBSTER, HIGHLAND Beck (1973) 283; Gregor (1885) 55; Ó Súilleabháin (1967) 31
MARWICK HEAD, ORKNEY Garrett (1987) 91–106; McCormick (1959) *passim*; Power (1926) *passim*; *Reader's Digest* (1977) 488–9
MILTON, WESTERN ISLES Dorson (1971) 148–53; Hogg (1903) 248–9; MacGregor (1882) *passim*; Vining (1967) 11–72
NICHOLSON'S LEAP, WESTERN ISLES Beck (1973) 45; Gerald of Wales (1978) 142–3; Lanier (1892) 33–8
NORTH RONA, WESTERN ISLES Campbell (1900) 220; Carmichael (1928–54) vol. 1, 126–7, vol. 2, 348; Marsden (1995) 145–8; Robson (1991) 2–4
NORTH RONALDSAY, ORKNEY Black (1903) 179–82; Gill (1929) 442; Thomson (1980) 139–49
PORT HENDERSON, HIGHLAND Dixon (1886) 162–3
PORTREE, HIGHLAND Campbell (1902) 44–5; Carew (1602) 31–2; Opie & Tatem (1992) 191–3
RABBIT ISLANDS, HIGHLAND Johnson (2009) 78; Opie & Tatem (1992) 191–3; Reader's Digest (1977) 73–4
ST KILDA, WESTERN ISLES Martin (1703) 280; Taylor (1969) *passim*; Watson (2005) 97–8
ST MOLVEG, WESTERN ISLES Anson (1965)

81; Campbell (1900) 244; Campbell (1902) 261–2; Carmichael (1928–54) vol. 1, 306–13; Hutton 1996) 369; Martin (1703) 28–9; Ross (1976) 162; Wood-Martin (1902) vol. 1, 376

SANDSIDE BAY, HIGHLAND Benwell & Waugh (1961) 111–13; Scott (1931) 653; *The Times* 8.9.1809, 3

SANDWOOD BAY, HIGHLAND Hutchins (1968) 26; Robertson (1948) 103–10

SCAPA FLOW, ORKNEY Baker (1979) 71; Ditmas (1973) *passim*; Hunt (1903) 231; Newbolt (1897) 11–12; Noyes (1963) 19–22; *The Times* 24.12.1964, 9

SHIANT ISLANDS, WESTERN ISLES Beck (1973) 241–2; Campbell (1900) 199–200

SOUND OF SLEAT, HIGHLAND Gould (1930) 141, 184; Heuvelmans (1968) 254–7; *Land and Water* vol. 14 (1872) 158; Olaus Magnus (1658) 235; Pontoppidan (1755) vol. 2, 195–202; Story (1895) 161–71

STROMNESS, ORKNEY Forbes (1909) 34–6; Lockhart (1837–8) vol. 2, 100–101

STRONSAY, ORKNEY Gould (1930) 239; Hamilton (1839) 313–26; Heuvelmans (1968) 114–42; WNHS vol. 1 (1811) 418–44. Information from Peter Drummond, Orkney Museum; Howie Firth, Orkney Science Museum; Clare McIntyre, National Museum of Scotland

SWILKIE, HIGHLAND Bassett (1885) 21; Lysaght *et al.*, (1999) 121–32; Morvan (1980) 67; Olaus Magnus (1658) 226; Poe (1948) 9; Thorpe (1851–2) vol. 1, 207–8

TARBAT NESS, HIGHLAND March (1998) 52–3; Miller (1835) 305–8

UIG, WESTERN ISLES MacDonald (1967) 120–23; PSAS vol. 4 (1863) 411–13; Robinson (2004) *passim*; Stratford (1997) 50–53; Wilson (1851) 567–8

VE SKERRIES, SHETLAND Hibbert (1822) 566–71; PSAS vol. 1 (1852) 87–8

WICK, HIGHLAND Fea (1912) 1–146; Johnson (1972) 358–69, 681–2; Scott (1822) vol. 1, i–vii; Watson (1978) *passim*

YELL, SHETLAND Hamilton (1839) 285–91; Hutchins (1968) 15–16

NORTHERN EIRE & NORTHERN IRELAND

ARRANMORE ISLAND, COUNTY DONEGAL Andrews (2003) 67; Hall (1841–3) vol. 3, 407–14; Lysaght *et al.* (1999) 223–45; ÓhEochaidh, Mac Neill & Ó Catháin (1977) 384; Thomson (1980) 180

BELFAST Baker (1979) 41–2; Foster (1999) *passim*; *Funk & Wagnalls* (1972) 868; Lockhart (1938) 171–9; MacDougall (1991) 11–22; Robertson (1900) *passim*, esp 1–3; Woodman & Stead (1922) 36. Information from Richard Parkinson, British Museum

RIVER BOYNE, COUNTY LOUTH Buckland (1860) 217–18; Gwynn (1903–35) vol. 4, 294–5, 454; Ó hÓgáin (2006) 151–3; Wood (1874) 326–7

CARLINGFORD, COUNTY LOUTH Gerald of Wales (1978) 65, 130–32; Ó hÓgáin (2006) 339

CASTLE GORE, COUNTY MAYO Bence-J ones (1988) 69–70; Gill (1932) 88–9; O'Sullivan (1966) 236–40, 283–4; Westropp (1918) 311

CLARE ISLAND, COUNTY MAYO AA (1932) 139, 203; Beck (1973) 332–6; O'Flaherty (1846) 376; Ó hÓgáin (2006) 374–6. Information from Julian Gaisford-St Lawrence, Howth Castle

DOWNPATRICK, COUNTY DOWN AA (1932) 97–8; Kennedy (1866) 331–2; Ó hÓgáin (1985) 34–5; Ó hÓgáin (2006) 92–3; O'Kelleher & Schoepperle (1918) 371–4

DROGHEDA, COUNTY LOUTH Kennedy (1866) 327–8; Ó hÓgáin (2006) 418–20

DRUMKIRK, COUNTY DOWN Norman (1931) 430–31

DUNDRUM BAY, COUNTY DOWN Ó hÓgáin (2006) 215–17

ERRIS HEAD, COUNTY MAYO Benwell & Waugh (1961) 17; Henderson (1910) 265–6; Lysaght *et al.* (1999) 223–45; Ó Catháin (1983) 63; Ó hÓgáin (2006) 343, 497; Thomson (1980) 77–93, 179

FAIR HEAD, COUNTY ANTRIM Foster (1951) 188; Hall (1841–3) vol. 3, 144–8

GALWAY, COUNTY GALWAY Thomson (1980) 191–6

GIANT'S CAUSEWAY, COUNTY ANTRIM Andrews (2003) 76; Hall (1841–3) vol. 3, 164–9; Kennedy (1866) 203–5

GREGORY'S SOUND, COUNTY GALWAY Ó hÓgáin (2006) 134, 279–80

HORN HEAD, COUNTY DONEGAL Ó hEochaidh, Mac Neill & Ó Catháin (1977) 240–41

INISHBOFIN, COUNTY GALWAY Beck (1973) 343–4; Gerald of Wales (1978) 66; Westropp (1912–13) 250–51; Wood-Martin (1895) 304

INISHGLORA, COUNTY MAYO Joyce (1894) x, 1–36; Kennedy (1866) 277–9; Ó hÓgáin (2006) 309–11

INISHMAAN, COUNTY GALWAY Gill (1929)

432–43; Gregory (1920) vol. 1, 7–25; O'Grady (1892) vol. 2, 72; Waldron (1865) 52–3

INISHMORE, COUNTY GALWAY Gregory (1920) vol. 1, 16, 23; Lysaght *et al.* (1999) 247–60; O'Flaherty (1846) 68–73; Ó hÓgáin (1999) 104–5; Southey (1815) vol. 1, 268–71; Westropp (1912–13) *passim*

INISHMURRAY, COUNTY SLIGO Mason (1936) 27–9; Wood-Martin (1895) 162; Wood-Martin (1902) vol. 2, 104–5

KID ISLAND, COUNTY MAYO Ó Catháin (1983) 52–6; O'Sullivan (1966) 188, 275

KINVARRA, COUNTY GALWAY Ó Catháin (1982) 107–8

LARNE WATER, COUNTY ANTRIM *Annals of the Kingdom of Ireland* (1851) vol. 1, 201–2; Benwell & Waugh (1961) 61; Joyce (1894) 97–105; O'Grady (1892) vol. 2, 265–9, vol. 2, 265–9; Ó hÓgáin (2006) 342–5

MULROY BAY, COUNTY DONEGAL Beck (1973) 363; Ó hÓgáin (1985) 32; O'Kelleher & Schoepperle (1918) 108; Severin (1996) 99–100

PORTACLOY, COUNTY MAYO Ó Catháin (1983) 46–7, 59–62

RATHLIN ISLAND, COUNTY ANTRIM Adamnan (1874) 251; *Cormac's Glossary* (1868) 41–2; Gwynn (1903–35) vol. 4, 80–87; Wilde (1887) vol. 1, 161, vol. 2, 126–7; Watson (2005) 94, 187

RATHLIN O'BIRNE ISLAND, COUNTY DONEGAL Lysaght *et al.* (1999) 247–60; Ó hEochaidh, Mac Neill & Ó Catháin (1977) 214–19

ST MACDARA'S ISLAND, COUNTY GALWAY AA (1932) 185; O'Flaherty (1846) 97–101

SCURMORE, COUNTY SLIGO Wood-Martin (1902) vol. 2, 212–13

SKERD ROCKS, COUNTY GALWAY Macpherson (1771); O'Flaherty (1846) 68–73; Southey (1815) vol. 1, 276

TEELIN, COUNTY DONEGAL Ó hEochaidh, Mac Neill & Ó Catháin (1977) 187–9, 209, 219, 383; Ó hÓgáin (2006) 206

TELTOWN, COUNTY MEATH Hone (1830) vol. 2, 278–81; Lysaght *et al.* (1999) 214; O'Flaherty (1846) 31–2; Ó hÓgáin (2006) 337–9; Wordsworth (1853) 115–16

TORY ISLAND, COUNTY DONEGAL Beck (1973) 14; Mason (1936) 16–18; Ó hEochaidh, Mac Neill & Ó Catháin (1977) 246–7, 385

SOUTHERN EIRE

ARDMORE, COUNTY WATERFORD Kenney (1993) 310–13; Ó hÓgáin (1985) 47; Ó hÓgáin (2006) 160–62, 444; Power (1914) xvi–xvii, 19–28; Stokes (1905) 201; Wood-Martin (1902) vol. 2, 233

BALLINSKELLIGS BAY, COUNTY KERRY Croker (1825–8) vol. 2, 3–20, 67–82. Information from Tadhg Ó hIfearnáin

BALLYHEIGE BAY, COUNTY KERRY Croker (1825–8) vol. 2, 21–9; Smith (1774) 210; Westropp (1912–13) 249–50

BANNOW BAY, COUNTY WEXFORD Armstrong (1970) 225–37; Gerald of Wales (1982) 41–2; Gerard (1597) vol. 2, 1391–2; Hall (1841–3) vol. 2, 154–5; Heron-Allen (1928) *passim*, esp 71, 105; Müller (1871) vol. 2, 583–604; Williamson (1983) 68

BANTRY BAY, COUNTY CORK Hall (1841–3) vol. 1, 143–4

BRANDON CREEK, COUNTY KERRY Baring-Gould (1899) vol. 1, 132–3; Barron & Burgess (2002) 1–64; Kennedy (1866) 337–40; Marsden (1995) 169–76; Ó hÓgáin (2006) 41–2; Severin (1996) *passim*

THE BULL, COUNTY KERRY Gwynn (1903–35) vol. 4, 311, 459–60; Joyce (1894) 29–30; Lysaght et al. (1999) 247–60; Ó hÓgáin (1985) 55, 152–4; Ó hÓgáin (1999) 57; Ó hÓgáin (2006) 178–80

CAHERDANIEL, COUNTY KERRY Lysaght *et al.* (1999) 37–49; Thomson (1980) 177–9

CLEAR ISLAND, COUNTY CORK Lysaght *et al.* (1999) 285–98

CORK, COUNTY CORK Jones (1880) 56; Meyer & Nutt (1895–7) vol. 1, 16–20; Ó hÓgáin (2006) 32–3, 85–6; Stokes (1905) 41

DINGLE, COUNTY KERRY *Béaloideas* vol. 2 (1930) 146–7; Lysaght *et al.* (1999) 133–48

DOONBEG BAY, COUNTY CLARE Croker (1825–8) vol. 2, 30–58; Joyce (1894) vi–vii; Kennedy (1866) 121–2

DUNKERRON CASTLE, COUNTY KERRY Benwell & Waugh (1961) 157–8; Croker (1825–8) vol. 2, 59–66; Smith (1774) 88–9

DURSEY ISLAND, COUNTY CORK Gill (1932) 281; Kennedy (1866) 121–2; Ó hÓgáin (2006) 206; Simpson (2004) 109; Wilde (1887) vol. 2, 41–5

FASTNET ROCK, COUNTY CORK Bywater (1932) 21–9; Hardwick (1967) 82–9; *en.wikipedia.org/wiki/UB65* (accessed 6.4.2012)

GALLARUS, COUNTY KERRY Croker (1825–8) vol. 2, 3–20

GLANDORE, COUNTY CORK Joyce (1870–1913) vol. 1, 188; Ó hÓgáin (2006) 85–6, 195; Wood-Martin (1895) 132–3; Wood-Martin (1902) vol. 1, 370–2

GREAT BLASKET ISLAND, COUNTY KERRY Beck (1973) 217; Thomson (1980) 26–8, 118–19

LAHINCH, COUNTY CLARE Griffin (1926) 106; Lysaght *et al.* (1999) 247–60; N&Q series 1, vol. 8 (1853) 145; Westropp (1912–13) 250–51

LAMBAY ISLAND, COUNTY DUBLIN Greenoak (1979) 28; Mason (1936) 6

MALAHIDE, COUNTY DUBLIN Bassett (1885) 352; Bassett (1917) 26–33; O'Curry (1861) 383–423

MOORESTOWN, COUNTY KERRY McKay (1940–60) vol. 1, 92–103; Ó hÓgáin (2006) 343–4, 354–7; O'Sullivan (1966) 21–37

ROCKABILL, COUNTY DUBLIN Gwynn (1903–35) vol. 3, 26–33; Meyer & Nutt (1895–7) vol. 1, 214–5; Ó hÓgáin (2006) 28–30

ROSSBEIGH STRAND, COUNTY KERRY Hogan (1993) 642; Jones (1880) 25; Lysaght *et al.* (1999) 247–60; Ó hÓgáin (2006) 195. Information from Micheál Ó Coileáin, Kerry County Councilá

SCATTERY ISLAND, COUNTY CLARE Kenney (1993) 364–6; N&Q series 4, vol. 8 (1871) 265–7; Ó hÓgáin (2006) 449–50; Ó Súilleabháin (1967) 31; Severin (1996) 75–6

TRAFRASK, COUNTY CORK Gwynn (1903–35) vol. 2, 26–35; Ó hÓgáin (2006) 76, 344

VICARSTOWN, COUNTY KERRY Beck (1971) 50–56; Beck (1973) 371–3; Smith (1774) 187–91. Information from Micheál Ó Coileáin, Kerry County Council

WEXFORD, COUNTY WEXFORD Kennedy (1867) 364; O'Reilly (1878) 97–103; Ó Súilleabháin (1967) 27, 70–71

YOUGHAL, COUNTY CORK Brewer (1825–6) vol. 2, 387–8; Westropp (1912–13) 254–60; Wood-Martin (1902) vol. 1, 217–18

ESSAY REFERENCES

ATLANTIS Bramwell (1937) *passim*; Donnelly (1882) *passim*; Friedrich (2000) 1–2, 79, 83–4; Kircher (1678) 82; Masefield (1932) 72–4; Plato (1971) *passim*; Westwood (1999) *passim*

CAULS Beck (1973) 231–2, 246–7; Forbes (1953) *passim*; Hill (1968) 56; Jones (1880) 111–13; Killip (1975) 71–2; LFWI (1971) 49–50; Lovett (1925) 52–3; N&Q series 9, vol. 3, (1899) 26, 77–8; Porter (1974) 17–18; Simpson & Roud (2000) 50–51. Information from Sara Brown, Cambridge and County Folk Museum

DOLPHINS AND PORPOISES Anderson (1959) 93; Anson (1965) 108; Bassett (1885) 61–3, 132–4, 241–5; Gervase of Tilbury (2002) 678–81; Jones (1880) 14–15; Kemp (1988) 258; West (1960) 64

FIDDLER'S GREEN AND DAVY JONES'S LOCKER Baker (1979) 184; Bassett (1885) 94; Bowen (1929) 34, 47; Cooper (1932) 34; Granville (1950) 76, 95; Kemp (1988) 232, 300–301; Marryat (1897) 47; Page & Ingpen (1985) 105; Part ridge (1973) 244–5, 314; Smollett (1831) vol. 1, 78–9; Smyth (2005) 293

FIGUREHEADS Campbell (1956) 6; Hardwick (1967) 245–6; Kemp (1988) 302–4; Laughton (1925) 63–101; *Reader's Digest* (1977) 75; Stammers (2005) 7–9

THE *FLYING DUTCHMAN* Bassett (1885) 343–9; Bassett (1917) 37–63; Dalton (1886) vol. 1, 551; Hadfield (1937) 1–15, 55; Lockhart (1924) 45–62; Thorpe (1851–2) vol. 3, 294–6

THE HEADLESS SAILOR Barham (1870) vol. 2, 2; Barham (1894) vol. 1, 160–65; Lane (1967) 56–61, 147–8

HERRING *Brewer's* (1981) 549; Defoe (1738) vol. 1, 70–71; Hammond (2005) 20; MacTaggart (1824) 430; Monger (2004) 99–100; Partridge (1973) 440; Saul (1981) *passim*; Simpson (2009) 142

HUNTING THE WREN Armstrong (1970) 141–66; Bullock (1816) 370–71; Hole (1976) 110–12; MacTaggart (1824) 157; Moore (1891) 133–40; Ó hÓgáin (2006) 37; O'Keeffe (1856) 171–2; Waldron (1865) 75–6; Wood-Martin (1902) vol. 2, 146–9

JONAH AND THE WHALE Brockie (1886) 121; Gregor (1881) 199; Marsden (1995) 169

KINGS CAN'T DROWN Fuller (1662) vol. 1, 356–7; Grose (1811) 69–70; Harland & Wilkinson (1882) 213; Jones (1880) 118–19; McPherson (1929) 188

THE *MARY CELESTE* Bryan (1942) *passim*; Conan Doyle (1892) 37–83; Hornibrook (1904) *passim*; Lockhart (1924) 171–202

MOTHER CAREY'S CHICKENS Armstrong (1970) 213; Bassett (1885) 127–8; *Brewer's* (1981) 759; Dampier (1703) 97; *EB* (1910–11) vol. 21, 315 16; Greenoak (1979)

29–30; Jonsson (1978) 32; Marryat (1841) 230–35; Trevelyan (1909) 119; Waterton (1839) 78

THE NAMING OF SHIPS Baker (1979) 24-5; Beck (1973) 20; Boase (1976) 162–3; Hardwick (1967) 244; Igglesden (1932) 117–19; Opie & Tatem (1992) 33; Stevenson (1883) 85; Trevelyan (1909) 4

THE NINTH WAVE Bassett (1885) 24–7; Hall (1841–3) vol. 3, 148; Morvan (1980) 58–9; Severin (1996) 1; Southey (1815) vol. 1, 231, 259; Tennyson (1972) 1479–80

PIRATES, AND PRIVATEERS Cordingly (2002) *passim*; Exquemelin (1969) 107–8; Johnson (1972) 71–94; Rogozinski (1997) *passim*

THE POWER OF SEAGULLS Anson (1965) 109, 150, 154; Armstrong (1970) 169; Bassett (1885) 127, 272; Campbell (1928) 216; Dyer (1880) 100; Foster (1951) 133; Greenoak (1979) 156–8; Jones (1880) 12–13; Owen (1896) 329–30; Sinclair (1978) vol. 13, 23; Woodford (1968) 60–61; Wright (1913) 216

SEA-SERPENTS Brainard (1841) 63; Hamilton (1839) 317–20; Hardwick (1967) 126–32; Harrison (2001) *passim*; Heuvelmans (1968) *passim*

SEA-SONGS Bone (1931) *passim*; Campbell (1956) 42; Healy (1976) 32–3; Hugill (1969) *passim*; Munnelly (1980–81) *passim*; Rogers (1929) 117–18; Whall (1927) 135–6

SIRENS, SEA-NYMPHS, AND MERMAIDS Anderson (1971) 249; Bennett (1953) 54–7; Benwell & Waugh (1961) *passim* ; Gervase of Tilbury (2002) 680–83; Gill (1932) 364; Lucian (1820) vol. 2, 440–41

SMUGGLERS AND WRECKERS Courtney (1890) 133; Coxe (1984) 1–20; *Daily Telegraph* 23.1.2007, 1, 4–5; *Eastern Daily Press* 28.2.2007, 2–3; Kemp (1988) 812; MacTaggart (1824) 455–6; Trevelyan (1909) 1–2. Information from Alison Kentuck, Receiver of Wrecks

STRANGE FISH AND JENNY HANIVERS Bassett (1885) 213–14; Benwell & Waugh (1961) 121–6; Buckland (1860) 326–7; Carrington (1961) 58–63; Chambers (1863–4) vol. 2, 310–11; Churchyarde (c. 1578) [26–7 – n. p.]; Dance (1976) 17–18; Hutchins (1968) 30–31; Jones (1871) 265–6; Partridge (1973) 424; Rondeletius (1554) 494

SUPERSTITIONS Bassett (1885) 474; Blakeborough (1898) 147–9; Brockie (1886) 215; Campbell (1900) 236–7; Gutch (1901) 51; Kemp (1988) 847; Martin (1703) 17–18; *Reader's Digest* (1977) 74; Simpkins (1914) 123–4

SURVIVAL CANNIBALISM Cowdell (2011) *passim*; Gibbings (1935) *passim*; Gilbert (1869) 85–9; Heaume (1976) 14–15; Poe (1838) 105–9; Simpson (1984) *passim*; Thackeray (1885) 127–8; Vaughan Williams & Lloyd (1959) 96

WONDER VOYAGES AND LOST LANDS Fritze (1998) 29–30, 42–3, 168; *Funk & Wagnalls* (1972) 515; Joyce (1894) 112–76, 236; Meyer & Nutt (1895–7) vol. 1, 2–34, 161–73, 236–7; Ó hÓgáin (2006) 333–4; Ramsay (1973) 61–74, 178–86

INDEX